AF697814

WITHDRAWN

UNVEILING THE COSMIC INFRARED BACKGROUND

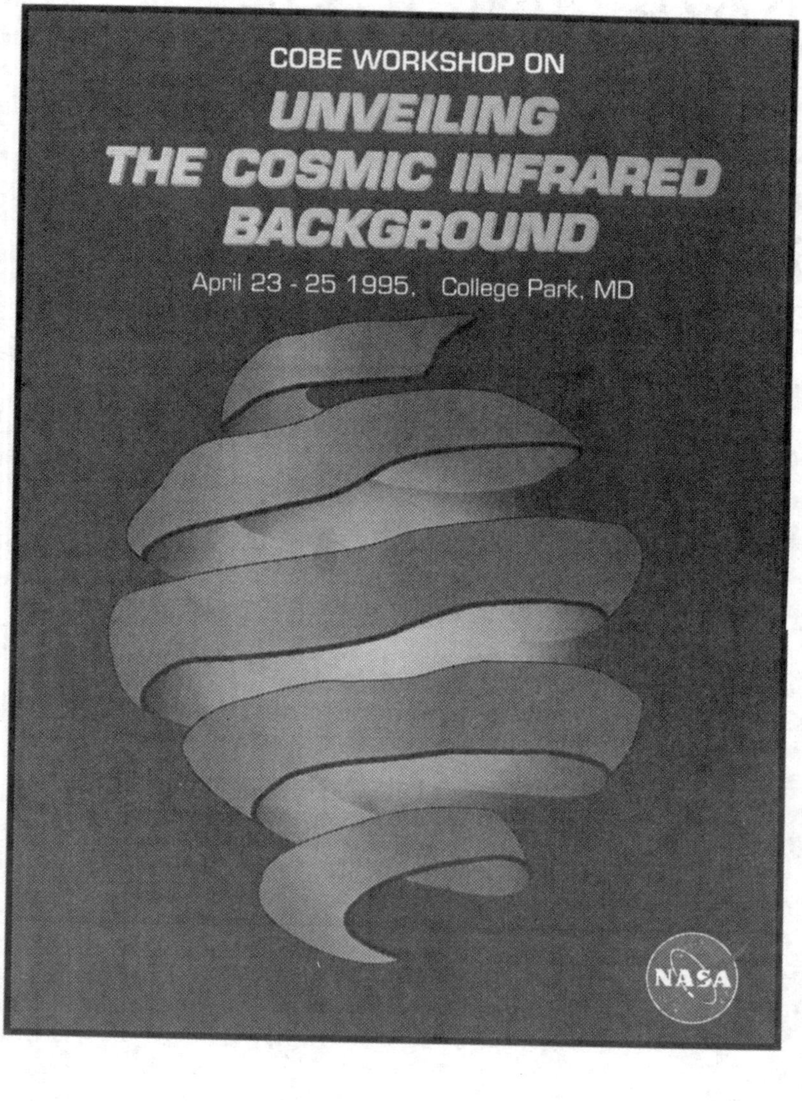
COBE WORKSHOP ON
UNVEILING
THE COSMIC INFRARED
BACKGROUND
April 23 - 25 1995, College Park, MD
NASA

UNVEILING THE COSMIC INFRARED BACKGROUND

College Park, MD April 1995

EDITOR
Eli Dwek
*NASA/Goddard
Space Flight Center*

AIP PRESS

AIP CONFERENCE PROCEEDINGS 348

American Institute of Physics

Woodbury, New York

QB
991
.C64
U58
1996

Authorization to photocopy items for internal or personal use, beyond the free copying permitted under the 1978 U.S. Copyright Law (see statement below), is granted by the American Institute of Physics for users registered with the Copyright Clearance Center (CCC) Transactional Reporting Service, provided that the base fee of $6.00 per copy is paid directly to CCC, 222 Rosewood Drive, Danvers, MA 01923. For those organizations that have been granted a photocopy license by CCC, a separate system of payment has been arranged. The fee code for users of the Transactional Reporting Service is: 1-56396-508-9/96 /$6.00.

© 1996 American Institute of Physics

Individual readers of this volume and nonprofit libraries, acting for them, are permitted to make fair use of the material in it, such as copying an article for use in teaching or research. Permission is granted to quote from this volume in scientific work with the customary acknowledgment of the source. To reprint a figure, table, or other excerpt requires the consent of one of the original authors and notification to AIP. Republication or systematic or multiple reproduction of any material in this volume is permitted only under license from AIP. Address inquiries to Office of Rights and Permissions, 500 Sunnyside Boulevard, Woodbury, NY 11797-2999; phone 516-576-2268; fax: 516-576-2499; e-mail: rights@aip.org.

L.C. Catalog Card No. 95-83477
ISBN 1-56396-508-9
DOE CONF- 9504215

Printed in the United States of America

GROUP PHOTO

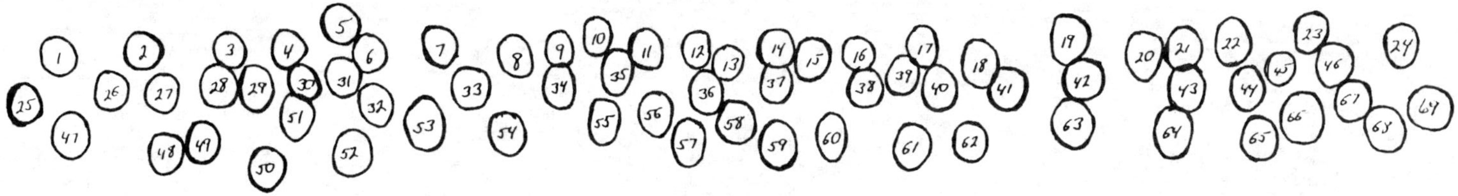

1. Franceschini
2. Giard
3. Salamon
4. Wishnow
5. Mather
6. Hanany
7. Richards
8. Granato
9. Reach
10. Peebles
11. Newmark
12. Comer
13. Sodroski
14. Crenshaw
15. Berriman
16. Banday
17. Kashlinsky
18. Toller
19. Talcott
20. Delabrouille
21. Leinert
22. Felten
23. Bernard
24. Gorski
25. Bartlett
26. Puget
27. Varosi
28. Hauser
29. Gardner
30. Leisawitz
31. Fall
32. Epchtein
33. Rowan-Robinson
34. Moseley
35. Kutyrev
36. Silk
37. Mamon
38. Odenwald
39. De Zotti
40. Kin-Wing
41. Guideroni
42. Bouchet
43. Desert
44. Chiu
45. Ruphy
46. Werner
47. Abergel
48. Rocca-Volmerange
49. Cohen
50. Jayaraman
51. Kelsall
52. Gordon
53. Ozernoy
54. Smith
55. Witt
56. Franz
57. Dwek
58. Vrtilek
59. Dermott
60. Aghanim
61. Stecker
62. Falvella
63. Boulanger
64. Arendt
65. Weiland
66. Ake
67. Lonsdale
68. Feldman
69. Heap

UNIVERSITY COLLEGE • CENTER OF ADULT EDUCATION •

Contents

4 44510 NAU CLINE:F-BKS
40 NAU 03/06/00 5033

OBSERVATIONAL LIMITS ON THE EXTRAGALACTIC INFRARED BACKGROUND

THEORIES ON THE ORIGIN OF THE COSMIC INFRARED BACKGROUND

FUTURE DIRECTIONS

POSTER CONTRIBUTIONS

PREFACE

On November 18 at 1434 UT, Delta Rocket No. 189 lifted into orbit a satellite that was soon to revolutionize the field of cosmology. Known as the Cosmic Background Explorer (*COBE*), it carried three experiments on board: The Far Infrared Absolute Photospectrometer (FIRAS), the Differential Microwave Radiometer (DMR), and the Diffuse Infrared Background Explorer (IDIRBE). The primary cosmological objectives of the three experiments were to measure the spectrum (FIRAS) and isotropy (DMR) of the cosmic microwave background and to detect the accumulated light from pregalactic objects, protogalaxies and evolving galaxies throughout the history of the universe (DIRBE). The light from these objects is expected to be redshifted or absorbed and reradiated by dust in the 1–1000 μm wavelength region that constitutes the cosmic infrared background (CIB).

The detection of the CIB has turned out to be the most challenging of the three *COBE* experiments. Its signal is hidden behind a veil of infrared emission from interplanetary dust, and from galactic stellar and dust emission. Peeling off these layers of foreground emissions, leaving the cosmic signal unscathed, is a formidable task. It requires detailed insights into the nature and origin of the interplanetary dust cloud, knowledge of the inventory of Galactic stellar objects and gaseous components, and of the abundance and properties of interstellar dust particles.

Moreover, the interpretation of any residual cosmological signal requires further expertise in understanding the physical phenomena that can contribute to this background. These include energy releases associated with galaxy formation, decaying dark matter, the synthesis of elements, active galactic nuclei, brown dwarfs and other dark baryonic matter, and any unknown phenomena that await discovery.

The *COBE* Workshop: Unveiling the Cosmic Infrared Background, held in College Park, Maryland, from April 23–25, 1995, was the first forum dedicated solely to the study of the CIB. It brought together the world's leading experts on subjects ranging from the Solar System to the far reaches of the cosmos—to discuss their implications on the unveiling and interpretation of the CIB.

The success of the Workshop reflects the enthusiastic response to the subject by all who attended, and the high quality of the presentations and discussions. The Maryland Inn and Conference Center at College Park provided an intimate and professional atmosphere condusive to our scientific thought and dialogue.

The Workshop's format and organization reflects the congenial spirit and collective thinking of the Scientific Organizing Committee:

Martin Cohen
Mike Hauser
Dave Leisawitz
Carol Lonsdale
Gary Mamon
John Mather
Jim Peebles
Joe Silk
Virginia Trimble
Jan Vrtilek

The session chairs

Tom Kelsall
Joe Silk
Jan Vrtilek
Jean Loup Puget
Gary Mamon
Carol Lonsdale
Jim Felten

are to be commended for keeping the speakers to their alloted time, and ensuring the smooth flow of the proceedings

The Workshop owes much of its success to the professional experience, attention to detail, and personal touch of

Denise Dunn and *Paula Webber*

and to the early guidance from *John Trasco*.

Finally, we acknowledge the support of the NASA Cosmic Background Explorer (*COBE*) Project, without which, this Workshop would not have been possible.

Eli Dwek

BANQUET SPEECH

HOW SHOULD WE TEACH SCIENCE AND WHAT SHOULD WE TEACH?

Martin Harwit
Laboratory for Astrophysics
National Air and Space Museum
Washington, DC 20560

When I was nine and at the English High School for Boys in Istanbul, Mrs. Peach taught us "General Knowledge" on Friday afternoons. I think it was the best class I ever attended. Here were natural history, geography, physical science – real knowledge of our world – all thrown together, so you could see how they fitted together. It was just fun to think about it. Much more exciting than the courses in physics or chemistry that came in later school years.

What made this "General Knowledge" class so memorable? Why was it a way in which science might best be taught?

I want to talk about teaching science to the public and to children in our schools. And to distinguish that right away from teaching youngsters to become scientists. Given that we already have more scientists and engineers than can currently find jobs, and given that the caliber of the work our scientists do is high, I don't think that teaching prospective scientists is a major national problem.

The task that seems to be much harder is having the public and their representatives in Congress understand what science and engineering are all about, what kind of problems can be solved by technology and what kind require a political resolution.

The trouble is that this is the very place where we scientists are least well equipped to offer ideas. This is partly because our own conceptions of what we do are so loosely based on lore, perhaps even myth, and because we have not sufficiently supported, or have paid too little attention, to the type of inquiry that would clarify the issues.

Let me give you one example. In the early 1930s two different philosophies of science and scientific progress emerged. One was by Karl Popper, whose name many scientists recognize, and whose ideas are widely accepted by them. The other was Ludwig Fleck, of whom few have heard.

Popper proposed that any advance in science comes about through a three step process. The scientist formulates a hypothesis to be tested. This hypothesis has to have measurable consequences. The second step is to make the measurement. And third, from this to conclude whether the original hypothesis is consistent with that measurement.

If the experiment or observation yields a result that differs from prediction, then the hypothesis is judged to be wrong. Agreement between the prediction and the observation, on the other hand, means merely that the hypothesis is viable, and may be correct. I say "May be correct" because there are always further tests that could be proposed that ultimately might disprove it.

What Popper was proposing is that science advances by successively eliminating all alternatives, until we are left with only one viable theory, which we accept as correct.

This is basically, the science that we now teach in schools. We tell the pupils that science is a logical process of measurement followed by conclusion.

Unfortunately, we also confuse the process by telling the students they'll need a lot of mathematics to become scientists, and this hides the rather simple concepts that underlie Poppers's widely accepted ideas of science. Sure– mathematics is important, and like any sophisticated tool, it may be hard to operate. But it is only a tool of science and engineering – not the essence.

It would be much better if students, particularly those who like to work with their hands on machines, could be told that the principles just enunciated, are exactly those which a proficient auto-mechanic applies in trouble shooting a balky engine. The mechanic also takes measurements – of spark plug gaps, or cylinder head pressure, or engine timing, to analyze the problem and discover what is wrong. Each measurement is made in response to a hunch or hypothesis about what might be wrong, and either confirms or contradicts that hunch.

Couched in such terms, science is stripped of much of its mystery and becomes far more mundane and comprehensible. Looking at it in this fashion, also permits us to understand far more – namely why you might take your car to one garage, which doesn't manage to fix the problem to your satisfaction, even though the bill you are asked to pay is high, and why a second garage does succeed?

Both may have competent mechanics, both may have carried out significant measurements in trying to locate the trouble, but they may have come up with different conclusions.

Why?

In part, it may be because they start out from quite different perspectives, and therefore trust different kinds of measurements, or interpret their meanings in different ways.

This is where the second of the 1930s view comes in.

Ludwig Fleck was a Polish physician. The translation of his groundbreaking book has the English title, "The Genesis and Evolution of a Scientific Fact". Both Popper and Fleck wrote in German, perhaps the most influential language of science at the time, but while Popper's work was readily accepted by scientists and philosophers of science, nobody paid any attention to Fleck until recent decades. Thomas Kuhn came across Fleck's work and mentions him in his celebrated book on scientific revolutions, but his indebtedness to Fleck does not fully come through.

What Fleck pointed out was that science is far messier than normally accepted by usscientists. So much so, that progress is appreciably more dependent on consensus than on measurement.

Why is that?

Partly it is because, as we all know, measurements can be faulty. If a measurement does not agree with a favorite hypothesis, we tend to discard it as possibly suspect – an experiment badly carried out, or perhaps influenced by a subtle effect, which we do not properly understand, but which could bring about a misleading result. The interpretation of measurements simply is not straightforward.

Where have we seen this before? Well, neutrino measurements on the sun, for two decades were held suspect, because they didn't agree with our theories about nuclear processes in the solar interior. When that wouldn't go away, we started telling ourselves that perhaps neutrinos have mass and oscillate. And perhaps they do. But perhaps the theory of stellar interiors also is wrong. Or perhaps the observations all are in error, and the theories have been right all along. We just don't know, and may not know for some time to come.

In the meantime we continue to work as though the theory of stellar interiors and thermonuclear reactions in the center of the sun are correct, and continue doing astronomy as though everything was OK. In other words, we have a consensus that permits us to continue working, exchanging ideas, writing text books, teaching students about stars. Life goes on despite that little caveat that we only dust off on special occasions, and which hardly influences what we as scientists do or teach.

But Fleck saw the issue as much deeper still. He pointed out that in many cases the defenders of different points of view are so far apart, that they do not even speak the same language. Fleck's book on the genesis of a fact gives a biological example. For present purposes, however, let me give a more familiar recent example.

One of the great disputes in geology concerned Wegener's hypothesis of continental drift, proposed in the 1930s, which found no acceptance at all among colleagues, despite his standing as a professional among them. His peers dismissed his ideas as ludicrous. Not until the 1960s did Wegener's teachings find acceptance, and now of course everyone is convinced that the continents do move.

Why then was this such an unacceptable idea in the 1930s? Certainly not for lack of publicity. Wegener lucidly described his thoughts and reasoning in a widely published book. There was the clear evidence of continental shapes where the continents had torn apart, and the supporting biological evidence that comparable species, not found elsewhere in the world, often were found in regions which now are an ocean apart, but would have previously fit together, as judged by the shape of the jig-saw puzzle.

Wegener's ideas may even have had popular appeal. When I was in that school in Istanbul, our Turkish geography teacher, whose name none of us boys ever was told, but whom we all called "Molotov" because he bore an uncanny resemblance to the Soviet foreign minister, asked us one day, at the end of a class, to think about why it was that all the continents and subcontinents were triangular in shape, facing South. He was going to tell us at the next class. But he must have forgotten, and never came back to the subject. I, on the other hand, having been asked about it, couldn't help wondering from time to time. And it wasn't until I heard about continental drift, decades later, that I realized that Molotov might have been aware of the continental drift idea and meant to tell us boys about it – but then never did.

Think also of the many contentious debates have we gone through over the millennia, before establishing the fact that Earth is a minor planet, circling a commonplace star, in the outskirts of a run-of-the mill galaxy in a universe beyond whose event horizons there might be untold numbers of similar regions separated by similar horizons.

In establishing this fact, we've had disputes where antagonists spoke right past each other, because they were so far apart – like in the confrontation between Church and Giordano Bruno and the Church and Galileo Galilei. For the Church, the earth-centered universe was at stake.

But the disputes are not all historical. In this century, we've had a milder, though similar Curtis-Shapley debate, about whether the universe is our own galaxy, or whether the Galaxy is just one of many.

And now, most recently, we are asking whether the universe we observe is what we should call the Universe, or whether it is only the observable part of an infinite array of similar universes going on forever. Surprisingly, there hasn't been much discussion about that, though there should be, because here we come to a real dilemma. Science is supposed to proceed through measurement and observation. Since all those other universes are in principle unobservable, because everything in them recedes from us faster than the speed of light, how can any statement about them even be made in the realm of science.

So, here we have one of the most important issues concerning our place in the scheme of things, and as scientists we are precluded from making statements, because science is based on measurement and our answer lies in something immeasurable. Where do we come off in making any statement? Does it have any further value than statements made by people with no science background at all?

What does this tell us about science? How does the process work? What can we say to children possibly wishing to choose careers? What can we say to Congressional leaders?

If Popper's interpretation is correct, and science is just a matter of turning a crank, then how do we account for differences of scientific opinion on matters where all the facts are equally available to everyone. Do we tell the Congressmen that atmospheric carbon dioxide is on the rise, or do we tell them it is relatively stable? Is there global warming or is there not? Is a glass a day of red wine good for you, as some physicians claim, or is it bad?

On the other hand, surely some opinions are better than others. We would like to think that professional scientists, engineers and physicians have greater insight, in their own fields that the average lay person. So, Popper can't be discounted by any means.

What we need to teach youngsters, and tell governmental leaders is that science does provide a means of gauging whether we are on the right track. But it is not infallible. A good deal of intuition is still required in assessing which pieces of evidence are more credible than others. If we know that a given observation is easy and unlikely to be subject to error, then we give it a higher credence than we give a very difficult experiment conducted for the first time. Our experience tells us that the latter have a history of often being wrong, even when done by a careful experimenter.

One way of teaching all this is by way of the history of science. It gives examples of true starts as well as false, of great achievements as well as mistakes. Those false starts and mistakes are not to be ridiculed as the work of ignoramuses, but rather of the vagaries of science, the true difficulties of doing science well. It isn't the sophisticated math. that gets you. The real difficulty often is the judgement and interpretation of sets of data that many others may see as well, but in a quite different light.

Perhaps the most fascinating aspect of science is that approach to judgement: The selection of problems that a given scientist considers worth solving. The evidence called on to define the problem. The theoretical or experimental methods chosen to tackle its resolution. The interpretation of new data gathered in the course of settling the issue. All this is what we tend to call "style" in science.

Many of us can look at a scientific paper and, by the style of the argument presented, recognize who the author must have been. Such differences of style and perspective are key to the conduct of the scientific enterprise. For, if we all thought exactly along identical lines, gave identical levels of credence to a spate of different experimental results, and reasoned in one and the same way on all issues, then science would be the poorer for it. It would have become a matter of turning the Popperian crank.

What we need to teach school children, is that there is plenty of room for many different approaches to science. There is room for people who derive pleasure in the orderly compilation of catalogues. Others love to speculate on the large picture. Others again, obtain satisfaction from sophisticated mathematical modeling and the solution of complicated theoretical problems, while a fourth group, likes to build sophisticated measuring equipment and apply it to obtaining increasingly refined observations. All these careers may be available to children showing different aptitudes.

All this, also needs to be explained to our law makers, who have to make difficult decisions involving public funding. They don't need to know scientific facts. They don't need to know how to evaluate the promise of a particular project. But they do need to understand the scientific process well enough to know how to ask the right kinds of questions. And they have to know that a whole range of different schools of opinion will exist on any given issue and how to figure out what level of credence to assign to each.

All that may be difficult to teach, but if we don't help those who need to make decisions on behalf of the public, we are not taking science and technology that extra step that we responsibly must – the step of helping the public understand how to make sensible decisions on behalf of us all. Science at current levels of expenditure, no longer is the domain of a few enthusiasts and practitioners. It involves a significant share of every citizens tax payments. And we owe it to our fellow citizens to help them decide – not in order to increase the amounts Congress appropriates to science research, but simply because it is the right thing to do.

Thank you for letting me talk about this tonight.

INTRODUCTORY OVERVIEWS

COSMOLOGY AND THE COSMIC INFRARED BACKGROUND RADIATION

P. J. E. Peebles
Joseph Henry Laboratories
Princeton University
Princeton NJ 08544

1. RADIATION BACKGROUNDS

The first thing to consider is the distinction between local and cosmic radiation backgrounds. We are in the plane of the Milky Way galaxy and near one side of the luminous part; the surface brightness of the galaxy accordingly is quite anisotropic. The same applies to the zodiacal light. The distribution of the nearby galaxies is quite irregular too, but the fluctuations average out to near isotropy in observations at great depth. Examples are the integrated X-ray background, which is isotropic to better than a few percent on angular scales larger than a few degrees (Boldt 1987), deep radio source catalogues (Gregory & Condon 1991), and deep galaxy catalogs (Maddox et al. 1990). Thus a defining property of the cosmic infrared background radiation (the CIBR) is its near isotropy.

If the distribution of matter in the universe were a scale-invariant fractal with dimension less than $D = 2$ there would be no "cosmic" background, only the "local" contributions from the levels of the hierarchy of clustering. It is obvious from the measurements of the backgrounds at other wavelengths that the universe is not constructed this way, however. Instead, to account for the observed extragalactic backgrounds we have to assume either that the universe is quite close to isotropic only about a position quite close to us, which seems unlikely, or else that the large-scale fractal dimension is quite close to $D = 3$. We might note also that the cosmological inflation scenario has taught us the universe may be close to homogeneous and isotropic out to the distances we are able to observe yet chaotic and without form on very much larger scales. But we can still distinguish cosmic radiation by its near homogeneity within the section of the universe we can observe.

A second signal of the CIBR may be the residual small angular scale departures from isotropy due to the spatial clustering of the sources. The rms fluctuation in the count of galaxies found within a randomly placed sphere is $\delta N/N = 1$ when the sphere radius is $R = 10$ Mpc (for Hubble constant $H_o = 75$ km s^{-1} Mpc^{-1}). That is, a characteristic scale of nonlinear clustering is 10 Mpc. If correlated or anticorrelated fluctuations on larger scales may be neglected the rms fluctuation in galaxy counts integrated to the Hubble length $c/H_o = 4000$ Mpc is $\delta n/n \sim (H_oR/c)^{1/2} \sim 0.05$ on angular scale $\theta \sim H_oR/c \sim 10$ arc minutes. If the CIBR sources clustered like the galaxies and the comoving luminosity density were not a rapidly varying function of redshift then the CIBR surface brightness would similarly fluctuate by about 5% on angular scales of about 10 arc minutes. Shectman (1974) used this signal to measure the cosmic background radiation in the optical, and Martin & Bowyer (1989) used it to detect the cosmic ultraviolet background. In these Proceedings Kashlinksy describes a similar approach to a measurement of the CIBR.

Figure 1 is a summary of the measurements and bounds on the spectrum

© 1996 American Institute of Physics

of the cosmic background radiation. As is now standard practice I plot the product $\nu i_\nu = \lambda i_\lambda$, where $i_\nu d\nu$ is the energy flux per unit area and solid angle in the frequency band ν to $\nu + d\nu$. The energy flux density integrated over the frequency band ν_1 to ν_2 is

$$i = \int_{\nu_1}^{\nu_2} \nu i_\nu d \ln \nu. \tag{1}$$

Thus if the vertical axis in the figure were stretched to linear, leaving a logarithmic horizontal axis, the area under the curve would be proportional to the cosmic energy density in the radiation. In the left hand axis I have expressed the energy density in terms of the mass density per logarithmic interval of wavelength (or logarithmic interval of frequency or photon energy) relative the critical Einstein-de Sitter mass density (for $H_o = 75$ km s^{-1} Mpc^{-1}). Here and below the symbol Ω means the ratio of a mass density to the critical Einstein-de Sitter value.

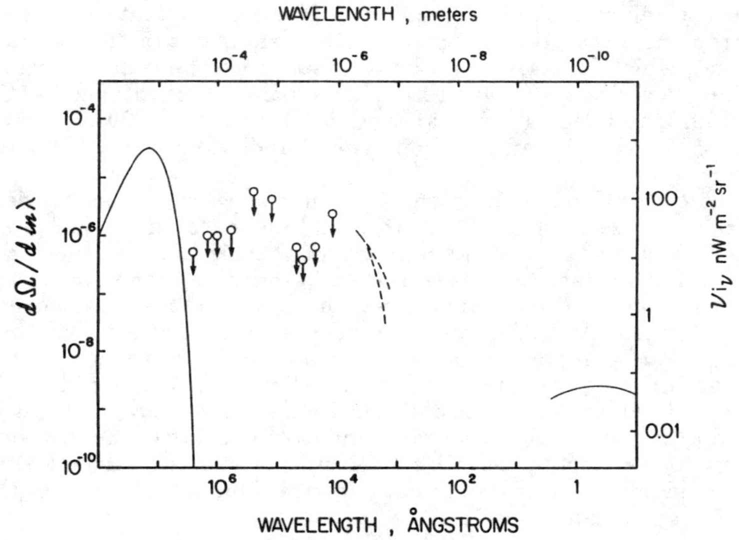

Figure 1: The cosmic background radiation spectrum.

I think the first recognizable version of this figure was published 25 years ago (in Peebles 1971). The measurements from the near infrared to the ultraviolet are much better now, the gap from the ionizing ultraviolet to soft X-rays remains, and the big gap in the infrared is being closed by the DIRBE experiment on COBE.

A considerable fraction of the mass of the universe is observable by the starlight in the relatively narrow band of frequencies of visible light. Dust degrades starlight into infrared radiation, and the cosmological redshift does the same thing to starlight emitted well before the present epoch. Since both of these processes have been operating we know the energy in the CIBR is at least comparable to that in the optical. And we may hope that measurements of the CIBR spectrum and angular distribution will constrain and guide attempts to understand what happened as the galaxies formed and evolved.

The open circles in the figure are minimum values of the residuals as a function of position on the sky for the DIRBE sky brightness measurements corrected for known local contributions (Hauser 1994 and in these Proceedings). Hauser cautions us that these numbers may go up as well as down with improvements in the models for the local components. The important point is that these preliminary numbers are comparable to the measured optical energy density, meaning DIRBE has reached cosmologically interesting surface brightnesses.

The energy in the X- and γ-ray background is well below the energy density of starlight, and may be of interest to this meeting mainly as a measure of the expected contribution of quasars and AGNs to the CIBR, if one has an opinion on the ratio of infrared to energetic photon luminosities in these objects.

2. COSMIC SOURCES

Table 1 lists commonly discussed sources of the CIBR entered in order of increasing speculation. The last entry is meant to be a reminder that Nature is quite capable of surprising us: maybe something new will be found in the broad band of the CIBR.

The idea that there might be detectable infrared radiation from decaying nonbaryonic matter is promoted to the fourth entry because there is a very real problem in the accounting of mass outside the bright central parts of galaxies: on larger scales the mass implied by the motions of the stars and gas generally is larger than the mass present in the visible stars. There is the chance that this is because we are using the wrong gravity theory, but I think it is appropriate that most of us spend most of our time working on the implications of conventional gravity physics until driven away from it. It certainly is easy to imagine that Nature chose to put most of the mass in the outer parts of the galaxies in forms that are not readily observable by us. The most carefully discussed picture in which the dark mass is nonbaryonic and unstable to electromagnetic radiative decay is that of Sciama (1994).

There is staright in the massive halos of galaxies, and reason to think there may be a considerable amount of star mass. This is based on the standard (and so far successful) model for the origin of the light elements as remnants of the rapid expansion and cooling of the early universe. The model requires that the density parameter in baryons (that is, the mean mass density relative to the Einstein-de Sitter value, as in the left-hand axis of Figure 1) is $\Omega \gtrsim 0.01$ (Walker *et al.* 1991). Since this exceeds the baryonic mass present in the bright parts of the galaxies the model predicts there is a significant amount of dark or dim baryons, likely forms being intragroup plasma and stars on and off the main sequence. And a reasonable place to look for the stars is in the massive halos of the galaxies.

A good example is the relatively nearby giant elliptical galaxy M87 in the Virgo cluster, at distance ~ 17 Mpc. The X-ray image is close to circular (but slightly elongated along the same position angle as the starlight) and extends to at least $1.5°$ or 500 kpc from the center, much larger than the optical half-light radius $R_e = 8$ kpc (de Vaucouleurs and Nieto 1978). The X-rays are thermal bremsstrahlung radiation from a hot plasma, identified by the emission lines from common highly ionized heavy elements. It is reasonable to assume that this plasma is gravitationally bound to the galaxy around which it is smoothly distributed and concentrated. The plasma pressure gradient indicates the mass

Table 1
Sources of the Cosmic Infrared Background Radiation

Source	Density Parameter
1. Nucleosynthesis	
Discs	$\Omega \sim 10^{-7}$
Spheroids	
2. Starlight absorbed by dust in low redshift galaxies	$\Omega \sim 10^{-6}$
3. Binding energies	
Stars	$\Omega \sim 10^{-6}\Omega_{\rm stars}/(1+z_f)$
White dwarfs	$\Omega \sim 10^{-2}\Omega_{\rm wd}/(1+z_f)$
AGNs and quasars	
4. Dim matter	
Black/brown dwarfs	
Decaying dark matter	
5. Other	

of the galaxy increases about linearly with distance to at least 500 kpc from the center of the galaxy (Fabricant & Gorenstein 1983). The light detected in the halo likely is starlight, because it is a smooth continuation of the starlight from the center of the galaxy (Boughn & Saulson 1983). If the star mass function (the distribution of star masses) is independent of radius the stars are a small fraction of the halo mass, but there always is the chance the mass fuction varies with radius. The observations of gravitational lensing of stars in the Magellanic Clouds by stellar mass objects in the halo of the Milky Way do not encourage the assumption that the dark mass halos are entirely baryonic, but the issue is still under discussion (Alcock et al. 1995). And whatever the baryonic mass fraction in galaxy halos we have a fascinating puzzle: how did stars form or get placed in smooth distributions in the exceedingly low densities and long crossing times characteristic of halos of galaxies? Further observations in the infrared may offer some hints.

The third entry refers to the release of binding energy in the formation of stars and other compact objects. The conversion from the energy released to the energy now in the background radiation depends on the redshift factor, as follows.

The cosmological redshift z is defined by the ratio of the wavelength $\lambda_{\rm obs}$ of a spectral feature measured at the present epoch by a comoving observer (that is, an observer moving so the large-scale pattern of the galaxy redshifts is isotropic) to the wavelength $\lambda_{\rm em}$ measured by a comoving observer at the source,

$$1+z=\frac{\lambda_{\rm obs}}{\lambda_{\rm em}}=\frac{a_{\rm obs}}{a_{\rm em}}. \qquad (2)$$

As indicated in the second part of the equation, the wavelength is stretched in proportion to the factor by which the universe has expanded between the epochs of emission and detection. It has become common practice to label epochs in the very early universe by the expansion factor expressed by the redshift z even when there is no way one could detect radiation freely propagating from that epoch to the present.

The stretching of the wavelength diminishes the energy of each photon by the factor $1+z$, and the expansion of the universe dilutes the photon number density by the factor $(1+z)^3$, so the relation between the mean surface brightness νi_ν of the cosmic background at emission and detection is

$$\nu i_\nu(\text{obs}) = \nu i_\nu(\text{em})/(1+z)^4, \qquad \nu(\text{obs}) = \nu(\text{em})/(1+z). \tag{3}$$

This relation also applies to the surface brightness of an object observed at redshift z in an inhomogeneous universe; the derivation from Liouville's theorem is an interesting exercise for the student.

We might pause to note two points that are sometimes confusing. First, the wavelength stretching in equation (2) is not the result of some cosmic force that is stretching all lengths, and in particular there is nothing forcing galaxies or people to expand with the general expansion of the universe. A better way to think of the cosmological redshift is to imagine a string of comoving observers placed along the path of a packet of photons. In an expanding universe each observer sees that the packet is overtaking the next observer, who is moving away. The result is a sequence of first-order Doppler shifts to the red. Second, because of the redshift the total electromagnetic energy in a fixed comoving volume is smaller at detection than at emission by the factor $1+z$. Energy is conserved locally: the radiation pressure is doing work on the expanding comoving volume. The PdV work does not go into kinetic energy of expansion, however: in a homogeneous universe there is no pressure gradient force, and in general relativity theory no general energy conservation law.

The gravitational binding energy per unit mass in a star with mass in the range 0.1 to $1M_\odot$ is close to constant at

$$\epsilon_{\text{stars}} \sim 2 \times 10^{-6}. \tag{4}$$

As indicated in the table, if baryonic mass Ω_{stars} were converted to stars it would release a fraction ϵ_{stars} in electromagnetic radiation, and if this happened at redshift z_f the present energy density would be down by the factor $1+z_f$. We see that even if the critical mass density $\Omega = 1$ were in baryons sequestered in low mass stars at redshift $z_f \sim 10$ the energy released would be an order of magnitude below the preliminary DIRBE bound.

Baryons sequestered in white dwarfs have to be burned to helium or heavier elements, to prevent the star from exploding, so the dominant binding energy is nuclear, $\epsilon \sim 0.01$. Thus we see that if $\Omega = 0.1$ were sequestered in white dwarfs at redshift $z_f \sim 10$ it would produce a radiation background $\Omega \sim 10^{-4}$. We can rule this out: there is no place to put the radiation.

The gravitational binding energy per unit mass in a galaxy is $\epsilon \sim (v/c)^2$, where v is the escape velocity. In the Milky Way the escape velocity is $v \sim 600$ km s^{-1}, so $\epsilon \sim 4 \times 10^{-6}$. The dark mass component may be bound by the primeval potential energy fluctuations; it need not have dissipated any energy (and perhaps cannot have dissipated energy except by gravitational interactions). If the baryonic component $\Omega \sim 0.01$ settled to the central parts of the galaxies at $z_f \sim 3$, about the minimum allowed by the observations of high redshift galaxies, it would contribute to the electromagnetic background $\Omega \sim 1 \times 10^{-8}$. Baryons settling in a galaxy may release the binding energy as

X-rays, and it is interesting that the energy is comparable to what is present in the X-ray background. But this is a subject for another conference.

The second entry in the table is known to be a significant source of the CIBR: an appreciable fraction of starlight has been absorbed by dust and the energy reradiated in the far infrared. The amount depends on the highly uncertain properties of the dust, and on what was happening at high redshift.

Dust in present-day galaxies is one of the topics for this conference, and an important problem for astronomy and cosmology. For example, obscuration of starlight by dust affects the observation of quasars behind galaxies, and hence the rate of detection of gravitational lensing (Fukugita & Turner 1991) and the rate of detection of absorption lines from foreground galaxies in quasar spectra (as discussed in these Proceedings by Michael Fall). The source strength for the CIBR depends on the distribution as well as amount of the dust. If dust distributed like the stars in galaxies had large optical depth for absorption of starlight it would have made Ω in the CIBR considerably larger than in the optical; DIRBE has shown us that this cannot be so. A thin opaque sheet of dust running through the plane of each spiral galaxy would absorb about half the starlight, which DIRBE would allow. Peletier and Willner (1992) favor dust distributed like the disc stars and with optical depth $\tau \sim 1$ perpendicular to the plane, which again would mean that about about half the starlight has been deposited in the CIBR.

The first entry in the table is the energy released in the production of the observed heavy elements. We can be sure the heavy elements were produced in stars out of matter that initially was almost pure hydrogen and helium, for that is the composition of the oldest stars. The discussion of the history of heavy element formation seems to naturally divide into two parts: the ellipticals and spheroid components of the spirals, and the thin disc components of the spiral galaxies. Silk addresses the former in these Proceedings; I discuss some recent ideas on the history of the thin discs.

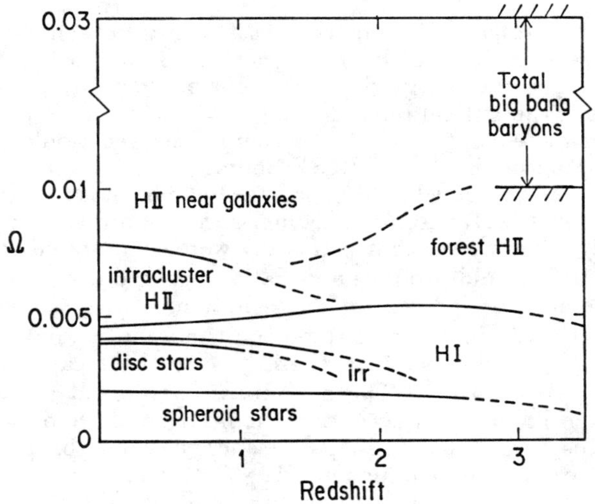

Figure 2: A history of the baryons.

Figure 2 shows the proposed history of the baryons. (The figure is from Fukugita, Hogan, & Peebles 1995. It would be inappropriate to present here a full list of the authors whose work inspired the figure. Instead I give a few examples and for further details refer the reader to the references in this paper and in recent proceedings of conferences on galaxies and the intergalactic medium at high redshifts.)

As in the left-hand axis of Figure 1, the amount of mass in each baryonic component is represented by its contribution to the cosmological density parameter (the ratio of the mean mass density in the component to the Einstein-de Sitter value). The numbers are stacked so that if we had a full accounting of all baryons the top would be level at the net density parameter in baryons. We indicate two main stellar components: discs and spheroids. We show little evolution in these components at $z < 1$ because observations of large galaxies at $z \sim 1$ suggest they are much the same as their present-day counterparts, with similar comoving number densities and luminosities. (A striking example of the evidence is given by Steidel, Dickinson, and Persson 1994.) Evolution at $z < 1$ is observed, to be sure, but it seems to involve relatively small mass fractions. For example, the evolution of luminosities of galaxies in the ultraviolet may be the effect of massive short-lived stars, the production of which seems to be decreasing as the large galaxies exhaust the supply of gas for the formation of new generations of stars. There is rapid evolution in a population of relatively low luminosity blue galaxies; they are indicated by the thin wedge labeled "irr."

The spectra of ellipticals observed at redshift $z \sim 1$ resemble that of a relatively old star population, suggesting the stars in ellipticals were formed well before $z = 1$. Consistent with this, galaxies at $1 \lesssim z \lesssim 3$ identified as radio sources have luminosities in the optical at emission comparable to the luminosities of present-day giant galaxies (McCarthy 1993). We accordingly show nothing much happening to the ellipticals since $z \sim 3$. We assume the same is true of the spheroid components of spiral galaxies because they look a lot like ellipticals.

The absorption lines in quasar spectra yield a remarkably detailed picture of the gas clouds that lie along the lines of sight. The Lyman-α forest is named for the dense set of narrow atomic hydrogen resonance absorption lines in quasar spectra blueward of the Lyα emission line of the quasar. The lines are produced by gas clouds that at $z \sim 3$ come close to filling space, that is, the cloud sizes are comparable to the mean distance between clouds. Most of the forest baryons are in clouds with neutral column densities on the order of 10^{13} cm^{-3}, and most of the cloud material is ionized. The density parameter in baryons in the forest seems to decrease significantly after redshift $z \sim 2$; a likely guess is that these baryons end up in plasma in and around groups and clusters of galaxies.

Most of the neutral atomic hydrogen, HI, at $z \sim 3$ is in the damped Lyman-α clouds (so called because the high column densities in neutral atomic hydrogen broaden the Lyα absorption line well into the radiation-damped wings of the transition). The density parameter in HI is roughly constant at $\Omega \sim 0.003$ at $2 \lesssim z \lesssim 4$, and decreases to a much smaller value at $z = 0$ (Lanzetta, Wolfe, & Turnshek 1995).

There is evidence that the HI in the damped Lyα clouds is being converted to the stars in the thin disc components of the spiral galaxies: the density parameter in the clouds at $z = 3$ is comparable to the density parameter in the stars now in the discs of the spiral galaxies; the largest HI column densities found in the clouds, $\sim 10^{22}$ cm^{-2}, are comparable to the baryonic column densities in the thin stellar discs of spirals, and the damped Lyα clouds tend to rotate about as rapidly as do the thin discs (Wolfe et al. 1994). In this picture the clouds would be rotationally supported. They could not be directly collapsing to thin stellar discs because the heavy element abundances are at least an order of magnitude below what is thought to be typical of thin disc stars. They could be forming the thick disc stars, and in the process accumulating heavy elements prior to a final collapse normal to the disc. It would be easy to imagine that a spiral galaxy that grows by accretion of star clusters or massive gas clouds acquires several generations of discs, only the last surviving the disruption of a major accretion event. There is enough material in the HI in the damped Lyα clouds to account for the discs in the spirals we see around us.

In this picture the heavy element radiation background from the formation of the thin disc population is

$$\Omega_{\rm HERB} \sim \frac{\epsilon Z \Omega_{\rm disc}}{(1+z_f)} \sim 10^{-7}, \tag{5}$$

where ϵ is the binding energy per unit mass in the heavy elements and the heavy element abundance $Z \sim 0.02$ was produced in the thin disc mass $\Omega_{\rm disc} \sim 0.002$ at mean redshift $z_f \sim 2$.

ACKNOWLEDGEMENTS

This work was supported in part by the US National Science Foundation.

REFERENCES

Alcock, C. et al. 1995, ApJ, 449
Boldt, E. 1987, Physics Reports, 146, 215
Boughn, S. P. & Saulson, P. S. 1983, ApJ, 265, L55
de Vaucouleurs, G. & Nieto, J.-L. 1978, ApJ, 220, 449
Fabricant, D. & Gorenstein, P. 1983, ApJ, 267, 535
Fukugita, M., Hogan, J. C., & Peebles, P. J. E. 1995, submitted to Nature
Fukugita, M. & Turner, E. L. 1991, MNRAS, 253, 99
Gregory, P. C. & Condon. J. J. 1991, ApJS, 75, 1011
Hauser, M. G. 1994, in the Proceedings of IAU Symposium 168
Lanzetta, K. M, Wolfe, A. M., & Turnshek, D. A. 1995, ApJ, 440, 435
Maddox, S. J., Efstathiou, G., Sutherland, W. J., & Loveday, J. 1990, MNRAS, 242, 43P
Martin, C. & Bowyer, S. 1989, ApJ, 338, 677
McCarthy, P. J. 1993, ARA&A, 31, 639
Peebles, P. J. E. 1971, Physical Cosmology (Princeton: Princeton University Press)
Peletier, R. F. & Willner, S. P. 1992, AJ, 103, 1761
Sciama, D. W. 1994, ApJ, 426, L65
Shectman, S. A. 1974, ApJ, 188, 233
Steidel, C. C., Dickinson, M., & Persson, S. E. 1994, ApJ, 437, L75
Walker, T. P. et al. 1991, ApJ, 376, 51
Wolfe, A. M. et al. 1994, ApJ, 435, L101

SEARCHING FOR THE COSMIC INFRARED BACKGROUND

Michael G. Hauser
Code 680
NASA Goddard Space Flight Center
Greenbelt, MD 20771

ABSTRACT

The cumulative emissions of pregalactic, protogalactic, and evolving galactic systems are expected to be recorded in a cosmic infrared background (CIB). Measurement of such a uniform background in the infrared requires absolute sky brightness measurements from space. The Diffuse Infrared Background Experiment (*DIRBE*) on the Cosmic Background Explorer (*COBE*) spacecraft is the first space experiment designed to conduct a systematic search for the CIB from 1.25 to 240 μm, and the *COBE* Far Infrared Absolute Spectrophotometer (*FIRAS*) provides additional absolutely calibrated all-sky measurements from 100 μm to 5 mm wavelengths. With the data from these two experiments in hand, the challenge now becomes discrimination of the bright foreground emissions from interplanetary dust and Galactic sources to reveal the extragalactic background. Conservative limits on the isotropic infrared background are given by the minimum observed sky brightness at each wavelength. Extensive modeling of the foregrounds is under way to isolate or strongly limit the extragalactic infrared component. The current approach to these modeling efforts is described and representative present residuals are reported. Results are compared with those from recent rocket flights and with the extragalactic light estimated from galaxy counts.

1. INTRODUCTION

The search for cosmic infrared background (CIB) radiation is a relatively new field of observational cosmology. Measurement of this distinct radiative background, expected to arise from the cumulative emissions of pregalactic, protogalactic, and galactic systems, would provide new insight into the cosmic 'dark ages' following the decoupling of matter from the cosmic microwave background (CMB) radiation (Partridge & Peebles 1967; Low & Tucker 1968; Peebles 1969; Harwit 1970; Kaufman 1976; Bond, Carr, & Hogan 1986, 1991; Franceschini *et al.* 1991; Franceschini *et al.* 1994; and presentations by many speakers at this Workshop).

The status of direct attempts to detect the CIB has recently been reviewed by Hauser (1995a; 1995b). Though the integrated energy density in the CIB might, on the basis of pre-COBE observations, be comparable to that of the CMB (Ressell & Turner 1990), there have been no corroborated direct detections of the CIB. Possible evidence for an isotropic infrared background, or at least limits on emission in excess of galactic foregrounds, has been reported on the basis of very limited data from rocket experiments (Matsumoto *et al.* 1988; Matsumoto 1990; Noda *et al.* 1992; Kawada *et al.* 1994). Indirect upper limits, and even possible lower limits, on the extragalactic infrared background have been inferred from measurements of TeV γ-rays (De Jager, Stecker, & Salamon 1994; Dwek & Slavin 1994; Biller *et al.* 1995). Since the status of the γ-ray

© 1996 American Institute of Physics

analyses is reviewed by Stecker at this Workshop (Stecker 1995), those results are not considered further here.

The *COBE* Diffuse Infrared Background Experiment (DIRBE) is the first satellite instrument designed specifically to carry out a systematic search for the CIB in the 1.25–240 μm range. The *COBE* FIRAS instrument, designed primarily to make a precise measurement of the spectrum of the cosmic microwave background radiation, is also a powerful instrument for the CIB search in the 140–2000 μm range. The search for the CIB is the most exploratory of the *COBE* objectives. Even from a spaceborne instrument, this cosmic fossil is far more difficult to discern than the CMB radiation. Whereas the CMB is the dominant celestial radiation over most of the sky at millimeter wavelengths, the local infrared foregrounds from interplanetary dust and the Galaxy are far brighter than the CIB.

Since the *COBE* experiments are such a major advance in this field, this review will necessarily focus on current results from them. Conservative limits based upon the darkest sky measurements from *COBE* will be presented. The status of our efforts to discriminate and remove solar system and Galactic foregrounds, a major theme of this Workshop, will also be described. Results will be compared with some theoretical predictions, with other diffuse infrared measurements, and with lower limits based upon integrated galaxy counts.

2. THE *COBE* EXPERIMENTS

2.1 The Observational Challenge

Since the term "cosmic infrared background" has been used with various meanings in the literature, let me begin by adopting this to mean all diffuse infrared radiation arising external to the Milky Way Galaxy. The search for the CIB is impeded by two fundamental challenges: there is no unique spectral signature of such a background, and there are many local contributors to the infrared sky brightness at all wavelengths, often quite bright. The lack of a distinct spectral signature arises in part because so many different sources of primordial luminosity are possible (e.g., Bond, Carr, & Hogan 1986), in part because the radiant characteristics of evolving galaxies are imperfectly known, and in part because the primary emissions at any epoch are then shifted into the infrared by the cosmic red-shift and possibly by dust absorption and re-emission. Hence, the present spectrum depends in a complex way on the characteristics of the luminosity sources, on their cosmic history, and on the dust formation history of the Universe.

Setting aside the difficult possibility of recognizing the CIB by its angular fluctuation spectrum (Bond, Carr, & Hogan 1991; Kashlinsky 1995; Veeraraghavan 1995), the only identifying CIB characteristic for which one can search is an isotropic glow. One must solve the formidable observational problem of making absolute brightness measurements in the infrared. One must then discriminate and remove the strong signals from foregrounds arising from one's instrument or observing environment, the terrestrial atmosphere, the solar system, and the Galaxy. Particular attention must be given, of course, to possible isotropic contributions from any of these foreground sources. Finally, if an extragalactic isotropic residual remains, one must evaluate the contribution from galaxies over their luminous lifetime to distinguish their light from that of pregalactic or protogalactic sources. The *COBE* investigations are presently at the stage of modeling the known solar system and Galactic foregrounds and assessing the uncertainties in the models and measurements so as to determine whether evidence for an isotropic residual remains.

2.2 The COBE Diffuse Infrared Background Experiment

A detailed description of the *COBE* mission has been given by Boggess *et al.* (1992), and the DIRBE instrument has been described by Silverberg *et al.* (1993). The DIRBE observational approach was to obtain absolute brightness maps of the full sky in 10 photometric bands at 1.25, 2.2, 3.5, 4.9, 12, 25, 60, 100, 140, and 240 μm. In order to facilitate discrimination of the bright foreground contribution from interplanetary dust, linear polarization was also measured at 1.25, 2.2, and 3.5 μm. Because of the Earth's motion within the interplanetary dust cloud, the diffuse infrared brightness of the entire sky varies over the course of a year. To monitor this variation, the DIRBE field of view was offset from the *COBE* spacecraft spin axis by 30°. This produced a helical scan of the sky during each *COBE* orbit. Over the course of six months this yielded observations of each celestial direction hundreds of times at all accessible solar elongation angles (depending upon ecliptic latitude) in the range 64°to 124°. All spectral bands viewed the same instantaneous field-of-view, and the helical scan allowed the DIRBE to map 50% of the celestial sphere each day. The instrument was designed to achieve sensitivities (1σ, 1 year) for each $0.7° \times 0.7°$ field of view of approximately $\nu I_\nu = 1$ nW m^{-2} sr^{-1} (1.25–100 μm) and 10 nW m^{-2} sr^{-1} (140–240 μm). Actual performance during the 10 months of operation with cryogen in the COBE dewar was close to these values, except for an increase to about 2 nW m^{-2} sr^{-1} at 1.25 and 2.2 μm because of source confusion, and to about 30 nW m^{-2} sr^{-1} at 140 μm. These single field-of-view values are generally below the actual sky brightness, and below many of the predictions for the CIB. Averaging over substantial sky areas, once foregrounds are removed, increases the sensitivity for an isotropic signal.

The DIRBE instrument is an absolute radiometer, utilizing an off-axis folded Gregorian telescope with a 19-cm diameter primary mirror. The optical configuration (Magner 1987) was carefully designed for strong rejection of stray light from the Sun, Earth limb, Moon or other off-axis celestial radiation, or parts of the *COBE* payload (Evans 1983). The instrument, which was maintained at a temperature below 2 K within the *COBE* superfluid helium dewar, measured absolute brightness by chopping between the sky signal and a zero-flux internal reference at 32 Hz. Instrumental offsets were measured about five times per orbit by closing a cold shutter located at the prime focus. A radiative offset signal in the long wavelength detectors arising from JFETs (operating at about 70 K) used to amplify the detector signals was identified and measured in this fashion and removed from the DIRBE data. Because the offset signal was stable over the course of the mission, it would appear as an isotropic signal if left uncorrected. Uncertainty in the offset correction was well below 0.1 nW m^{-2} sr^{-1} at wavelengths shortward of 60 μm, approximately 1 nW m^{-2} sr^{-1} at 60 and 100 μm, and 8 (3) nW m^{-2} sr^{-1} at 140 (240) μm. Internal radiative reference sources were used to stimulate all detectors when the shutter was closed to monitor the stability and linearity of the instrument response. The highly redundant sky sampling and frequent response checks provided precise photometric closure over the sky and reproducible photometry to $\sim 1\%$ or better for the duration of the mission. Calibration of the photometric scale was obtained from observations of isolated bright celestial sources.

2.3 The COBE Far Infrared Absolute Spectrophotometer

The FIRAS instrument is a polarizing Michelson interferometer, providing spectral coverage from about 100 to 10,000 μm wavelength. The instrument has an instantaneous field of view of 7°FWHM aligned with the *COBE* spin axis, so as to map the sky at an elongation angle of 94°. The instrument operated in a fully differential mode, continuously measuring the difference between the spectrum of the sky and that of an ultrastable internal calibrator. Absolute calibration was provided by inserting a full-beam precision black-body source into the sky horn. The FIRAS was most sensitive in its 500–10,000 μm range, but had sensitivity per beam in its sky maps better than $\nu I_\nu = 10$ nW m^{-2} sr^{-1} at wavelengths down to about 200 μm. Its systematic offset calibration uncertainty was well below 1 nW m^{-2} sr^{-1} at wavelengths longer than 200 μm.

Comparison of the DIRBE and FIRAS absolute calibrations at 100, 140, and 240 μm shows excellent agreement between these two independent instruments on *COBE*. The *COBE* instruments have provided an extensive, consistent data set with which to search for evidence of the CIB.

3. THE CHARACTER OF THE INFRARED SKY

The DIRBE sky brightness maps show the dominant anticipated features of Galactic starlight and zodiacal light at short wavelengths, emission from the interplanetary dust dominating at 12 and 25 μm, and emission from the interstellar medium dominating at longer wavelengths (Hauser 1993; DIRBE Sky Maps at Elongation 90°, available from the National Space Science Data Center (NSSDC): see Leisawitz 1995). The brightness of the sky in any fixed direction varies roughly sinusoidally over the year, though the detailed time dependence and amplitude of variation depends upon direction in the sky (Kelsall *et al.* 1993), indicating rather complex features of the interplanetary dust (IPD) cloud. The DIRBE polarization maps reveal the expected strong linear polarization arising from the zodiacal light (Berriman *et al.* 1994; DIRBE Weekly Sky Maps, NSSDC: see Leisawitz 1995). The FIRAS provides maps (FIRAS Spectral Sky Maps, NSSDC: see Leisawitz 1995) from 100 μm to wavelengths well beyond where the CMB dominates the sky brightness, showing Galactic emission from dust and gas in the interstellar medium (Wright *et al.* 1991; Bennett *et al.* 1994; Reach *et al.* 1995). The high-frequency FIRAS data also include detectable IPD emission. The *COBE* data dramatically illustrate the challenge of measuring the CIB: the diffuse sky brightness is definitely not isotropic in the infrared, and is variable over the year in all directions due to our motion in the IPD cloud.

The most conservative direct observational limits on the CIB are the minimum observed sky brightnesses. In each DIRBE weekly sky map, the faintest direction has been determined for each wavelength. The smallest of these values at each wavelength over the duration of the mission provides a 'dark sky' upper limit to the CIB. Table 1 lists these dark sky values. At wavelengths where interplanetary dust scattering or emission is strong, the sky is darkest near the ecliptic poles. At wavelengths where the interplanetary cloud signal is rather weak (i.e., longward of 100 μm), the sky is darkest in regions of low HI column density, which are generally at high galactic latitude. The error shown for each value is the absolute calibration uncertainty. A similar analysis of the FIRAS high frequency data, after removal of the CMB signal, yields the "FIRAS dark sky" values shown in Table 1 (Shafer *et al.* 1995), which are 95% confidence limits on the total brightness due to the solar system, Galaxy, and CIB.

TABLE 1. UPPER LIMITS ON THE COSMIC INFRARED BACKGROUND

λ μm	νI_ν nW m^{-2} sr^{-1}	Reference
1.25	393±13	DIRBE dark sky
2.2	150±5	"
3.5	63±3	"
4.9	192±7	"
12	2660±310	"
25	2160±330	"
60	261±22	"
100	74±10	"
140	57±6	"
240	22±2	"
111	103	FIRAS dark sky
143	63	"
250	30	"
500	6	"
500-5000	$680/\lambda(\mu m)$	Mather *et al.* (1994)

As described by Mather *et al.* (1994), the CMB spectrum in the wavelength range 0.5–5 mm deviates from a 2.726 K blackbody shape by less than 0.03% of the peak intensity. Taking this as an upper limit to an additional cosmic infrared background implies $\nu I_\nu < 340/\lambda(\mu m)$ nW m^{-2} sr^{-1}. To make some allowance for systematic errors in separating the Galactic signal from the FIRAS CMB signal, the above limit has been doubled in Table 1.

4. FOREGROUND MODELING

In order to determine whether an isotropic extragalactic signal is present, one must address the problem of discriminating the various contributions to the measured sky brightness. The DIRBE team is presently approaching this in a sequential process: (1) find a parameterized model of the interplanetary dust (IPD) cloud; (2) integrate this IPD model along the line of sight to calculate the sky brightness for each celestial direction for each week of the DIRBE mission; (3) subtract the calculated IPD map from each weekly DIRBE map and average the residual weekly maps; (4) blank bright sources discernible in the residual averaged maps, both stellar and more extended sources; (5) use a detailed model of discrete Galactic sources to calculate a map of the integrated sky brightness due to sources fainter than those blanked in the previous step, and subtract this map from the DIRBE residual map; (6) determine an interstellar medium emission map at each wavelength using the DIRBE 100 μm map as a template, and subtract this map from the residual map; (7) test the final residual maps for isotropy.

A brief summary of an early version of this approach was presented by Hauser (1995b). Since rather detailed expositions of these steps are presented by other DIRBE team members at this Workshop (Reach *et al.* 1995b; Franz *et al.* 1995; Arendt *et al.* 1995; Weiland *et al.* 1995), we here proceed to present representative results after removal of foreground signals determined from current models used by the DIRBE team.

5. STATUS OF THE CIB SEARCH

After removing from the measured sky brightness the contributions from interplanetary dust, bright and faint discrete galactic sources, and the interstellar medium, the residual signal at high galactic and ecliptic latitudes is positive and generally rather featureless, though low level artifacts from systematic errors in the models are clearly present. To illustrate the magnitude of the foreground signals, Figure 1 shows the DIRBE specrum of the total observed sky brightness averaged over a $10° \times 10°$ region at the north Galactic pole and the individual contributions from the foreground sources. Scattering and emission from the interplanetary dust dominates all other signals from 1.25 to 100 μm at high galactic latitude. This is true even at 3.5 μm, the spectral 'window' between the maxima of the scattered and emitted IPD signal. Only at 140 and 240 μm does some other foreground signal, that from the interstellar medium (infrared cirrus), become dominant.

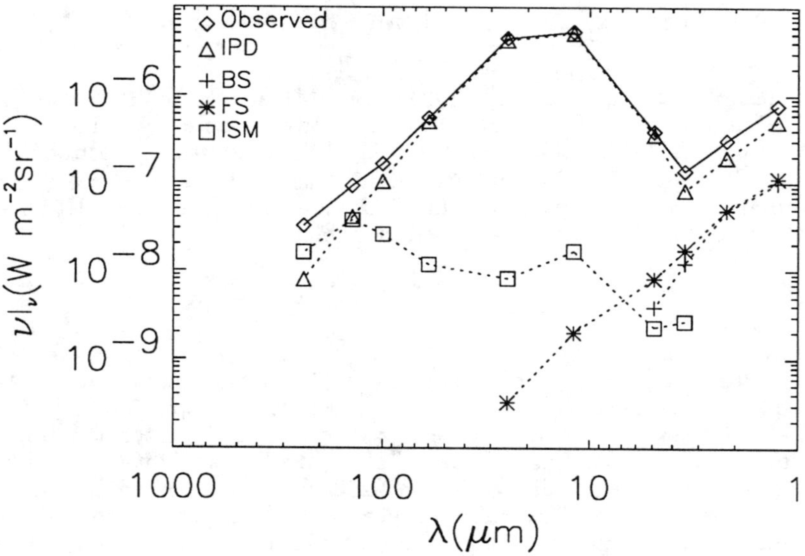

Figure 1: DIRBE sky brightness toward the North Galactic Pole (Observed) and foreground contributions from the interplanetary dust (IPD), bright galactic sources (BS), faint galactic sources (FS), and interstellar medium (ISM).

The next critical step in the search for the CIB is to test the residual maps for isotropy. The process of assessing the uncertainties in the foreground models and the DIRBE measurements, and using this knowledge to test whether the residuals show significant isotropy, is not yet complete. Hence, no definitive conclusions regarding DIRBE detection of, or limits upon, the CIB more restrictive than the dark sky limits in Table 1 can yet be stated. The FIRAS team is also analyzing their residual signals at wavelengths shorter than 500 μm to reach definitive conclusions regarding the CIB (Shafer *et al.* 1995).

However, some insight into the challenge and present status is provided by looking at the residual brightnesses in several high latitude regions. For this purpose, the DIRBE team has examined the residuals in $10° \times 10°$ fields at the Galactic poles and ecliptic poles, and a $5° \times 5°$ field in the 'Lockman Hole' (Lockman, Jahoda, & McCammon 1986; Jahoda, Lockman, & McCammon 1990), the region of minimum HI column density at $(\ell, b) \sim (148°, +53°)$. Table 2 lists the minimum and maximum mean residual brightnesses for these 5 patches after all of the foreground removal steps listed above. These values differ modestly from those presented by Hauser (1995b, Table 2) because of some improvements made in the foreground models since that earlier report. While the range of values at each wavelength is still substantial, typically a factor of 2 or more, comparison with Table 1 shows that we are now dealing with residuals which are small fractions, approaching 10% at wavelengths shortward of 100 μm, of the 'dark sky' limits. However, the fact that the residuals are brightest in the region of peak IPD thermal emission, 12 to 25 μm, strongly suggests that significant foreground emission still remains, at least in the middle of the DIRBE spectral range. The DIRBE team plans to report their conclusions regarding CIB limits based upon such analyses in the near future.

TABLE 2. RANGE OF DIRBE SKY BRIGHTNESS RESIDUALS AT HIGH LATITUDE

λ μm	νI_ν nW m^{-2} sr^{-1}
1.25	50 - 104
2.2	15 - 26
3.5	15 - 24
4.9	9 - 22
12	102 - 164
25	136 - 210
60	31 - 42
100	20 - 35
140	12 - 63
240	8 - 33

The current state of direct infrared background measurements is summarized in Figure 2. Figure 2a shows the *COBE* DIRBE (diamonds) and FIRAS (triangles) dark sky limits from Table 1, the range of DIRBE residuals after foreground removal at high latitude from Table 2 (+), and the FIRAS residual upper limit in the 500–1000 μm range from Table 1 (dashed line). The range of residuals from recent near-infrared rocket measurements of two sky regions by

Noda *et al.* (1992) is bounded by dotted lines. The residual upper limit at 154 μm from the rocket measurement of Kawada *et al.* (1994) is also shown.

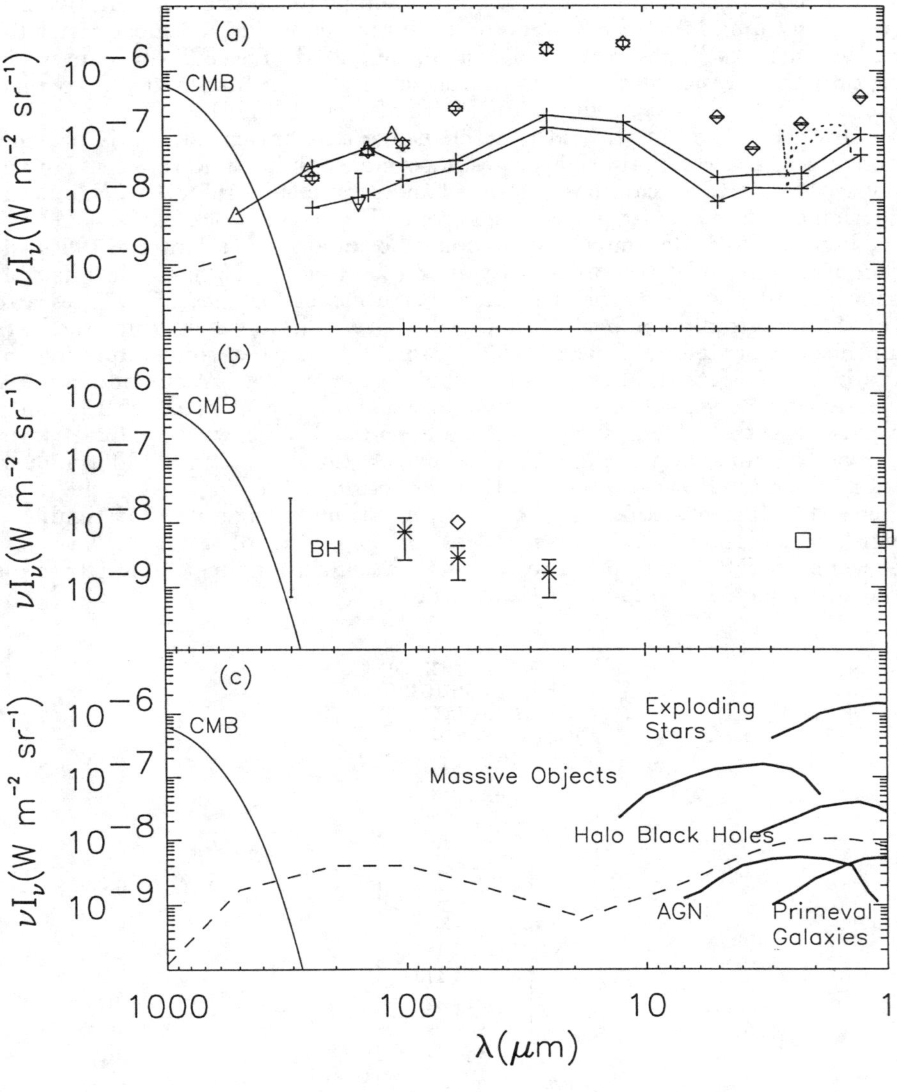

Figure 2: a) Recent direct measurements of the diffuse infrared background (see text). b) Extragalactic infrared background estimated from galaxy counts (see text). c) Representative predicted contributions to the CIB from pregalactic sources in a dust-free Universe and evolving galaxies (see text).

Lower limits to the extragalactic infrared background can be obtained by integrating the brightness of observed galaxies. Figure 2b shows such results from the near-infrared galaxy counts of Cowie *et al.* (1990) (squares), and from the IRAS survey by Hacking and Soifer (1991) (asterisks) and Beichman and Helou (1991) (BH). The IRAS results are shown as a range to encompass the various galaxy luminosity or density evolution models considered in those studies. The results of a recent analysis of previously unreduced IRAS "filler" mode observations at 60 μm by Gregorich *et al.* (1995) is also shown (diamond). It is comforting to see that the integrated discrete source estimates still lie below the diffuse sky brightness residuals, though the gap is not large at some wavelengths. For example, the result of Gregorich *et al.* is about a factor of two brighter than that of Hacking and Soifer, and is only about a factor of two below the lower end of the DIRBE residual range at 60 μm. The estimated integrated galaxy far-infrared background contribution should become less uncertain as deeper counts from future space infrared missions such as *ISO*, *WIRE*, and *SIRTF* are obtained. Even the present estimated integrated galaxy background may be discernible at some wavelengths using *COBE* data if the foreground contributions to the *COBE* measurements can be modeled to about the 1% level, a difficult but perhaps achievable goal.

To provide some perspective on the implications of residuals at these levels, though there is much more input on that subject at this Workshop, Figure 2c presents a few representative theoretical predictions from the literature. Bold solid curves show estimates of contributions from some possible pregalactic and protogalactic sources in a dust-free universe by Bond, Carr, & Hogan (BCH) (1986), and the dashed curves are estimated contributions from evolving galaxies by Franceschini *et al.* (1994). Few of the BCH estimates are consistent with the present residuals if one regards the residuals as upper limits, and some are even ruled out by the dark sky limits in the near infrared. The estimated evolving galaxy contributions are still consistent with the dark sky limits and residuals.

Finally, we can note the implications of the results presented here for the integrated energy density of the CIB, an overall constraint on the integrated cosmic luminosity. Denoting the integrated infrared background energy density in units of the critical closure energy density by Ω_{IR} and the corresponding quantity for the CMB by Ω_{CMB}, one finds that (for $T_{CMB} = 2.726$ K, Mather *et al.* 1994) $\Omega_{IR}/\Omega_{CMB} = 1 \times 10^{-3} \times I_{IR}$ (nW m^{-2} sr^{-1}), where I_{IR} is the sky brightness integrated over the infrared spectrum. Taking the range of integration for the infrared to be 1–400 μm, the dark sky upper limits of Table 2 give $\Omega_{IR}/\Omega_{CMB} < 4.3$, not a very restrictive limit. If the maximum values of the DIRBE residuals shown in Table 2 are taken as upper limits, one finds $\Omega_{IR}/\Omega_{CMB} < 0.5$. To provide substantially more stringent limits on the integrated infrared background, the peak in the limits over the thermal infrared range ($\sim$ 5–60 μm), which is presumably largely due to the IPD signal, must be substantially reduced. It is hoped that further advances in modeling will provide modest steps in this direction, but ultimately measurements from well beyond the bulk of the IPD cloud would be most effective.

6. CONCLUSIONS

Measurement of the extragalactic infrared background radiation will advance our understanding of cosmic evolution since decoupling. The high quality and extensive new measurements of absolute infrared sky brightness obtained with the DIRBE and FIRAS instruments on the *COBE* mission have already set significantly improved limits upon this elusive background. When foreground models are completed and uncertainties carefully assessed, the *COBE* data will certainly provide cosmologically valuable constraints upon, or perhaps even direct detection of, the extragalactic infrared background to the limits imposed by our bright astrophysical environment.

ACKNOWLEDGMENTS

The author gratefully acknowledges the contributions to this report by the many participants in the *COBE* Project, to his colleagues on the *COBE* Science Working Group, and especially the many scientists, analysts, and programmers engaged in the DIRBE investigation. Those particularly involved in the work reported here include E. Dwek, T. Kelsall, H. Moseley, R. Silverberg, E. Wright, R. Arendt, G. Berriman, B. Franz, H. Freudenreich, C. Lisse, M. Mitra, N. Odegard, W. Reach, J. A. Skard, T. Sodroski, S. Stemwedel, G. Toller, and J. Weiland. The National Aeronautics and Space Administration/Goddard Space Flight Center (NASA/GSFC) is responsible for the design, development, and operation of the *COBE*. Scientific guidance is provided by the *COBE* Science Working Group. GSFC is also responsible for the development of the analysis software and for the production of the mission data sets.

REFERENCES

Arendt, R. G. *et al.*, 1995, "Subtraction of Galactic Stellar Emission from *COBE*/DIRBE Observations", this Workshop

Beichman, C. A. & Helou, G., 1991, ApJ, 370, L1

Bennett, C. L., *et al.*, 1994, ApJ, 434, 587

Berriman, G. B., *et al.*, 1994, ApJ, 431, L63

Biller, S. D., *et al.*, 1995, ApJ, 445, 227

Boggess, N. *et al.*, 1992, ApJ, 397, 420

Bond, J. R., Carr, B. J., & Hogan, C. J., 1986, ApJ, 306, 428

Bond, J. R., Carr, B. J., & Hogan, C. J., 1991, ApJ, 367, 420

Cowie, L. L., *et al.*, 1990, ApJ, 360, L1

De Jager, O. C., Stecker, F. W. & Salamon, M. H., 1994, Nature, 369, 294

Dwek, E. & Slavin, J., 1994, ApJ, 436, 696

Evans, D. C., 1983, SPIE Proc., 384, 82

Franceschini, A. *et al.*, 1991, A&A SupplSer, 89, 285

Franceschini, A. *et al.*, 1994, ApJ, 427, 140

Franz, B. A. *et al.*, 1995, "Inversion of the DIRBE Photometry to Derive the Three Dimensional Distribution of the Interplanetary Dust", this Workshop

Gregorich, D. T. *et al.*, 1995, AJ, 110, 259

Hacking, P. B. & Soifer, B. T., 1991, ApJ, 367, L49

Harwit, M., 1970, Rivista del Nuovo Cimento, II, 253

Hauser, M. G., 1993, in *Back to the Galaxy*, AIP Conf. Proc. 278, ed. S. S. Holt & F. Verter (New York: AIP), 201

Hauser, M. G., 1995a, in *Extragalactic Background Radiation*, Space Telescope Sci. Inst. Symp. Ser. 7, ed. D. Calzetti, M. Livio, & P. Madau, (Cambridge: Cambridge Univ. Press), 135

Hauser, M. G., 1995b, to be published in Proc. IAU Symposium 168, *Examining the Big Bang and Diffuse Background Radiations*, ed. M. Kafatos & Y. Kondo, (Dordrecht: Kluwer)

Jahoda, K., Lockman, F. J., & McCammon, D., 1990, ApJ, 354, 184
Kaufman, M., 1976, ApSpSci, 40, 369
Kawada, M., *et al.*, 1994, ApJ, 425, L89
Kashlinsky, A. 1995, this Workshop
Kelsall, T., *et al.*, 1993, Proc. SPIE Conf. 2019, *Infrared Spaceborne Remote Sensing*, ed. M. S. Scholl (Bellingham: SPIE), 190
Leisawitz, D., 1995, this Workshop
Lockman, F. J., Jahoda, K., & McCammon, D., 1986, ApJ, 302, 432
Low, F. J. & Tucker, W. H., 1968, PRL, 22, 1538
Magner, T. J., 1987, OptEng, 26, 264
Mather, J.C., *et al.*, 1994, ApJ, 420, 439
Matsumoto, T., 1990, in *The Galactic and Extragalactic Background Radiation*, IAU Symposium 139, ed. S. Bowyer & C. Leinert (Dordrecht: Kluwer), 317
Matsumoto, T., Akiba, M., & Murakami, H., 1988, ApJ, 332, 575
Noda, M. *et al.*, 1992, ApJ, 391, 456
Partridge, R. B. & Peebles, P. J. E., 1967, ApJ, 148, 377
Peebles, P. J. E., 1969, Phil. Trans. Royal Soc. London, A, 264, 279
Reach, W. T. *et al.*, 1995a, ApJ, to be publ.
Reach, W. T. *et al.*, 1995b, "DIRBE Observations of the Zodiacal Light", this Workshop
Ressell, M. T. & Turner, M. S., 1990, Comments on Astrophysics, 14, 323
Shafer, R. A. *et al.*, 1995, ApJ, in prep.
Silverberg, R. F. *et al.*, 1993, Proc. SPIE Conf. 2019, *Infrared Spaceborne Remote Sensing*, ed. M. S. Scholl (Bellingham: SPIE), 180
Stecker, F. W., 1995, this Workshop
Veeraraghavan, S. 1995, this Workshop
Weiland, J. L. *et al.*, 1995, "The Interstellar Dust Contribution to the Diffuse Infrared Sky Brightness", this Workshop
Wright, E.L. *et al.*, 1991, ApJ, 381, 200

DIFFUSE INFRARED EMISSION FROM THE SOLAR SYSTEM

THE ORIGIN AND DYNAMICS OF THE INTERPLANETARY DUST CLOUD

S. F. Dermott, S. Jayaraman, Y. L. Xu, K. Grogan and B. A. S. Gustafson
Dept. of Astronomy University of Florida, Gainesville, FL 32611-2055
Email ID
dermott@astro.ufl.edu

ABSTRACT

Obtaining a model of the zodiacal cloud to predict the flux in a given waveband, in a given direction, at a given time of the year, to one per cent of the peak brightness, or better, will require a detailed understanding of the dynamics of the particles that originate from each of the various sources. Each component of the cloud (asteroidal, cometary and interstellar) gives rise to characteristic features and by resolving the cloud into its various components and determining the amplitudes of the signals associated with each, we are able to determine the fractional composition of the cloud.

We can expect to have a good understanding of the dynamics of the asteroidal particles that give rise to the solar system dust bands and get trapped in the Earth's circumsolar resonant ring. Our interpretations of the dust band observations and of the cloud's trailing/leading asymmetry indicate that about one third of the particles in the cloud are asteroidal, implying that about two thirds of the particles originate from comets. Some progress has been made with understanding the dynamics of cometary particles, but this remains a challenging problem. In this paper, to illustrate the quality of the DIRBE data and to show how the observed structure of the cloud is amenable to a dynamical interpretation, we analyse the asymmetry of the cloud observed in the DIRBE 12 μm waveband.

1. INTRODUCTION

Recent observations by the IRAS, COBE and ULYSSES spacecraft have shown that the zodiacal cloud probably consists of four (or five) distinct and significant components. These are:

1. The asteroidal dust bands discovered by IRAS (Low et al., 1984) that appear to be associated with some of the prominent asteroidal families (Dermott et al., 1984, 1994a).
2. The Earth's circumsolar resonant ring (Dermott et al., 1994b).
3. Dust associated with background asteroids (Dermott et al., 1994a).
4. Dust associated with cometary debris (Sykes & Walker, 1992; Liou & Dermott, 1995a).
5. And, possibly, interstellar dust (Grun et al., 1994).

Given the existence of these distinct components, it is not meaningful to talk about the properties of the zodiacal cloud as whole. Cometary and asteroidal particles have, respectively, high and low orbital eccentricities and the orbital dynamics of these two particle populations and the types of cloud feature that they are associated with are quite different. Similarly, given that the particles in the cloud have a range of sizes

© 1996 American Institute of Physics

and compositions, it is not meaningful to talk about particle properties without distinguishing between particles from the various sources and without allowing for the effects of particle size. The challenge that we now face is that of untangling or resolving the zodiacal cloud into its various components.

We consider that the key to understanding the structure of the zodiacal cloud lies in the interpretation of the dust particle dynamics. Our approach is to start with a postulated source of particles, either asteroidal, cometary or interstellar, and then model the features of the zodiacal cloud with which they are associated, taking into account:

1. The orbital evolution of the particles due to Poynting-Robertson light drag, using equations of motion that include, solar wind drag, light pressure and gravitational perturbations due to the planets (Dermott et al., 1992).
2. Resonant trapping (Dermott et al., 1994b).
3. The size-frequency distribution of the particles.
4. The thermal and optical properties of the particles and their variation with particle size (Gustafson, 1994).

We have made a start on this program using the IRAS data, but many of the important questions can only be tackled using the DIRBE data and much work remains to be done.

2. ASTEROIDAL DUST BANDS

In our previous work on the structure of the dust bands, we have shown that:

1. The central dust bands are associated with the Themis and Koronis asteroid families. However, the ten-degree bands cannot be accounted for by the Eos family alone unless the dispersion of the inclinations of the dust particle orbits is very large, $\sim 2°$. This is not impossible, but it may be unlikely (Dermott et al., 1993, 1994a).
2. The observed forced inclinations of the bands place their source firmly in the asteroid belt, but the observed variation of the shape of the bands with elongation angle shows that they extend to the inner solar system (Dermott et al., 1992).
3. Only asteroidal particles can produce a dust band because the orbital inclinations of the particles in the observed bands must have very small dispersions, $\leq 1°$. Dispersions as small as this are characteristic of those produced by asteroidal breakup. Particles in cometary-type orbits have high orbital eccentricities. Jovian gravitational perturbations produce large variations in these eccentricities and these variations are coupled to those in the inclinations. Even if a group of cometary-type orbits initially have identical inclinations, Jovian perturbations would disperse those inclinations over a wide-range on a timescale of a few precession periods showing that it is impossible for a comet to produce a well-defined dust band (Liou & Dermott, 1995a).
4. Although the observed amplitudes of the dust band signals are small, $\sim 1\%$ of the peak background, these signals are only the tip of the iceberg and the particles associated with the dust bands actually constitute $\sim 10\%$ of the whole zodiacal cloud (Dermott et al., 1994a).

The IRAS data have three limitations: (i) lack of reliable calibration, particularly in the 60 and 100 μm wavebands, (ii) limited coverage in elongation angle for most ecliptic longitudes of the Earth, and (iii) a limited number of wavebands. The DIRBE data compensate for these shortcomings and allow us to tackle three important questions.

1. The wide range of elongation angles used by DIRBE at all times of the year will enable us to model the dustbands from elongation angles of 64° to 124° and to determine how the effective area of the dust associated with a given dust band varies with heliocentric distance. Does it increase due to particle break-up, or does it decrease due to particle loss by radiation blow-out? Preliminary work with the IRAS data indicates that the area decreases with decreasing heliocentric distance.
2. Observations of the amplitudes of the dust bands at various wavelengths, but particularly in the scattering and IR regimes, will allow us to determine the albedos and colors of the particles in the dust bands.
3. To-date, all of our dust band models have been for single-size particles (Dermott et al., 1994a). The extensive and well-calibrated DIRBE data, and detailed modeling using a power-law particle size distribution, will allow us to produce sophisticated models of the dust bands and to determine the structure and extent of that fraction of the cloud that is asteroidal. Preliminary work with the IRAS data indicates that about one third of the particles in the cloud are asteroidal. This estimate is based on the observed amplitudes of the dust bands that indicate that about 10% of the particles in the cloud are associated with asteroidal families and from our estimate that the total amount of asteroidal dust is 3.4 times greater than that associated with the families alone (Dermott et al., 1994a).

3. EARTH'S CIRCUMSOLAR RESONANT RING

The Earth's resonant circumsolar ring may prove to be the most important feature of the zodiacal cloud because it will undoubtedly play a key role in resolving the cloud into its various components. The ring originates from asteroidal particles that spiral in towards the sun due to Poynting-Robertson light drag and get captured into outer mean motion resonances with the Earth (Dermott et al., 1994b). We are confident that the particles are asteroidal and not cometary because only particles with low orbital eccentricities have significant probabilities of capture into resonance. The particles are captured into a large range of overlapping resonant orbits that form a pattern that is near stationary in a frame centered on the Sun and corotating with the Earth's mean motion. The particles trapped in the resonances form a toroidal circumsolar ring that contains the Earth embedded in a cavity and a cloud or concentration of particles that trails the Earth in its orbit. The chief observational consequence of this trailing cloud is the observed trailing/leading asymmetry of the zodiacal cloud. At all times of the year, infrared observations of the zodiacal cloud in the trailing direction (opposite to the Earth's orbital motion) tend to be significantly greater than those in the leading direction. The IRAS data showed evidence of an asymmetry $\sim 3-4\%$ – in the 12, 25 and 60 μm wavebands (Dermott et al., 1994). Recently, the DIRBE data have provided much more accurate and extensive measurements of the infrared flux in 10 wavebands from 1.2 to 240 μm. Reach et al. (1995) subtracted empirical models of the background zodiacal cloud and the dust bands to obtain an image of the residual flux at 12 μm that is a striking match of the structure of the trailing dust cloud shown in our model (see Jayaraman & Dermott, this meeting).

Analysis of the structure of the ring will enable us to determine accurately the fraction of asteroidal particles in the zodiacal cloud. The number of particles in the ring and the leading/trailing asymmetry are directly dependent upon the number of asteroidal particles reaching the inner solar system, which in turn is dictated by the asteroidal dust production rate. Using models of the ring and an asteroidal background cloud made up of single-size particles, we have estimated that one third of the zodiacal cloud is asteroidal in origin (this agrees with our estimate based on the amplitudes of the dust bands). In order to improve on this estimate, it is essential to extend the models to include a realistic distribution of particle sizes. We have calculated the capture probabilities for asteroidal particles of density 2.7 g/cm^3 ranging in diameter from 9 to 45 μm and found that the total capture probability decreases steeply with decreasing size providing a lower cut-off for particle size in the ring of about 5 μm. There is also an effective cut-off at larger sizes despite the higher trapping probability, because although the larger particles have low drag rates and are easily captured into resonance, their resonant orbits are near-symmetric with respect to the mean Sun-Earth line and they do not contribute to the observed trailing/leading asymmetry of the cloud. Thus, the asymmetry is caused by a narrow range of particle sizes from 5–30 μm in diameter. We have constructed single-size particle rings for 9, 12, 15, 19, 25 and 36 μm diameter particles and we intend to combine these according to a power law size-frequency distribution, allowing for the variation of capture probability with size, to construct a realistic model of the ring that can then be compared with the observations of the ring in the various DIRBE wavebands.

4. MODELING THE ZODIACAL CLOUD

The dust bands and the resonant ring should be the easiest components of the zodiacal cloud to model. The background asteroidal component of the cloud will be modeled using information derived from the observed structures and amplitudes of those two components.

1. From the observed variations of the shapes and amplitudes of the dust bands with elongation angle, we should be able to determine the variation of the number density or, more precisely, the effective area per unit volume of the particles in the dust bands with heliocentric distance. We will assume that this variation also applies to the background asteroidal dust.
2. Observations of the mean amplitude of the trailing/leading asymmetry due to the resonant ring will be used to determine the absolute number density of asteroidal particles in the vicinity of the Earth.

Because the orbital elements of the background asteroidal particles and their variation with heliocentric distance are known (at least for a significant range of particle diameters), the above items are all that is needed (apart, perhaps, from information on the optical properties of the particles) to produce a model of the background asteroidal dust. Models of these three components (dust bands, resonant ring and background asteroidal dust) will be subtracted from the DIRBE data to reveal the structure of the remnant background cloud. This component we will ascribe, by default, to comets.

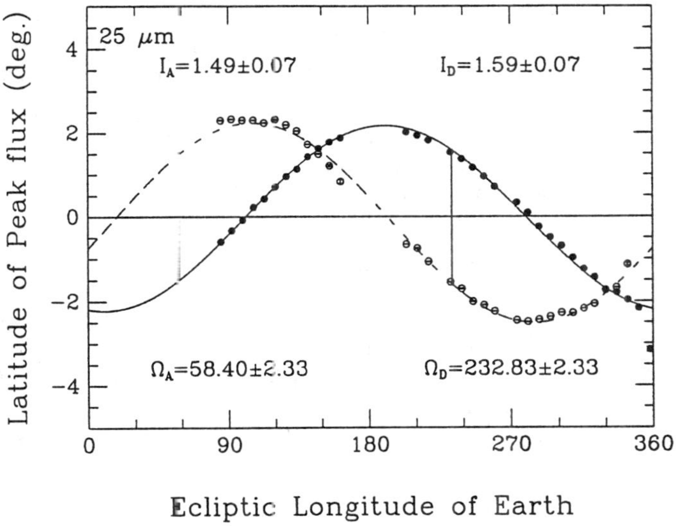
25 μm
I_A=1.49±0.07
I_D=1.59±0.07
Ω_A=58.40±2.33
Ω_D=232.83±2.33
Latitude of Peak flux (deg.)
Ecliptic Longitude of Earth

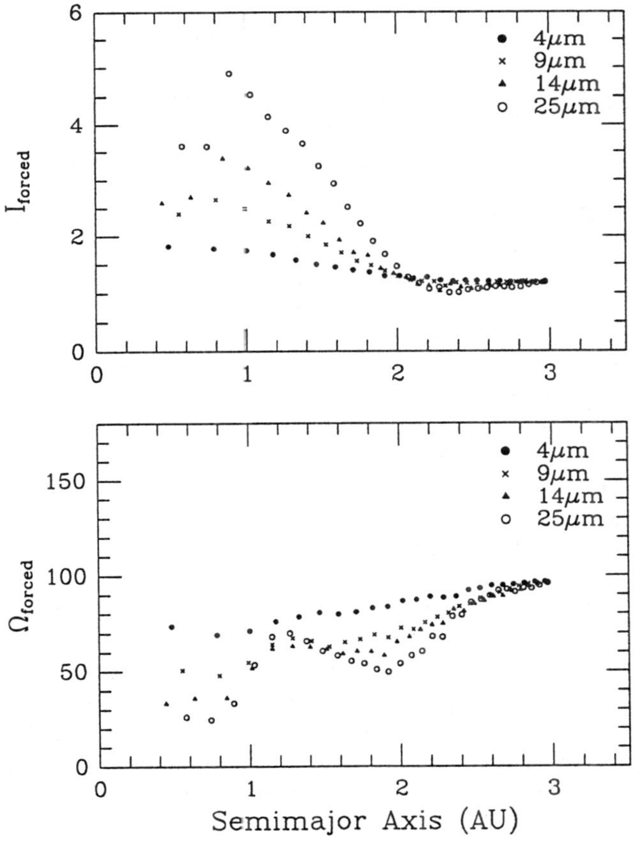
4μm
9μm
14μm
25μm
I_{forced}
Ω_{forced}
Semimajor Axis (AU)

Figure 1: (a) Variation of the ecliptic latitude of the peak background zodiacal emission with the ecliptic longitude of the Earth observed by DIRBE in the 25 μm waveband at an elongation angle of 90° in the leading (open circles) and trailing (full circles) directions. The numbers and vertical lines refer to the inclinations and the longitudes of the ascending and descending nodes of the cloud with respect to the ecliptic. The points on the curves have error bars that are too small to be seen in this figure. (b) Variation of the forced inclination and the longitude of the forced ascending node with heliocentric distance for asteroidal silicate particles of density $2.7g/cm^3$ and diameters 4, 9, 14 and 25 μm.

This does not imply that there is no direct evidence for a cometary component. It appears, for example, that a cometary component is needed to account for the large polar scale height of the cloud observed at $1AU$ (Liou & Dermott, 1995a). This large scale height cannot be accounted for by the asteroidal component alone. We are making some progress with modeling the dynamical structure of a cloud of cometary particles (Liou & Dermott, 1995a), but, at this stage, it may be simpler to model this part of the cloud empirically, unless the cloud also contains signatures of an interstellar component in which case a model of that component would also need to be subtracted.

That the DIRBE data is amenable to a dynamically driven modeling approach can be illustrated by an analysis of the variation of the peak brightness of the cloud observed at an elongation angle of 90° in the leading and trailing directions.

In Fig. 1(a) we show the observed variation of the ecliptic latitude of the peak background zodiacal emission with the ecliptic longitude of the Earth observed by DIRBE in the 25 μm waveband at an elongation angle of 90° in the leading and trailing directions. From these observations, we determine the inclination and orientation of the cloud with respect to the ecliptic. We will interpret these in terms of the forced inclinations and nodes of asteroidal dust particles (Fig. 1(b)). If the inclination and orientation of the plane of symmetry of the cloud are due to the forced inclinations of asteroidal dust particles, and this would appear to be the case (Dermott et al., 1992), then we should also expect these particles to have forced eccentricities, in which case the sun will not be at the center of symmetry of the cloud (Dermott et al., 1985). In Fig. 2 we show the variation of the forced eccentricity and pericenter of 9 μm diameter asteroidal particles with heliocentric distance. The forced eccentricity in the vicinity of the Earth is small, ~ 0.01 and smaller than the Earth's own orbital eccentricity ($= 0.0167$). The longitude of the forced pericenter $\approx 130°$, although there will, of course, be some variation with particle size and density.

In Fig. 3(a), we show the observed variation of the peak flux of the smooth background cloud (after filtering out the dust bands) with the ecliptic longitude of the Earth as seen by DIRBE in the 12 μm waveband. The observations in the trailing (filled circles) and leading (open circles) directions have been normalized to a solar elongation of 90°. The trailing (solid line) and leading (dashed line) data have been fitted with combinations of a sine and a double sine curve. In Fig. 3(b), we show the variation of the flux due to the resonant ring in the trailing (filled circles, upper curve) and leading (open circles, lower curve) directions obtained from a model of the ring, normalized such that the mean difference in the trailing and leading signals is the same as that observed. Fig. 3(c) shows the variation of the peak zodiacal flux due to the background cloud alone

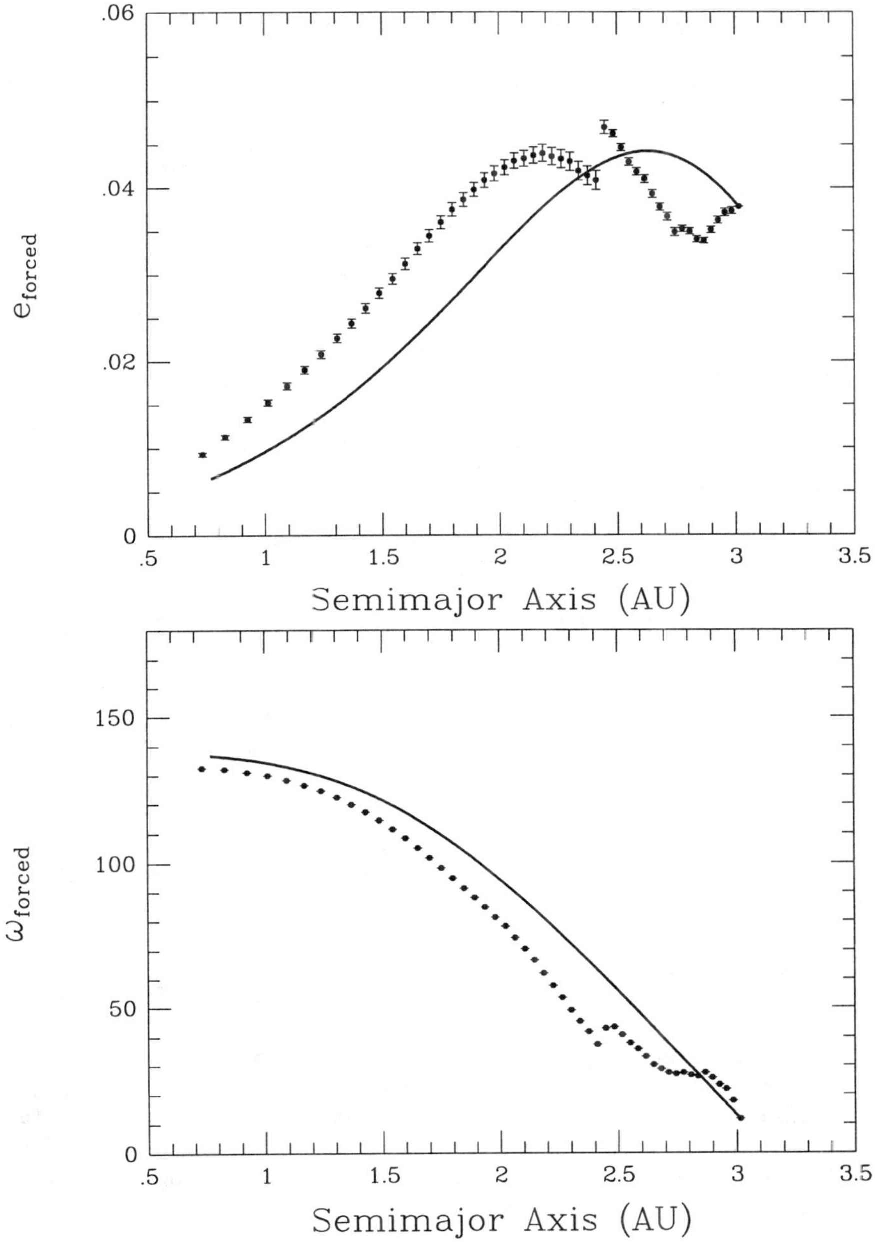

Figure 2: Variation of the forced eccentricity and the longitude of the forced pericenter with heliocentric distance of asteroidal silicate particles of density 2.7 g/cm^3 and diameter 9 μm. The points represent the results of pure numerical integrations that take account of the gravitational perturbations due to the four major planets but exclude the smaller terrestrial planets. The solid curve is the result of a similar calculation that is partially analytical (Dermott et al., 1992; Liou 1993).

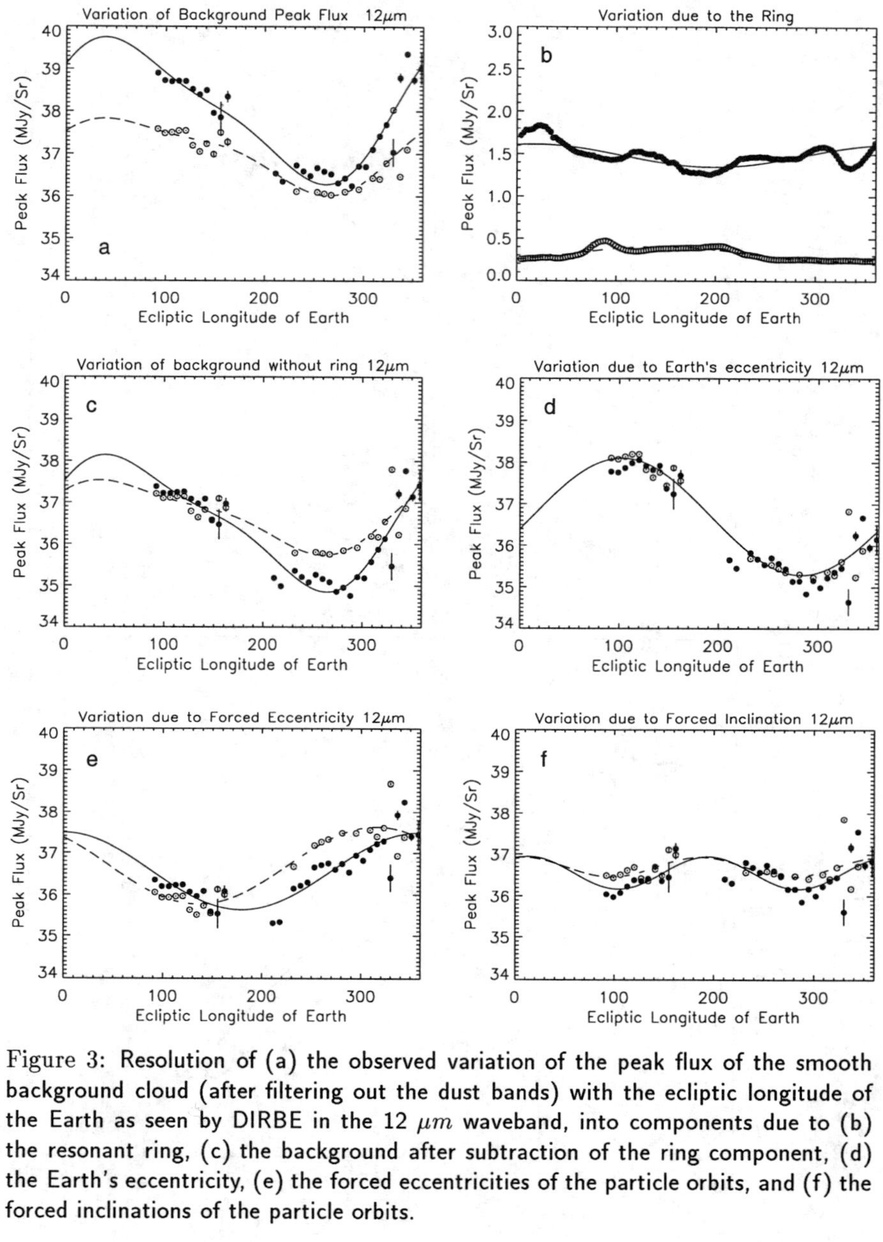

Figure 3: Resolution of (a) the observed variation of the peak flux of the smooth background cloud (after filtering out the dust bands) with the ecliptic longitude of the Earth as seen by DIRBE in the 12 μm waveband, into components due to (b) the resonant ring, (c) the background after subtraction of the ring component, (d) the Earth's eccentricity, (e) the forced eccentricities of the particle orbits, and (f) the forced inclinations of the particle orbits.

after subtracting the average trailing/leading asymmetry due to the resonant ring. The subsequent figures resolve this background flux variation into three different components that we have identified on the basis of dynamical arguments. These three components are due to (1) the eccentricity of the Earth's orbit (Fig. 3(d)), (2) the forced eccentricity of the zodiacal cloud (Fig. 3(e)), and (3) the forced inclination of the zodiacal cloud Fig. 3(f). In Fig. 3(e), the lower of the two points where the trailing and leading curves cross occurs at the observed longitude of the forced pericenter and this is in good agreement with the expected value shown in Fig. 2.

5. Particle Properties

All of our models of the various components of the zodiacal cloud will have a range of particle sizes that we will represent by a simple power law. The particles may also have different compositions, structures and porosities. In order to determine the particle properties (size, albedo, color) from the amplitudes of the signals in the various wavebands, we must use a physical model of the particles that describes their interaction with light. Many radiative transfer and scattering calculations are still based on Mie-theory. But in most practical cases, the scattering particles do not have high spherical symmetry and are not homogeneous. Chondritic particles often appear to be aggregates of smaller grains (Brownlee, 1985).

In addition to the ground-based observations of the polarization of the zodiacal light (Levasseur-Regourd and Dumont, 1980; Leinert et al., 1981), the diffuse sky was recently surveyed by DIRBE for the overall polarization characteristics (Berriman et al., 1995). To infer information about the dust particles from these observations, optical cross-sections are required before models can be computed for specific grain candidates. Several approaches have been followed to obtain absorption and scattering cross-sections for finite non-spherical particles. The T-matrix technique involves an exact expansion of the exterior radiation field of the particles in terms of spherical harmonics (Waterman, 1971; Mishchenko, 1991). The discrete dipole approximation (Purcell & Pennypacker, 1973) is also used to calculate the optical properties of particles of arbitrary shapes. Recently a comprehensive solution to the multisphere-scattering problem has been presented by Xu (1995). This multisphere-scattering theory provides a new approach to calculating the optical properties of aggregates of small grains. The agreement of the theoretical predictions with the experimental data is very good. Initially, though, we will still use Mie theory to calculate the particle properties, but we anticipate that theoretical and experimental work at the University of Florida will soon allow the use of more sophisticated and realistic dust models to interpret the data.

6. Interstellar Dust

Impact data from the ULYSSES dust detector at $5AU$ from the sun have been interpreted as a flux of 1.5 10^{-2} $m^{-2}s^{-1}$ submicron sized interstellar dust particles arriving from ecliptic longitude 252° and latitude 2.5° (Grun et al., 1994). This corresponds to a particle number density of 5.8 10^{-15} cm^{-3}, or equivalently, in terms of surface area per unit volume, 2.9 10^{-23} cm^2/cm^3. The contribution of any large particles to the IR emission is easiest to model, because gravity is the dominant force, and interaction with the solar wind, Lorentz forces, etc., can be neglected. In this case, the particles are focussed in the downstream direction, ecliptic longitude 72°, and the problem essentially reduces to that of Rutherford scattering (Fig. 4). Using a SIMUL type approach

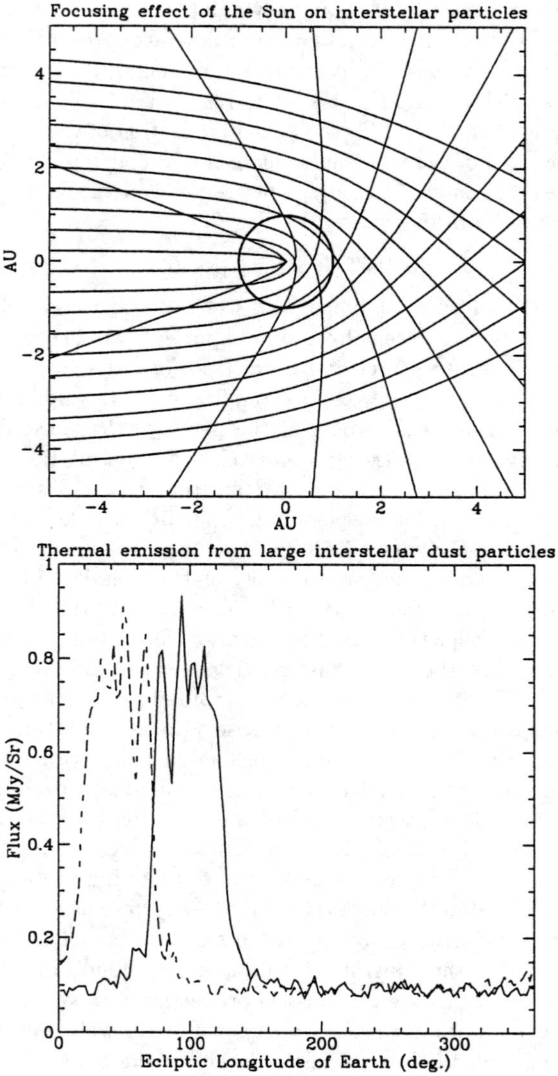

Figure 4: The predicted variation with ecliptic longitude of the Earth of the flux in the 25 μm waveband observed from a cloud of 'large' interstellar particles (incoming direction ecliptic longitude 252°, latitude 2.5°) observed at an elongation of 115° in the trailing (solid curve) and leading (dashed curve) directions. The surface area per unit volume of large particles at infinity in this calculation is set to $8.8\,10^{-24}\,cm^2/cm^3$ in order to produce an observable peak flux of around 1 MJy/sr. This area per unit volume is approximately 30% of the value for the smaller particles detected by ULYSSES at 5 AU.

to model the resulting particle cloud (Dermott et al., 1989), we are able to predict the possible signal present in the observations in both the trailing and leading directions (Fig. 4). Assuming that a flux of about 1 MJy/Sr is needed for a strong detection, we have estimated that the required effective area per unit volume of these "large" particles at infinity is 8.8 10^{-24} cm^2/cm^3, or about 30% that of the small particles detected by ULYSSES.

Although ULYSSES detected small particles for which gravity is not the dominant force, preliminary numerical investigations of the motion of these particles show that these small particles are also focussed to produce regions of enhanced number density, albeit in different locations to those in the purely gravitational treatment. These locations are partially determined by the interplanetary magnetic field and will vary with time according to the solar cycle. In addition, small particles may also radiate strongly in the IR: a 0.6 μm radius silicate particle at 1 AU will radiate twice as much thermal energy per unit surface area in the 10 μm waveband than a 10 μm or bigger sized particle, due to a combination of a $60K$ temperature excess and an increased emissivity (there are strong silicate emission features at wavelengths of 10 μm and 20 μm). These considerations suggest that we have a good chance of detecting interstellar dust particles in the DIRBE data. After removing the effects of the dust bands and the Earth's resonant circumsolar ring, we will look for characteristic signatures of interstellar dust in the residual DIRBE data in the 12 and 25 μm wavebands. An example of a characteristic signature is shown in Fig. 4 where we have plotted the expected flux from the putative interstellar component as seen in the 25 μm waveband in the trailing and leading directions at an elongation angle of 115°. The features are very broad and by suitable coadding of data we should be able to detect a signal considerably less than 1 MJy/Sr. A knowledge of the number density and size of interstellar dust particles is needed in order, for example, to determine whether cometary dust from the Kuiper belt can migrate to the inner solar system without being destroyed by collisions (Liou, Zook & Dermott, 1995b).

ACKNOWLEDGEMENTS

This research was supported by NASA grants NAGW 1923 and NAG5–2654.

REFERENCES

Berriman, G. B., Boggess, N. W., Hauser, M. G., Kelsall, T., Lisse, C. M., Mosely, S. H., Reach, W. T. and Silverberg, R. F. 1995. Ap. J. Lett. (in press).

Brownlee, D. E. 1978. In 'Cosmic Dust', (Ed. J. A. M. McDonnell), Wiley, Chichester, 275–336.

Dermott, S. F., Nicholson, P. D., Burns, J. A. and Houck, J. R. 1984. Nature, 312, 505-509.

Dermott, S. F., Nicholson, P. D., Burns, J. A. and Houck, J. R. 1985. In 'Properties and Interaction of Interplanetary Dust', (eds. R. H. Giese and P. Lamy) D. Reidel Pub. Co., 395-409.

Dermott, S. F., Nicholson, P. D., Kim, Y., Wolven, B. and Tedesco, E. 1988. In 'Comets to Cosmology' (ed. A. Lawrence), Berlin, Springer-Verlag, 3-18.

Dermott, S. F. and Nicholson, P. D. 1989. In 'Highlights of Astronomy', 8, 259-266.

Dermott, S. F., Gomes, R. S., Durda, D. D., Gustafson, B. A. S., Jayaraman, S. Xu, Y-L. and Nicholson, P. D. 1992. In 'Chaos, Resonance, and Collective Dynamical Phenomena in the Solar System', (ed. S. Ferraz-Mello), Kluwer Academic Publishers, Dordrecht, 333–347.

Dermott, S. F., Durda, D. D., Gomes, R. S., Gustafson, B. A. S., Jayaraman, S., Liou, J. C., Nicholson, P. D. and Xu, Y-L. 1993. In 'Meteoroids and Their Parent Bodies', (eds. J. Stohl and I. P. Williams), Astronomical Inst., Slovak. Acad. Sci., Bratislava, 357–366.

Dermott, S. F., Durda, D. D., Gustafson, B. A. S., Jayaraman, S., Liou, J. C., and Xu, Y-L. 1994a. In 'Asteroids, Comets, Meteors, 1993' (eds. A. Milani, M. Martini and A. Cellino), Kluwer, Dordrecht, 127–142.

Dermott, S. F., Jayaraman, S., Xu, Y-L. Gustafson, B. A. S. and Liou, J.C. 1994b. Nature, 369, 719–723.

Grun, E. et al., Astr. Ap., 286, 915 (1994).

Gustafson, B. A. S. 1994. Ann. Rev. Earth Planet Sci., 22, 553–595.

Leinert, C., Richter, I., Pitz, E. and Planck, B. 1981. Astron. Astrophys., 103, 177–188.

Levasseur-Regourd, A. C. and Dumont, R. 1980. Astron. Astrophys., 84, 277–289.

Liou, J. C. 1993. PhD thesis, University of Florida.

Liou, J. C. and Dermott, S. F. 1995a. Planet. Space Sci. (in press).

Liou, J. C., Zook, H., and Dermott, S. F. 1995b. Proc. Lunar and Plan. Sci. Conf. (in press).

Mishchenko, M. I. 1991. J. Opt. Soc. Am. A, 8, 871–882.

Purcell, E. M. and Pennypacker, C. R. 1973. Ap. J. 186, 705–714.

Reach, W. T., Franz, B. A., Weiland, J. L., Hauser, M., Kelsall, T. N., Wright, E. L., Rawley, G., Stemwede, S. W. and Spiesman, W. J. 1995. Nature, 374, 521–523.

Sykes, M. and Walker, R. 1992. Icarus, 95, 180–210.

Waterman, P. C. 1971. Phys. Rev. D, 3, 825–839.

Xu, Y. L. 1995. Applied Optics, (in press).

DIRBE Observations of the Zodiacal Light

William T. Reach
Universities Space Research Association
Code 685, NASA/GSFC, Greenbelt, MD 20771

Bryan A. Franz
Applied Research Corporation
Code 685.3, NASA/GSFC, Greenbelt, MD 20771

Thomas Kelsall
Code 685, NASA/GSFC, Greenbelt, MD 20771

Janet L. Weiland
General Sciences Corporation
Code 685.3, NASA/GSFC, Greenbelt, MD 20771

Abstract. The Diffuse Infrared Background Experiment (DIRBE) aboard the *Cosmic Background Explorer* (*COBE*) [1] mapped the entire sky redundantly in 10 wavebands at 1.25, 2.2, 3.5, 4.9, 12, 25, 60, 100, 140, & 240 μm. The scattering or thermal emission from interplanetary dust contributes significantly to the sky brightness in all 10 wavebands, dominating most. The sky brightness is modulated in time due to the changing viewing aspect of the DIRBE line of sight through the interplanetary dust cloud. A three-dimensional semi-physical model for the distribution, emission, and scattering of interplanetary dust was optimized to match the time-dependence of the sky brightness as observed by DIRBE. The method and results of this fitting procedure are described, as are the difficulties and some future prospects for disentangling the zodiacal light from other contributions to the diffuse infrared sky brightness.

INTRODUCTION

The zodiacal light is the dominant foreground at infrared wavelengths, and the DIRBE search for the cosmic infrared background radiation (CIBR) begins with its removal from the observed brightness. The infrared spectrum of the North Galactic Pole (NGP), together with the estimated contributions from zodiacal light, starlight, and the interstellar medium, is shown as a Figure in the contribution by Hauser (1995). The estimated fraction of the total brightness due to zodiacal light varies from about 2/3 in the near-infrared (1.25–3.5 μm) to more than 90% in the mid-infrared (4.9–60 μm). Even in the far-infrared the zodiacal contribution cannot be neglected; based on the results described below, 25% of the NGP brightness at 240 μm is zodiacal light. Based on current infrared background residuals, the limits on the total energy density of the extragalactic

[1] The National Aeronautics and Space Administration/ Goddard Space Flight Center (NASA/GSFC) is responsible for the design, development, and operation of the Cosmic Background Explorer (*COBE*). Scientific guidance is provided by the *COBE* Science Working Group. GSFC is also responsible for the development of the analysis software and for the production of the mission data sets.

© 1996 American Institute of Physics

infrared radiation field is dominated by the relatively poor limits in the mid infrared (Hauser 1995). Thus any cosmological processes that produce energy that is redshifted to this spectral region are only poorly constrained at present by infrared observations. It has been argued that limits to the extragalactic near- and mid-infrared radiation field from high-energy γ rays are much more stringent (cf. Stecker 1995). In order to limit the CIBR by direct observations, the zodiacal light has to be modeled to unprecedented precision.

This paper describes the observations and modeling of the zodiacal light by the Diffuse Infrared Background Experiment (DIRBE) science team. The goal of this work is to provide a method for subtracting the zodiacal light from DIRBE observations at 1.25 μm through 240 μm, preserving any isotropic component of the sky brightness and achieving $\sim$ 1% accuracy at high latitude. To this end, we created a three-dimensional model for the interplanetary dust cloud that predicts scattered sunlight and thermal emission, optimized parameters in the cloud model to match the modulated component of the observed brightness, and subtracted the model from the data to examine residuals for temporal and angular variations.

OBSERVATIONS

DIRBE is an absolutely calibrated photometer, with an accurate zero-brightness offset measurement in its 0.7° beam (Boggess *et al.* 1992). The gain calibration is stabilized using an internal reference source and non-variable celestial sources on short and long timescales, respectively (DIRBE Explanatory Supplement 1995; Mitchell *et al.* 1995). Absolute gain calibration was determined using observations of bright celestial sources. The absolute calibrator is Sirius from 1.25 to 12 μm, NGC 7027 for 25μm, Uranus for 60 and 100 μm, and Jupiter for 140 and 240 μm. Ideally, the same calibrator would be used at all wavelengths, but compromises were forced by the fact that Sirius becomes too faint at wavelengths longer than 12 μm and Jupiter is so bright that the 60 and 100μm detectors become nonlinear. For these reasons we have avoided using DIRBE as a spectrometer where possible, and our model includes free parameters for the gain in each waveband.

TABLE 1. Properties of the Diffuse Infrared Background Experiment

Band	1	2	3	4	5	6	7	8	9	10
Wavelength (μm)	1.25	2.2	3.5	4.9	12	25	60	100	140	240
Gain Uncertainty (%)	4	4	4	3	12	15	9	14	11	12
Offset Uncertainty(nW/m^{2}sr)	0.05	0.03	0.02	0.01	0.02	0.01	2	1	8	3
Gain Stability (%)	0.5	0.5	0.5	0.5	0.75	0.75	2	3	5	5

The combination of the orbit of the *COBE* satellite about Earth, spin of the DIRBE field of view about the satellite axis, and the orbit of the Earth around the Sun lead to a helical scan pattern on the sky. In one week, half the sky is fully sampled. The weekly sky maps are the basis for this work, as a compromise between using the full time-ordered data (which are too numerous) and using longer averaging times (which smear temporal variations of the zodiacal light). In formulating our goodness-of-fit criterion, we weighted each observation using a combination of a 'random' and 'systematic' error estimate. The random error estimate is based on the observed variation of brightness with time in each pixel during the week. The systematic error is based on the estimated gain stability (listed in Table 1), determined from observations of a network of point sources throughout the mission.

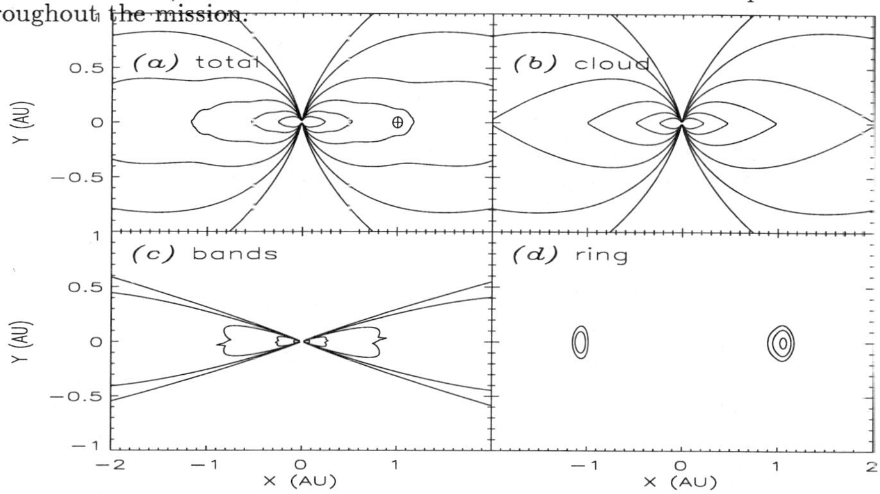

FIGURE 1. Density contours of our zodiacal cloud model. *(a)* total of all components; *(b)* 'smooth cloud'; *(c)* dust bands; *(d)* Earth's dust ring.

THREE-DIMENSIONAL MODEL FOR THE ZODIACAL CLOUD

The brightness of the zodiacal light is the integral along the line of sight of scattering and thermal contributions:

$$Z_\nu = \rho \int n(R, z, \theta)\{A_\nu F_\nu^\odot \Phi(\Theta) + (1 - A_\nu)E_\nu B_\nu(T)\}\,\mathrm{d}s, \qquad (1)$$

where ρ is the volumetric extinction cross-section at 1 AU, A_ν is the albedo at frequency ν, $F_\nu^\odot$ is the solar flux, $\Phi(\Theta)$ is the phase function at scattering angle Θ, E_ν is the emissivity at frequency ν, and T is the grain temperature. The phase function is taken from previously published fits to the visible zodiacal light (Hong 1988), which are consistent with determinations using DIRBE data (Berriman *et al.* 1995). We assume the temperature varies with distance from the Sun as $T(R) = T_0 R^{-\delta}$. Throughout this paper, R is the distance from the Sun, z is the distance from the ecliptic plane.

The treatment of the albedos and emissivities is important for this work. We hope to achieve $\sim$ 1% accuracy in removal of the zodiacal light from 1.25–240μm, but the accuracy of our calibration from waveband to waveband in the mid-infrared is relatively poor. Furthermore, it is unlikely that the assumption of a single grain temperature will apply with high accuracy. Models for the emission from interplanetary dust predict that a small range of temperatures contributes to the mid-infrared emission, expanding to a wider range at shorter wavelengths (Reach 1988). The wavelength-dependence of the albedo is essentially unknown, and a constant albedo is unlikely. In order to allow for the imprecision of the waveband-to-waveband calibration as well as the restrictive nature of our spectral model, we allow the emissivities and albedos to be free parameters. There are three albedos (at 1.25, 2.2, & 3.5 μm) and 8 emissivities (at 3.5, 12, 25, 60, 100, 140, & 240 μm).

The cloud density is composed of 4 components,

$$n(R, z, \theta) = n_C + n_{B1} + n_{B2} + n_R,$$

each of which is shown in Figure 1 and discussed in turn below.

Smooth cloud model. The smooth cloud model is a cylindrical fan model, similar to that fitted to the *IRAS* data in producing the *IRAS Sky Survey Atlas* (Good *et al.* 1986; Good 1994):

$$n_C(r, z, \theta) = R_{xy}^{-\alpha} \exp(-\beta |z_c / R_{xy}|^{\gamma}), \tag{2}$$

where R_{xy} is the distance from the cloud center projected into the cloud midplane. The geometry is defined as follows. Let (x, y, z) be heliocentric Cartesian coordinates. The center is shifted from the Sun by (x_0, y_0, z_0), leading to the coordinates $(x', y', z') \equiv (x - x_0, y - y_0, z - z_0)$. The vertical distance from the midplane is then

$$z_c = x' \sin\Omega \sin i - y' \cos\Omega \sin i + z' \cos i, \tag{3}$$

where i and Ω are the inclination and ascending node of the dust midplane with respect to the ecliptic plane.

Dust bands. The dust bands were discovered in the *IRAS* data (Low *et al.* 1984), and are believed to be asteroidal collisional debris (Dermott *et al.* 1984; Sykes *et al.* 1989). The dust bands have been studied using the DIRBE data (Spiesman *et al.* 1995) confirming the observational results from *IRAS* data and extending them to the near-infrared. In particular, the parallactic and spectroscopic distances to the bands are less than the distance to the asteroid belt, so that the material producing them is likely to be debris spiralling into the Sun under Poynting-Robertson drag. For this work, we used a dust band density based on the migrating model (Reach 1992), but with a simpler analytic formulation that is easier to evaluate and optimize:

$$n_{Bi}(r, \zeta_{Bi}) = n_{1Bi} R^{-1} \exp\left[-\left(\zeta_{Bi}/\delta_{\zeta_{Bi}}\right)^6\right] \left[v_{Bi} + \left(\zeta_{Bi}/\delta_{\zeta_{Bi}}\right)^{p_{Bi}}\right], \tag{4}$$

where n_{1Bi} is the density at 1 AU, relative to that of the smooth cloud, of band i, $\zeta_{Bi} \equiv z_{Bi}/R$, z_{Bi} is the vertical distance from midplane of band i, and $\delta_{\zeta_{Bi}}, v_{Bi}, p_{Bi}$ are adjustable shape parameters. We included two bands, which appear at ecliptic latitudes around $\pm 10°$ and $\pm 1.4°$ in the sky maps and have been attributed (Sykes *et al.* 1989) to the Eos asteroid family ($\pm 10°$) and a blend of the Themis and Koronic families ($\pm 1.4°$).

Earth's dust ring. The Earth temporarily traps particles migrating asteroidal particles if they are in low-eccentricity orbits such as expected for asteroidal debris (Dermott *et al.* 1994; Dermott 1995). We have confirmed the existence of the Earth's dust ring by subtracting a smooth cloud model from two weekly sky maps, revealing the signature of the ring in remarkable agreement with the predictions (Reach *et al.* 1995). For this work, an empirical ring density model was developed to emulate the numerical simulations of Dermott *et al.* (1994).

It consists of a Gaussian (radial and vertical) toroid, with an enhancement in a 3-dimensional Gaussian blob trailing the Earth.

FITTING THE MODEL TO THE DATA

The infrared sky brightness is a combination of foregrounds due to interplanetary dust, starlight, and interstellar dust, as well as the CIBR. We will operationally define the CIBR to be an isotropic component of the sky brightness not associated with the abovementioned foregrounds. The problem at hand is to determine the contribution from interplanetary dust without modifying any of the other contributions. To this end, we use the one distinct signature of the zodiacal light: it is the only component that is not fixed on the celestial sphere. The brightness observed along a given celestial direction depends on the observer's line of sight with respect to the Sun on that day, as well as the position of the Earth in its orbit. The apparent time variation of the brightness toward two different directions is shown in Figure 2. Toward the ecliptic pole, the main causes for variation of the brightness are the motion of the Earth vertically, with respect to the inclined midplane of the dust distribution, as well as the motion of the Earth radially, due to its orbital eccentricity. At lower latitudes, the apparent temporal variation is primarily due to the changing solar elongation of the line of sight.

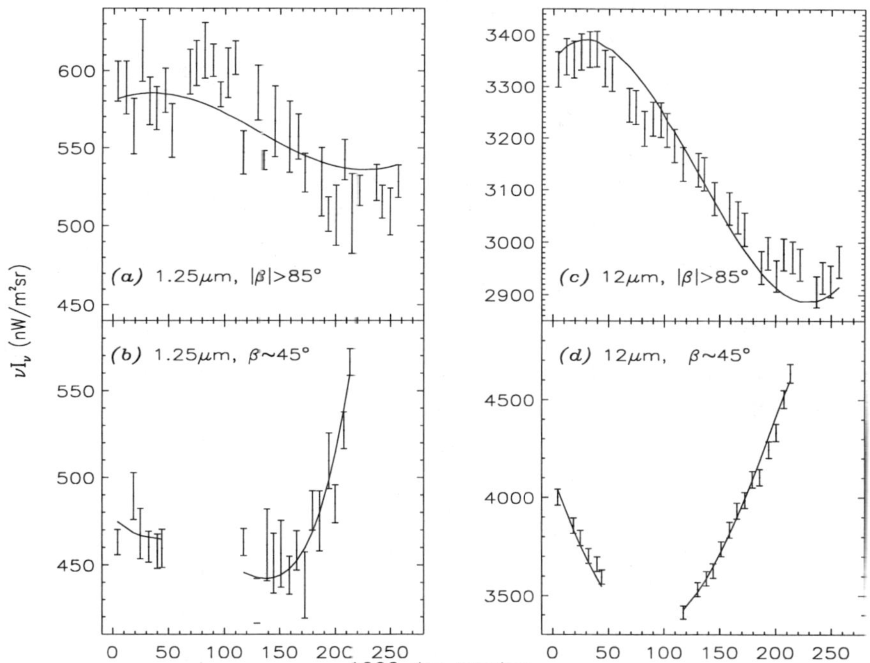

FIGURE 2. Observed brightness over the DIRBE mission for two directions in two wavebands. Error bars include random and systematic uncertainties.

We define the goodness of fit so that the model is optimized to match the modulation as follows. Let t be the observation time (an index over the weekly sky maps), p be the celestial position (an index over selected pixels), and b be the wavelength (an index over wavebands). The observed brightness is I_{tpb}, and the model evaluated for the same conditions is Z_{tpb}. Then we define the goodness of fit as

$$\chi^2[Z(t)] = \sum_{t,p,b} \left[(I_{tpb} - \langle I \rangle_{pb}) - (Z_{tpb} - \langle Z \rangle_{pb}) \right]^2 w_{tpb}. \quad (5)$$

In this equation, w_{tpb} is the weight given to an observation, which is based on a combination of random errors and the uncertainty in the temporal stability of the gain (Table 1). The mean brightness (over the mission) in each pixel is subtracted from both the model and the data. Therefore the sky brightnesses due to non-varying foregrounds and the CIBR are free parameters in this fitting method, with no assumptions about their morphology or spectrum.

The idea of using only the time variation to determine the zodiacal light was tested on simulated observations (including noise) of a model zodiacal cloud 'observed' by DIRBE. The above fitting procedure was able to recover the properties of the smooth cloud well. But for low-contrast features such as the dust bands and the Earth's dust ring, the time variation was inadequate to retrieve their shape. We therefore used a different method for part of the database. At 12 and 25 μm, and at absolute galactic latitude greater than 30°, the zodiacal light is the dominant sky brightness. If we assume all other contributions are isotropic, then we can define another goodness of fit,

$$\chi^2[Z + C] = \sum_{t,p,b} \left[(I_{tpb} - \langle I \rangle_{b}) - (Z_{tpb} - \langle Z \rangle_{b}) \right]^2 w_{tpb}, \quad (6)$$

where mean brightness *over the sky* was subtracted from each observation.

The total χ^2 for our fitting procedure is a combination of $\chi^2[Z + C]$ for the high-galactic latitude regions at 12 and 25 μm and $\chi^2[Z(t)]$ everywhere else. The parameters were determined using the Levenberg-Marquardt nonlinear least-squares optimization scheme (Bevington 1969). We selected 1920 lines of sight—one every $5° \times 5°$ patch of sky, and the total number of observations (including all wavebands and weekly samples) is 2×10^5. There are 30 model parameters as well as the free mean brightness in each pixel, which still leaves 1.8×10^5 degrees of freedom for the model fitting.

RESULTS

The fits to time strings are fairly 'good' in all wavebands, to within the accuracy of the DIRBE data as currently understood. The fit is shown together with the data in Figure 2 for selected individual pixels. The global goodness-of-fit χ^2/d.o.f. = 2.3 for the present model. While the fit appears excellent on a pixel-by-pixel basis, averages over patches large enough to reduce the random errors show some time variation in the residuals, which must be due to the zodiacal light. Maps of the total intensity, zodiacal light model, and residual intensity at 25 μm are shown in Figure 3. The residual maps at 1.25, 3.5, 12, & 60 μm are shown in Figure 2 of the contribution by Franz *et al.* (1995). Residual maps show that the bulk of angular variation that can be attributed to the zodiacal light is removed, but systematic residuals are clearly present

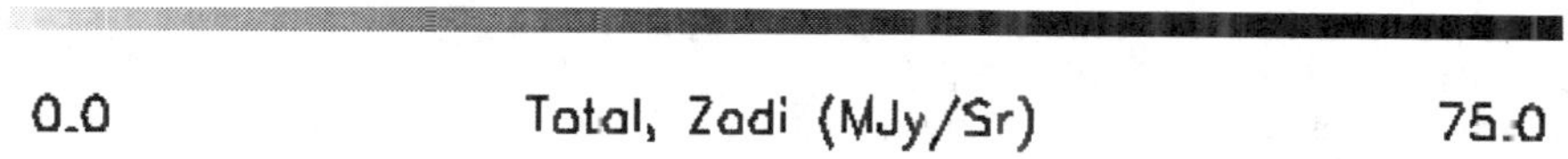

Figure 3. DIRBE annual average maps at 25μm.

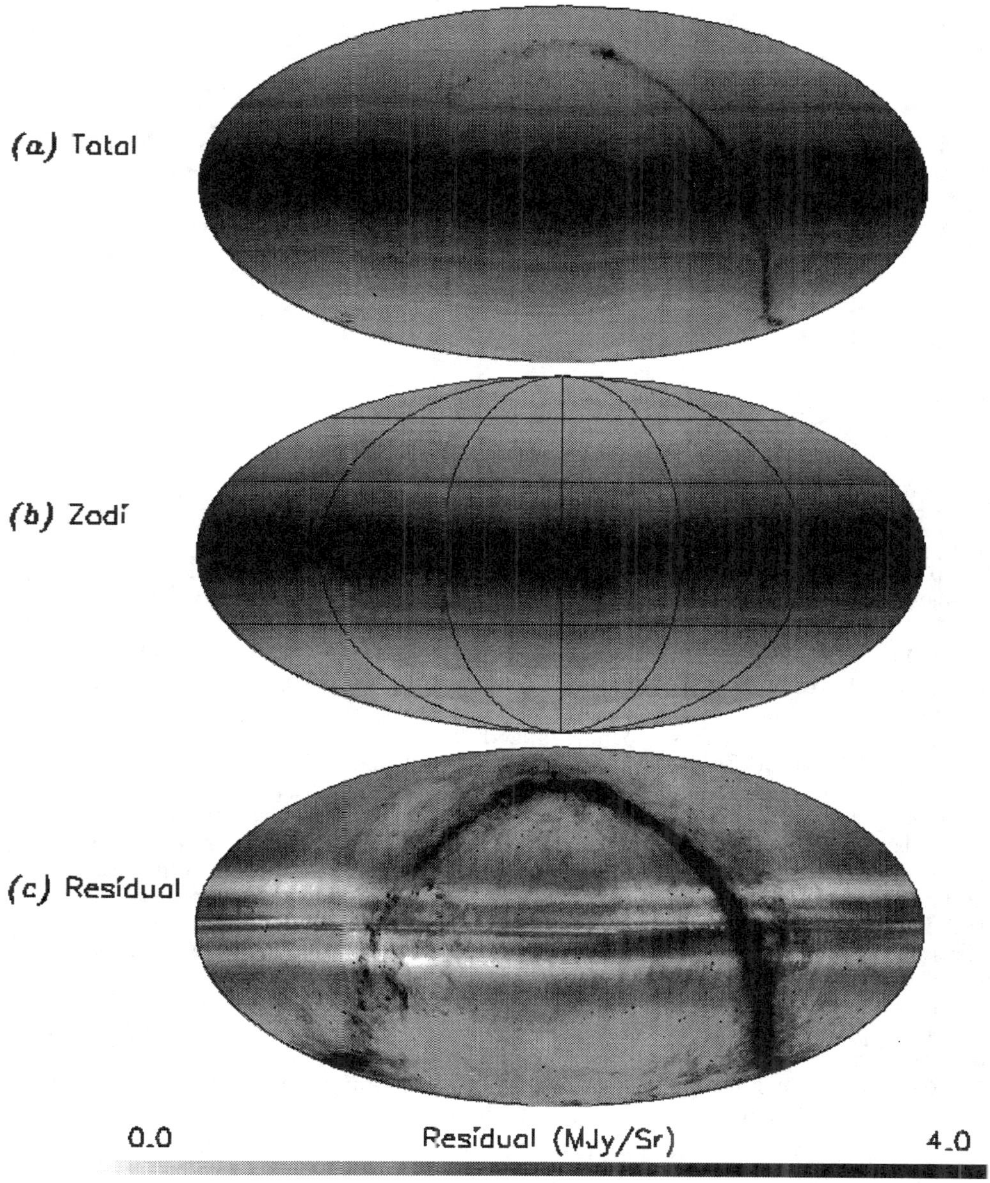

in nearly every waveband. Curiously, the residual signature is nearly always positive near the ecliptic, which is to say there is a band of enhanced brightness there. We are experimenting with alternative formulations for the density of the 'smooth cloud' in order to determine whether the particular model presented here systematically biases the results toward positive residuals at higher ecliptic latitude, where the best limits to the cosmic infrared background are obtained.

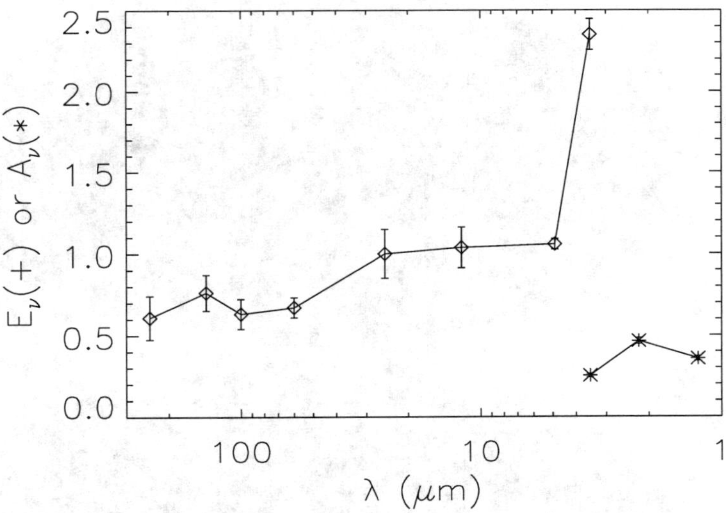

FIGURE 4. Emissivity (◇) and albedo (*) for each of the DIRBE wavebands for which they are determined. The error bars shown here include the absolute gain uncertainty.

Thermal emission from interplanetary dust was detected in 8 wavebands—including the first clear detection at 240 μm—and scattering was detected in 3 wavebands. The emissivities and albedos are shown in Figure 4. By our definition, the emissivities are normalized to unity at 25 μm. Ideal particles with radius a would be predicted to have an emissivity of unity at wavelengths $\lambda << 2\pi a$, and the emissivity would scale as a/λ at long wavelengths. We find that the emissivity is relatively flat through the mid-infrared and drops only modestly in the far-infrared. Based on the size distribution of interplanetary meteoroids (Grün *et al.* 1985), the particles that are expected to dominate the zodiacal light are large, with radii $\sim$ 30 μm (Reach 1988). Our observed slow rolloff of the emissivity in the far-infrared is consistent with such large particle sizes. On the other hand, the apparent emissivity enhancement at 3.5 μm is clearly not consistent with expectations based on physical optics and a single grain temperature. In principle, this feature could be due to a spectral line or other property of the grain material. A likely hypothesis is that the 3.5 μm emission, which is well into the Wien portion of the Planck function at the temperature (280 K)dominating the mid-infrared emission, is due to grains at temperatures $\gtrsim$ 400 K.

The fit we have obtained is already fairly 'good', despite systematic residuals, because we have removed the bulk of the time variation to within the systematic uncertainty in the DIRBE gain. Future improvements could arise from using a better functional form for the radial and vertical structure, using different fitting techniques, or relaxing some of the assumptions we have made about the interplanetary dust. Some of the assumptions that we will likely need to relax include the following:

1. The thermal emission is characterized by a greybody spectrum. This probably leads to the 'unphysical' 3.5 μm emissivity. Improvement of the spectral kernel, *e.g.* to include multiple temperatures, will not only make the emissivity variation with wavelength (Figure 4) more reasonable, but it may also lead to an improved fit to the angular dependence at 3.5 μm, which depends on the temperature more than in any other DIRBE waveband for which we have detected thermal emission.
2. The cloud is assumed to be plane-parallel. A warp is expected because the forced orbital elements depend on the semimajor axis of the orbit (Dermott 1995).
3. The phase function is assumed to be independent of wavelength. This is unlikely, as the scattering may be more isotropic at longer wavelengths. The albedos are also directly affected by the normalization of the phase function in equation (1). There has been an initial attempt to determine the phase function from the DIRBE data (Berriman *et al.* 1995).
4. Particle properties are assumed independent of location in the Solar System. The interplanetary dust consists of of asteroidal and cometary particles, which may have distinct properties. Further, structures such as the Earth's dust ring will have different size distributions due to size-dependent dynamical selection effects, and collision rates will vary throughout the Solar System. While these effects are hard to quantify, it may at least prove plausible to find different particle properties in the 'smooth cloud', dust bands, and Earth's ring.

Acknowledgements

For contributing ideas we acknowledge contributions from M. G. Hauser, C. M. Lisse, H. T. Freudenreich G. B. Berriman and the rest of the DIRBE science team.

References

Berriman, G. B., Weiland, J. L., Lisse, C. M., Reach, W. T., Hauser, M. G., & Kelsall, T. N. 1995, this workshop.

Bevington, P. R. 1969, *Data Reduction and Error Analysis for the Physical Sciences* (McGraw-Hill Inc.), pp 237-240.

Boggess, N. W., *et al.* 1992, ApJ, 397, 420

Dermott, S. F. 1995, this workshop.

Dermott, S. F., Nicholson, P. D., Burns, J. A., & Houck, J. R., 1984, Nature, 312, 505

Dermott, S. F., Jayraraman, S., Xu, Y. L., Gustafson, B. Å. S. & Liou, J. C. 1994, Nature, 369, 719

DIRBE Explanatory Supplement 1995, eds. M. G. Hauser *et al*, in preparation.

Franz, B. A., Reach, W. T., Kelsall, T., & Weiland, J. L. 1995, this workshop.

Good, J. C., Gautier, T. N. & Hauser, M. G. 1986, Adv. Space Res., 6, 84

Good, J. 1994, in *IRAS Sky Survey Atlas Explanatory Supplement*, ed. Wheelock, S. L. *et al.* JPL Publication 94-11 (Pasadena: JPL)

Grün, E., Zook, H. A., Fechtig, H., & Giese, R. H. 1985, Icarus, 62, 244

Hauser, M. G. 1995, this workshop.

Hong, S. S. 1988, A& Ap, 146, 67

Low, F. J. *et al.*, 1984, ApJ, 278, L19

Mitchell, K. J., Berriman, G. B., Kelsall, T., Richardson, D., Skard, J. A., & Stemwedel, S. W. 1995, this workshop.

Reach, W. T. 1988, ApJ, 335, 468

Reach, W. T. 1992, ApJ, 392, 289

Reach, W. T., Franz, B. A., Weiland, J. L., Hauser, M. G., Kelsall, T. N., Wright, E. L., Rawley, G., Stemwedel, S. W., & Spiesman, W. J. 1995, Nature, 374, 521

Spiesman, W. J., Hauser, M. G., Kelsall, T., Lisse, C. M., Moseley, S. H., Jr., Reach, W. T., Silverberg, R. F., Stemwedel, S. W. & Weiland, J. L. 1995, ApJ, 442, 662

Stecker, F. 1995, this workshop.

Sykes, M. V., Greenberg, R., Dermott, S. F., Nicholson, P. D., Burns, J. A., & Gautier, T. N., Jr. 1989, in *Asteroids II*, eds. R. P. Binzel, T. Gehrels, & Matthews (Tucson: U. Arizona), p. 336

COBE-DIRBE OBSERVATIONS OF THE EARTH'S RESONANT RING

Sumita Jayaraman and Stanley F. Dermott
Dept. of Astronomy University of Florida, Gainesville, FL 32611-2055
Email ID
sumita@astro.ufl.edu

ABSTRACT

The marked trailing/leading asymmetry of the zodiacal cloud, first detected in the IRAS (Infrared Astronomical Satellite) data by Dermott et al. (1988), has since been confirmed by DIRBE (Reach et al., 1995). Our analysis of the DIRBE data shows that the asymmetry exists in a range of wavebands from 4.5 to $60\mu m$ as well as at different solar elongations. We proposed (Dermott et al., 1994) that this asymmetry is caused by a trailing cloud of asteroidal dust particles that corotates with the Earth in a circumsolar ring of particles trapped in mean motion resonances. Here, we compare predictions derived from this model with the DIRBE observations and discuss the implications of the results for the sizes of the particles near the Earth, and their origin, and for the overall structure of the zodiacal cloud.

1. INTRODUCTION

The brightness of the infrared sky, especially at the mid infrared wavelengths, is dominated by the interplanetary dust complex in the solar system. Modeling the interplanetary dust cloud (also known as the zodiacal cloud) shows that a significant fraction of the signal comes from dust particles located within about 0.5 AU from the Earth. Therefore, it is important to study the dust distribution near the Earth especially in view of the recent discovery that the dust environment near the Earth is responsible for one of the most prominent features in the IRAS data, namely, the trailing/leading asymmetry. The flux observed when the line of sight of the telescope is in a direction opposite to the Earth's orbital motion (the trailing direction) is always 3–4% greater than that in the opposite direction (leading direction). This asymmetry was detected in the IRAS data (Dermott et al., 1988) and later studied by Reach (1991) who attributed it to the uncertainties in the IRAS flux calibration. Dermott et al. (1994) proposed that the flux asymmetry is real and caused by a cloud of dust particles that trails the Earth in its orbit. This trailing cloud is embedded in a circumsolar ring of asteroidal dust particles that are trapped in mean motion resonances. The presence of this asymmetry has been recently confirmed by Reach et al. (1995) using the well calibrated DIRBE data.

In this paper, we discuss the numerical model of the Earth's solar ring and compare the predictions of the model to the results from the DIRBE observations. We calculate the thermal emission from the asteroidal dust and estimate its overall contribution to the zodiacal cloud. A better understanding of the ring leads to a more accurate model of the zodiacal cloud and paves the way to obtaining stronger constraints on any isotropic component of infrared emission of cosmological origin.

© 1996 American Institute of Physics

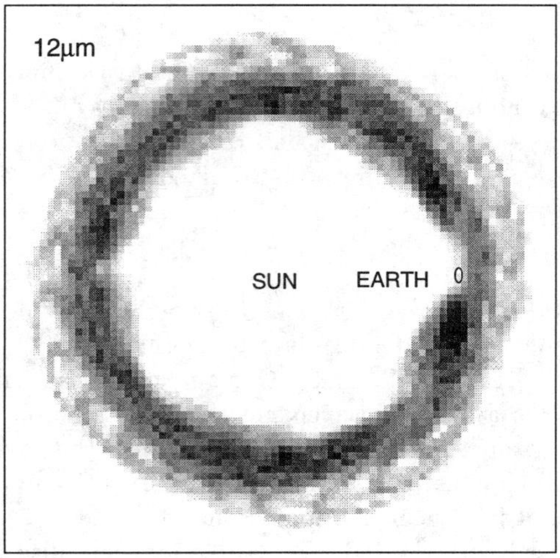

Figure 1: A numerical model of the Earth's resonant ring made up of 12 μm diameter asteroidal particles (assuming the material to be astronomical silicates with a density of 2.7gms/cm^3) showing the trailing clou of dust behind the Earth.

The origin and magnitude of the trailing/leading asymmetry can be understood by studying the formation and structure of the trailing dust cloud embedded in the resonant ring. The ring is formed in a frame rotating with the average orbital velocity of the Earth, by dust particles that spiral in towards the sun due to radiation and solar wind drag forces (Figure 1). The particles are trapped into the outer resonances of the Earth with the probability of capture into resonance increasing markedly for particles with decreasing eccentricity, thereby favoring an asteroidal origin for the ring. The distribution of the orbits in the resonances gives rise to an embedded cloud of dust trailing the Earth (Dermott et al., 1994). We have generated a geocentric view of the ring in Figure 2 which clearly shows the structure of the trailing cloud (the large blob on the left). This image is in very good agreement with that obtained from the DIRBE data after subtracting the smooth zodiacal background and the dustband contribution (Reach et al., 1995). The model also shows an offset of the cloud from the ecliptic (also present in the observations) due to the inclination of the ring to the ecliptic. Because the particles in the ring will have a forced inclination with respect to the ecliptic, the location of the trailing cloud will be above or below the ecliptic depending upon the longitude of the Earth with respect to the nodes of the plane of symmetry of the ring. The maximum displacement will occur at a longitude of 90° from the node, while the magnitude of the offset is determined by the distance of the Earth from the cloud and can be as high as 10. A view of the ring from the Earth with the trailing cloud on the left, when the Earth was at an ecliptic longitude of about 6°.

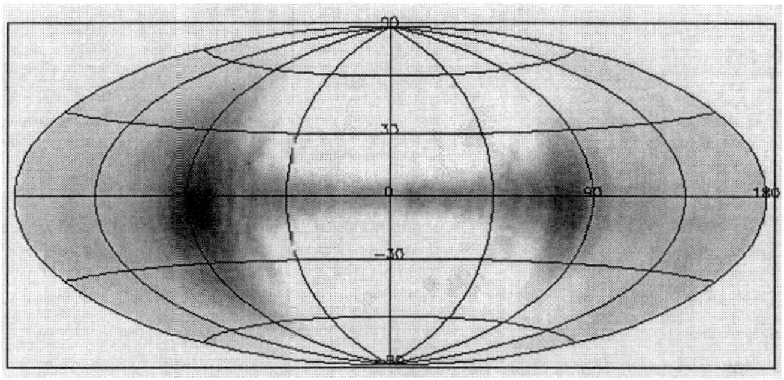

Figure 2: A view of the ring from the Earth with the trailing cloud on the left, when the Earth was at an ecliptic longitude of about 6°

We have extended our investigation of the ring to include a range of particle sizes because drag rates vary steeply with particle size (Gustafson, 1993). The capture probability of a particle into a resonance is function of the drag rate and decreases with increasing drag rates. For particles smaller than about $5\mu m$ in diameter, their high drag rates reduce the capture probability is near zero. As the particle size increases we calculate the total probability of capture to be, for example, 18% for $12\mu m$ particles (assuming astronomical silicate particles with a density 2.7gms/cm^3) and 40% for $19\mu m$ particles (Figure 3a). Particles larger than about $30\mu m$ are trapped in the ring but do not have any flux asymmetry (Jayaraman and Dermott, 1995) because of their low drag rates as shown in case of a ring formed by $36\mu m$ particles (Figure 2b). Therefore, the flux asymmetry is caused by particles in a limited size range from about 10 to $30\mu m$.

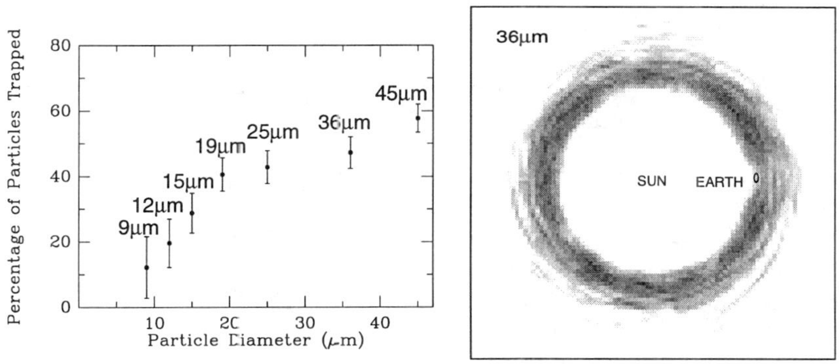

Figure 3: a) (left) The overall capture probability of asteroidal dust particles as a function of particle diameter. The lower cutoff for the particle size below which the probability of capture is zero is approximately $5\mu m$. b) (right)The Earth's resonant ring for $36\mu m$ diameter particles shows the presence of the leading as well as a trailing enhancement nearly symmetric with respect to the Earth. This explains the upper size limit for particles contributing to the flux asymmetry.

3. DIRBE OBSERVATIONS

The DIRBE instrument aboard the COBE spacecraft has provided extensive images of the infrared sky at ten wavebands, from 1.2 to 240μm. In comparison with the IRAS data, the DIRBE observations have absolute, and therefore more accurate, flux measurements that cover a wide range of solar elongation angles from 64° to 124°. We have used the DIRBE data products which are in the form of weekly averaged maps of the sky in 10 wavebands and cover forty one weeks of the mission (DIRBE Explanatory Supplement, 1995).

The following procedure was used for the data analysis. First, we separated the observations into trailing and leading directions. Then, scans at constant solar elongation (as done by IRAS) were simulated in the trailing and leading directions at different times of the year from the weekly all-sky maps. Applying a Fourier filter to remove the zodiacal dust band contribution, the peak flux of the zodiacal cloud was derived from polynomial fits to each scan. The peak flux was normalized to an elongation of 90° and plotted as a function of the ecliptic longitude of the Earth, for leading and trailing directions (Figure 4). Finally, we fit curves with a sine plus a double sine component to the observations as predicted from the model of the zodiacal cloud (Dermott et al., this proceeding). The ratio of the average trailing to leading flux for the year was calculated to be 1.024±0.008, 1.028±0.002, 1.029±0.001, 1.009±0.006 and 1.002±0.036 for 4.9, 12, 25, 60 and 100μm respectively. The large errorbars in the 100m waveband is due to the Galactic emission which will be subtracted in the future to obtain improved values. The wide disparity between the amplitude of the trailing and leading curves is not caused by the ring as the variations due to the Earth's epicycle and the forced inclination of the ring are less than 1%. The reason for this effect can be traced to dynamical effects in the background cloud (see Dermott et al. this proceeding). The decrease in the asymmetry at longer wavelengths is expected due to the upper limit to the particle size contributing to the asymmetry. No asymmetry was detected in the J, K and L bands (1.25, 2.2 and 3.5 microns respectively) that measure the scattered light from the zodiacal dust. The zodiacal signal is significantly weaker at these wavelengths and very noisy as compared to the bands in thermal emission. Therefore, further analyses is needed before the lack of asymmetry can be confirmed.

4. IMPLICATIONS FOR THE ZODIACAL CLOUD

We have been successful in modeling the structure of the Earth's resonant ring and more importantly, in accounting for the flux asymmetry due to the trailing dust cloud. This enables us to remove the variations produced by the trailing cloud and to set stronger constraints for the Cosmic Infrared Background Radiation.

In addition to the direct subtraction of the ring component, the significance of the ring model lies in the fact that we know the source of the particles, their orbital distribution and location and, most importantly, their sizes. This presents an unprecedented opportunity to determine the number density of asteroidal dust in the ring. The fraction of dust in the ring is a direct measure of the actual asteroidal dust in the zodiacal cloud. Using models of the ring and the background cloud, populated by single size particles of diameter 12μm, we have estimated that the asteroidal component is one third of the total cloud. Hence, extending these models to include a given size distribution of particles, using the magnitude of asymmetry from observations, and the numerically calculated probability of capture of dust into resonances, we will be able to determine accurately the fractional contribution of asteroidal dust to the zodiacal cloud.

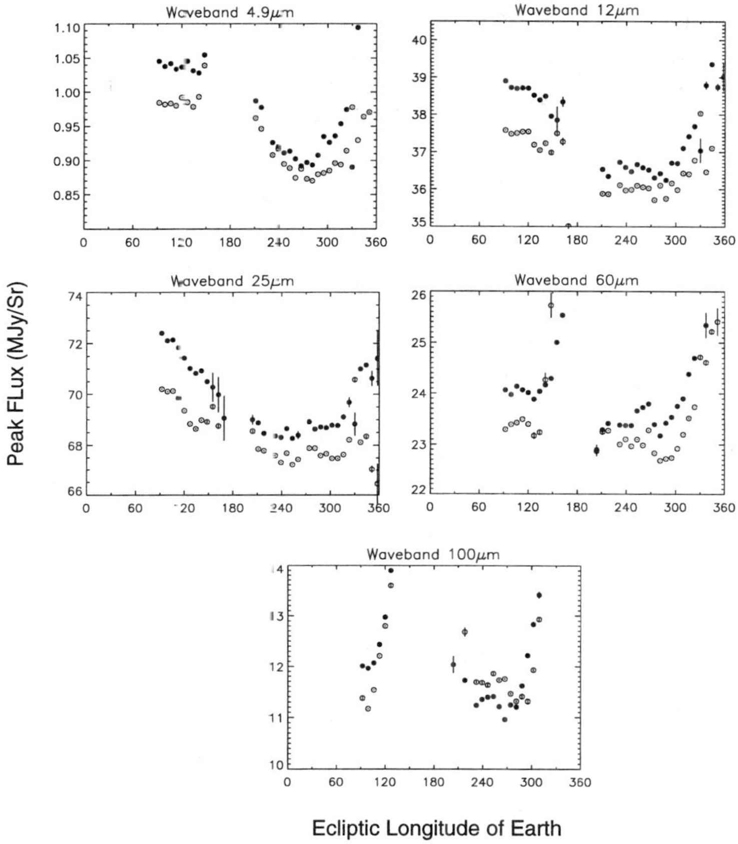

Figure 4: Variation with the ecliptic longitude of the Earth, of the peak flux of the zodiacal cloud at a solar elongation of 90°, as seen by DIRBE in 5 wavebands. The flux in the trailing direction (filled circles) is higher than the flux leading direction (open circles) at all wavebands except 100 microns

ACKNOWLEDGEMENTS

This research was supported by NASA grants NAGW 1923 and NAG5–2654.

REFERENCES

Dermott, S.F., Nicholson P., Kim, Y., Wolven and E. Tedesco 1988, In 'Comets to Cosmology', (ed. A. Lawrence), Berlin, Springer-Verlag, 3–18.

Dermott, S. F., Jayaraman, S., Xu, Y-L, Gustafson, B. A. S. and Liou, J.C. 1994, Nature, 369, 719–723.

Gustafson, B. A. S., 1994, Ann. Rev. Earth and Planet Sci., 22, 553–595.

Jayaraman, S. and Dermott, S. F. 1994. In the NATO-ASI 3rd Course : The Cosmic Dust Connection, (ed. J. M. Greenberg) (in press).

Reach, W. T., 1991, Ap. J, 369, 529–543.

Reach, W. T., Franz, B. A., Weiland, J. L., Hauser, M., Kelsall, T. N., Wright, E. L. Rawley, G., Stemwedel, S. W., and Spiesman, W. J. 1995, Nature, 374, 521–523.

SOLAR SYSTEM OBJECTS AND EXTRAGALACTIC BACKGROUND LIGHT

Christoph Leinert
Max-Planck-Institut für Astronomie, Königstuhl 17, D-69120 Heidelberg, Germany

E-mail ID
leinert@mpia-hd.mpg.de

ABSTRACT

The contribution of different solar system objects to the diffuse infrared sky brightness is discussed under the viewpoint whether and to what extent these contributions could mimic or hide true extragalactic background light emission. The objects considered range from interplanetary dust through a possible Kuiper belt population to the Oort cloud of comets. It is mainly uncertainties in the distribution of interplanetary dust which may compromise the detection of extragalactic background light, unnoticed contributions from the Kuiper belt or from interstellar particles streaming through the solar system being the next probable sources of uncertainty. A separation of the extragalactic component seems only possible if accurate sky brightness measurements over a wide range of wavelengths with largely different ratios of local to extragalactic contributions are available.

1. GENERAL REMARKS

There is no obvious relation between solar system objects and the extragalactic background light (EBL) except for the trivial fact that from the earth or from earth orbit we have to look through the former to see the latter. Or, posed as question: " can solar system objects by their infrared radiation mimic or hide a contribution of extragalactic background light?" Depending on whether the answer is "yes", "no" or "maybe" different degrees of difficulty will be encountered in the attempt to establish a credible value for the brightness of the EBL. I will approach this question in a qualitative way by using rough estimates for the optical depth (=fraction of the celestial sphere covered by a particular class of objects) and of their temperature to estimate their contribution to the diffuse sky brightness under the assumption of blackbody radiation. A sperical shape and specific densities of 0.6 g/cm^2 for comets and 2.6 g/cm^2 otherwise are used to transform masses to radii and cross-sections where necessary. Generally the uncertainties in the distributions of interplanetary and outer solar system bodies are large enough to justify such a crude back-of-the-envelope type calculation. The aim of this contribution is to present justifiable estimates, nothing more, and to give the literature which will allow the reader to do a more thorough evaluation, if needed.

The reference for the appraisal of such brightness estimates is given by the zodiacal light, which can be characterized by an optical depth of $\tau \approx 10^{-7}$ and T(1 AU) $\approx$ 280 K (Dumont and Levasseur-Regourd 1988). I assume that no effects smaller than 1 % of the zodiacal light brightness in the ecliptic at elongation $\epsilon = 90°$ can be observed (for wavelengths $\geq$ 100 μm this estimate may be too optimistic). This leads to the limits given in Table 1, below which contributions to the infrared sky brightness are to be considered negligible. For

© 1996 American Institute of Physics

comparison, the darkest spots on the sky as presented at this conference from the measurements of the DIRBE experiment at the same wavelengths correspond to 2400 ± 480 nW/m^{2}sr, 300 ± 60 nW/m^{2}sr, 90 ± 18 nW/m^{2}sr and 140 ± 28 nW/m^{2}sr, respectively to 20 MJy/sr, 6.0 MJy/sr, 3.0 MJy/sr and 6.5 MJy/sr.

Table 1: Limits of 1% of zodiacal light as function of wavelength

Unit	25 μm	60 μm	100 μm	140 μm
F_ν (MJysr^{-1})	0.70	0.23	0.12	0.050
λF_λ (nWm^{-2}sr^{-1})	84	11	3.6	1.1

2. INVENTORY OF SOLAR SYSTEM CONTRIBUTIONS

I will survey the following classes of objects in turn: comets in the Oort cloud, in the inner Oort cloud, and in the Kuiper belt; dust in the Kuiper belt, asteroidal bodies, meteors and meteorites, and two classes closely related experimentally to the detection of interplanetary dust – a possible spherical halo of dust particles at Jupiter to Uranus distances, and a stream of interstellar particles penetrating the solar system. A sketch of the spatial distribution of these objects is given in Figure 1, where the main component of interplanetary dust is assumed to extend out to ≈ 3 AU from the sun in a lenticular distribution. Similarly, mass range and flux densities for the different kind of objects are shown in Figure 2.

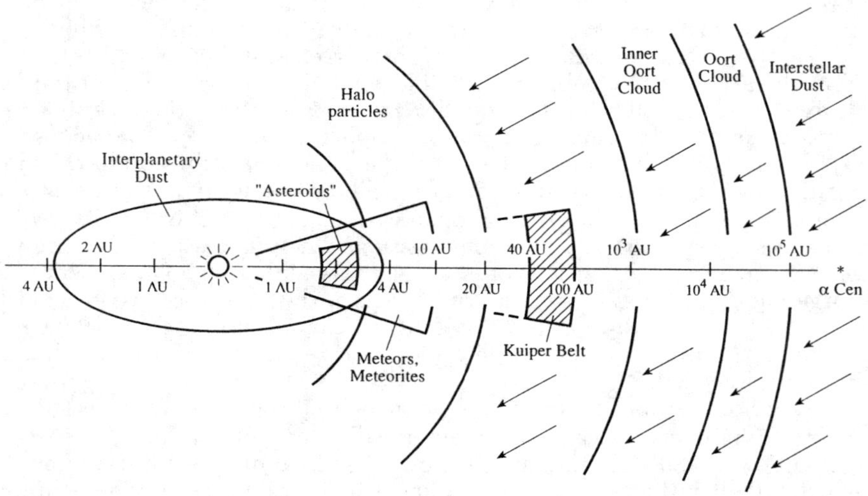

Figure 1: Geometry: the regions of interplanetary space occupied by different kinds of solar system objects. The horizontal line represents the ecliptic.

The parameters describing the individual populations are taken from that literature known and appearing relevant to me, with no attempt to homogenize partly conflicting data between different distributions. Conversion from fluxes to spatial densities was done assuming a particle velocity of 20 km/s. The brightness data are given for a wavelength of $\lambda = 60\ \mu$m, outside the main emission of zodiacal light and well representing the heliocentric distance range of 10 - 100 AU. A summary of estimated brightness contribution (Table 2) including discussion is given in the following chapter.

2.1 Oort cloud [$2 \cdot 10^4$ - 10^5 AU]

Observed as source region of long period comets.
Parameters(Weissman 1990,1991):
- $N \approx 1.0 \cdot 10^{12}$ comets, isotropically distributed
- radial dependence $n(a)da \sim a^{-2}da$ (Marochnik et al. 1988)
- mean radius $R = 2.5$ km (from mean mass of $3.8 \cdot 10^{13}$ kg)
- albedo = black (typical values of albedo = 0.07 - 0.3 (Hammel et al. 1987))
- $T \approx 5$ K (rough estimate)

Resulting brightness contribution negligible.

2.2 Inner Oort cloud [$\approx 10^3$ AU - $2 \cdot 10^4$ AU]

Postulated source for replenishment of Oort cloud, probably created from comets having formed in the Uranus/Neptun region.
Parameters(Weissman 1990,1991):
- $N \approx 6 \cdot 10^{12}$ comets, approximately in isotropic distribution
- radial dependence, radius, albedo as above
- excentricity $< e^2 > \approx 0.5$, i.e. $< 1/r^2 > = < 1 / a^2 > / \sqrt{1 - 0.5}$
- $T \approx 10$ K (rough estimate)

Resulting brightness contribution negligible.

2.3 Kuiper belt of comets [40 - 100 AU]

Postulated among other reasons for being a source providing short-period comets with correct inclination of $<|i|> \approx 15°$ by Duncan et al. (1988); a number of possible Kuiper belt comets have been detected (Luu 1994), and more candidates are being found from ground and by the HST Kuiper belt search (Cochran et al. 1995).
Parameters (Weismann 1990, Luu 1994):
- $N = 10^8$ - 10^{10} comets, flattened distribution with $<|i|> \approx 15°$
- mean radius $R = 6$ km (to lead to upper mass limit of 1 earth mass for 10^{10} comets)
- radial dependence unknown, assumed constant
- $T \approx 40$ K (blackbody temperature at 50 AU)

Resulting brightness contribution negligible.

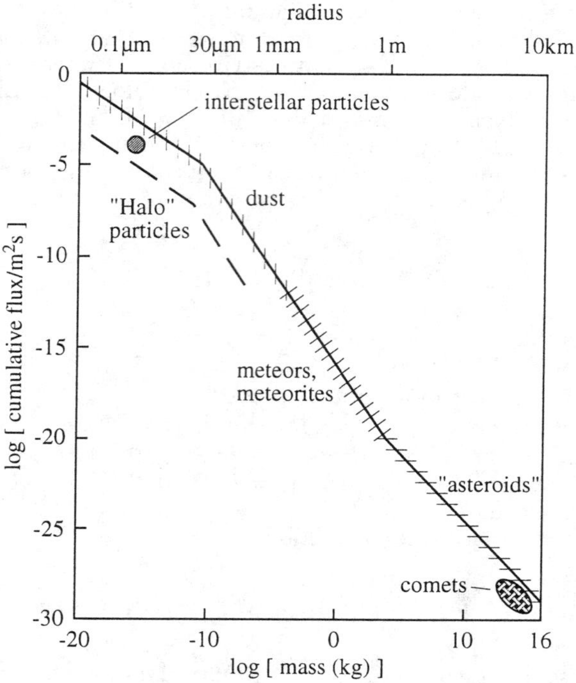

Figure 2: Masses: cumulative flux curve at 1 AU giving masses and fluxes for different interplanetary populations. (Adapted from Dohnanyi 1978).

2.4 Dust in the Kuiper belt [40 - 100 AU]

Existence probable (e.g. produced in collisions between comets). Parameters assumed to be the same as for interplanetary dust (Leinert and Grün 1990), e.g. projected area = 5 m^2/kg, but lower temperature of T $\approx$ 40 K. Resulting optical thickness and brightness contribution then are proportional to dust mass M and approximately scaling with zodiacal light brightness F_ν as

$$\frac{F_\nu(KB)}{F_\nu(ZL)} = \frac{M_{KB}}{M_{ZL}} \cdot \frac{a^2_{eff}(ZL)}{a^2_{eff}(KB)} \cdot \frac{B_\nu(T_{eff}(KB))}{B_\nu(T_{eff}(ZL))},$$

where a is the average distance from the observer and the indices KB and ZL refer to the Kuiper belt and the zodiacal light, respectively. The last two fractions on the right hand side have approximately the values of $10^{-3} \cdot 4 \text{x}\ 10^{-3}$. For a dust mass in the Kuiper belt corresponding to the complete dispersal of two large comets in a collision ($2 \cdot 10^{15}$kg $\approx$ 0.1 M(ZL)) the brightness contribution is negligible. Since the dynamical upper limit for the Kuiper belt mass of 1 earth mass is about 10^8 times the mass of the interplanetary dust cloud, much more than needed to explain the observed sky brightness at 60 μm, at most 10^{-3}

earth masses of dust can be present in the Kuiper belt, and speculations on a substantial brightness contribution by Kuiper belt dust cannot be forbidden. However, in steady state particles would be removed from the Kuiper belt by the Poynting-Robertson-effect on a time scale of a few million years, with a resulting spatial density distribution of $n(r) \sim r^{-1}$ at smaller heliocentric distances, and with the result that at most a few percent of the 60 μm emission due to such particles would arise outside 10 AU. Probably there is no substantial brightness contribution due to dust in the Kuiper belt.

2.5 Asteroidal bodies [1.8 AU - 3.8 AU]

Here I do not talk of known individual asteroids but of smaller asteroidal bodies, assumed to populate the asteroid belt according to the extrapolated mass distribution given by Dohnanyi (1978).
Parameters:
- $n(m)dm = 1.24 \cdot 10^{-19} m^{-1.837} dm$ per m^3
- shape flattened, with $<|i|> \approx 10°$
- radial range 1.8 AU - 3.8 AU (Kessler 1971, reproduced in Dohnanyi 1978)
- upper mass limit 10^{16} kg (radius = 10 km)
- lower mass limit (daring extrapolation in number densities) 10^3 kg (Poynting-Robertson life time = $3 \cdot 10^9$ years); smaller asteroidal bodies are taken care of under the populations of meteoroids, meteorites and interplanetary dust.
- albedo ≈ 0.1, black (see tabulation by Tedesco(1989))

Resulting brightness contribution still negligible.

2.6 Meteors and meteorites [0.1 AU to ≈ 10 AU]

Observed, spatial distribution of meteors(Southworth 1967, Southworth and Sekanina 1973) includes massive extrapolation for unobservable orbits.
Parameters:
- mass range 10^{-5} - 10^3 kg
- $N(m)dm = 1.1 \cdot 10^{-22} m^{-1.69}$ dm per m^3 (Ceplecha 1988,1992)
- radial distribution assumed to be that of the meteors
- shape flattened, with $<|i|> \approx 17°$
- albedo assumed ≈ 0.08, black (Dumont and Levasseur-Regourd 1988)like for interplanetary dust

Resulting brightness contribution negligible.

2.7 The "Halo" dust population [3AU - 20 AU]

Custom tailored to explain the Pioneer 10/11 penetration detector results (Divine 1993)
Parameters:
- $N(m)$ dm = $5.42 \cdot 10^{-30}$ $m^{-2.6}$dm at 3.4 AU, distribution like for interplanetary dust, less steep at small masses ($m < 10^{-9}$ kg)
- shape isotropic
- radial dependence rather flat from 3 - 20 AU, maximum at 3.4 AU
- albedo = 0.02 (black), as modeled by Divine(1993)

Predicted brightness contribution at 60 μm is F_ν = 8.1 MJy/sr or λF_λ = 400 nW/m^{2}sr, quite similar to the 33 % found by Staubacher et al. (1993) from more detailed calculations. Since the predictions for the "Halo" dust population exceed the dark sky measurements of COBE, there may be difficulties with Divine's postulation of this additional component of interplanetary dust. Actually, the preferred interpretation for the impacts observed in the solar system beyond the asteroid belt now has changed (see following section).

2.8 Interstellar dust particles [$\geq$ 2 AU]

Detected by Ulysses (Grün et al. 1993).
Parameters (Grün et al. 1994):
– mean mass ($3{\cdot}10^{-13}$ g) and observed flux ($1.5{\cdot}10^{-4}$m^{-2}s^{-1}) lead to a geometrical scattering cross section of $2.9{\cdot}10^{-21}$ m^2/m^3 ($\approx$ 1/170 of the value for interplanetary dust at 1AU)
– radial depence assumed to be constant, corresponding to an extinction of $\approx$ 0.1 mag/kpc; approach to $\leq$ 2 AU may or may be not inhibited
– shape isotropic, but gravitational focussing and interactions with solar wind and interplanetary magnetic field may be important inside a few AU
– albedo unknown, assumed $\approx$0.2
Resulting brightness contribution just negligible.

3. OVERVIEW ON THE DIFFERENT SOLAR SYSTEM POPULATIONS

The overview is given in tabular form in Table 2. The contributions of most populations are negligible with respect to the zodiacal light, with three possible exceptions:

Dust in the Kuiper belt could give a noticeable contribution, but it would require the equivalent of 400 000 comets of radius 6 km to produce 10 % of the zodiacal light brightness at 60 μm. In addition, as mentioned above, one would expect that large amounts of dust in the Kuiper belt would lead to a dust distribution increasing in spatial density towards the sun, with the result that material originating in the Kuiper belt would emit more than 95 % of its 60 μm radiation inside 10 AU. However likely or unlikely one judges such a contribution (which is not obvious how to decide), it would be flattened like the short-period comets, whose origin the Kuiper belt was supposed to explain, i.e with $<\mid i \mid> \approx 15°$.

"Halo" dust if present in the quantities postulated by Divine (1993) would be a major and isotropic contributor to zodiacal light. However, many if not all of the measurements in the outer solar system it was supposed to explain, now can be discussed in terms of an influx of interstellar particles, shown to exist by the Ulysses dust experiment (Grün et al 1993). Therefore I suggest not to seriously consider this contribution unless observational evidence, which probably had to be new evidence, is supporting the separate existence of this interplanetary population.

Interstellar dust particles finally have been shown to be part of the interplanetary dust complex. Whether they contribute more to the zodiacal light brightness than the amount of $\approx$ 1% estimated above, depends on unknown size and poorly known spatial distribution in the planetary system. Grün et al (1994) show a scattering contribution of the interstellar dust particles to the visual zodiacal light brightness of up to 60 % near the ecliptic poles. The simple but not necessarily incorrect model presented here would also result in a higher contribution in the visual than in the infrared, but only by 1.5 % in the ecliptic or 5% at the poles, if the interstellar particles, as assumed by Grün et al., are limited to the region outside 2 AU. The contribution by these particles is expected to be roughly isotropic.

With some reservation then, only the zodiacal light, representing the known interplanetary dust, has to be considered when discussing the possibility of a component of extragalactic background light.

Table 2: Summary of brightness contributions due to different solar system populations*

Object	Range [AU]	Typical temperature [K]	Total mass [kg]	τ	Brightness $F_\nu(60\ \mu m)$ [MJy/sr]
Oort cloud	$2\ 10^4$ - 10^5	5	$4\ 10^{25}$	$8\ 10^{-13}$	$< 10^{-20}$
Inner Oort cloud	10^3 - $2\ 10^4$	10	$2\ 10^{26}$	$5\ 10^{-11}$	$<10^{-10}$
Kuiper belt	40 - 100	40	$< 6\ 10^{24}$	$5\ 10^{-9}$	$\leq 10^{-3}$
Kuiper belt dust	40 - 100	40	e.g. $2\ 10^{15}$	$4\ 10^{-12}$	10^{-5}
			e.g. $6\ 10^{24}$	10^{-2}	10^3 - 10^4
Asteroidal bodies	≈ 1.8 - 3.8	180	$2\ 10^{19}$	$4\ 10^{-10}$	0.03
Meteors, meteorites	< 10	100 - 275	10^{16}	$2\ 10^{-11}$	0.002
"Halo" dust	≈ 3 - 20	60 - 160	$7\ 10^{17}$	$4\ 10^{-7}$	8
Interstellar dust	≥ 2	< 200	$2\ 10^{24}/m^3$	0.1/kpc	0.003 - 0.015
Zodiacal light	≤ 3.5	275	10^{16} - 10^{17}	$\approx 10^{-7}$	22

*all values subject to considerable uncertainties

4. ZODIACAL LIGHT AND EXTRAGALACTIC BACKGROUND LIGHT

4.1 Choice of a qualitatively acceptable zodiacal light model

The following parameters were taken from the literature:
- radial distribution $n(r) \sim r^{-1.0}$ (Levasseur-Regourd et al. (1991) give a value for the exponent of 0.90 ± 0.15)
- dust temperature at 1 AU: 275 ± 20 K (ibidem)

– temperature dependence $T(r) \sim r^{-0.35}$ (ibidem)
– out-of-ecliptic decrease of spatial density as function of heliocentric ecliptic latitude $\beta_\odot$: $n(\beta_\odot) \sim \exp(-2.7 \mid sin\beta_\odot \mid)$ according to Giese and Kneissel (1989), a so-called "fan" model.
Since only relative brightness distributions have to be discussed, no further characterization of particle properties is necessary, except that I assume blackbody emission. The calculations are performed for 25 μm and the great circle from the ecliptic to the poles at an elongation of 90°. Figure 3 shows that the model qualitatively reproduces the zodiacal light distribution. A good fit to the IRAS data , since not needed for our discussion, was not attempted. Note that in this model the spatial density of dust 1AU above the solar poles is only 6.7 % of the value at 1 AU in the ecliptic.

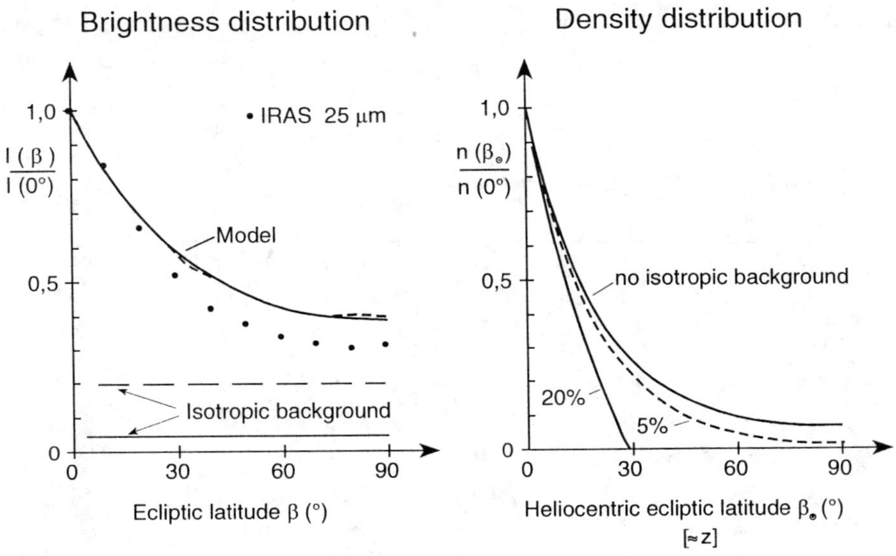

Figure 3: Right: out-of-ecliptic spatial density distributions: an interplanetary dust model (solid line) and two narrower distributions to be used in conjunction with an isotropic component of 5% (broken line) or 20 % (solid line). Left: Resulting out-of-ecliptic brightness distributions at a wavelength of 25 μm: the model and the combined brightness for 5 % isotropic component fall together (solid line), while the combined brightness for an isotropic contribution of 20 % shows small deviations (broken lines). The isotropic background components are also shown separately for comparison, as are the IRAS results.

4.2 Hiding or mimicking an isotropic contribution

Since the extragalactic background light has to be isotropically distributed (cosmological principle), it is important to know how well-defined the isotropic part of zodiacal light is. Unfortunately, from pure parametric modelling, this contribution is far from unique, as is shown in Figure 3. Easily the observed brightness can be decomposed into a stricly isotropic and a remaining zodiacal light component, as long as the brightness of the isotropic component is less than 7% of the zodiacal light in the ecliptic: simply by subtracting the corresponding constant from the out-of ecliptic distribution function. The resulting, somewhat steeper distribution function is shown in Figure 3 as dashed line. Even for an isotropic component of 20%, a steeper out-of-eclip[tic density distribution can still be found (solid line in Figure 3), for which the sum of isotropic and modified zodiacal light brightness fits the model brightness within $\pm$ 1%. We can look at this from two sides. Either the true distribution of interplanetary dust is more concentrated to the ecliptic than we think, and an isotropic contribution thus goes unnoticed, remains hidden. Or we assume the interplanetary dust to be more concentrated to the ecliptic than it actually is, and thus postulate an additional isotropic component in order to fit the measurements. Obviously, both possibilities have to be taken care of in an interpretation of actual data.

4.3 Destructive interference of isotropic and narrow distributions

As a special case for the foregoing discussion we ask whether an unnoticed contribution with the inclination distribution of short-period comets ($<|\, i\, |> \approx 15°$) could combine with an isotropic component to reproduce the assumed brightness distribution of the zodiacal light. Figure 4 shows that this is indeed the case if the isotropic contribution is chosen to provide 20% of the total brightness in the ecliptic. Among the two narrow spatial distributions, the first one was chosen as f($\beta_\odot$) $\sim cos^{30}\beta_\odot$ for convenience (since generally $cos^n\beta_\odot$ in out-of ecliptic distribution corresponds to f(i) $\sim sin(i)cos^n i$ in inclination distribution; substitute x = $cosi/cos\beta_\odot$ in the integral relating f(i) to f($\beta_\odot$) to see it (e.g. in Leinert et al.)). The second spatial distribution, $cos\beta_\odot - 1.5sin^{0.7}\beta_\odot$

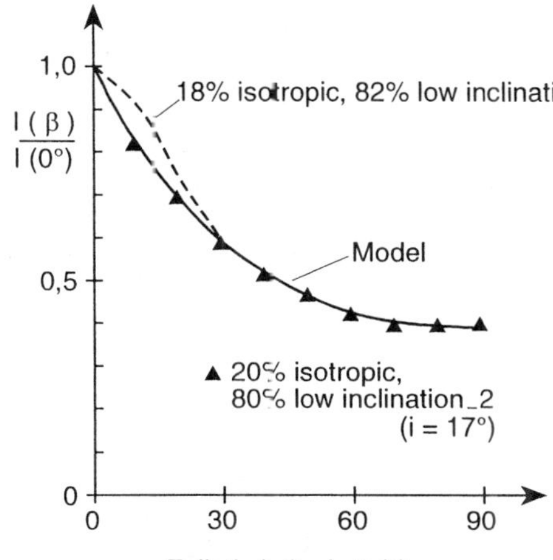

Figure 4: Two examples showing that it is easy to reproduce the assumed out-of-ecliptic distribution of zodiacal light by superposition of an isotropic and a low-inclination distribution, similar to what would be expected for a contribution from particles from the Kuiper belt. Wavelength is 25 μm.

(but setting negative values to zero) was chosen to make the discrepancy at small ecliptic latitudes disappear. Both of these distributions have the wanted average inclination, $<|i|> = 16°\text{-}17°$.

5. CONCLUSION

The only object of proven relevance to the infrared extragalactic background light is the zodiacal light. However, once the level of a few percent of the zodiacal light brightness is reached, the possible effects of interstellar particles or low-inclination objects have to be discussed also, and in more detail than done here.

To detect an isotropic component in the presence of zodiacal light requires both accurately (to $\approx \pm 1\%$) measured brightness profiles and reliable (therefore at least partly physical) models of dust properties and dust distribution.

In addition measurements over a wide range of wavelengths with very different ratios of extragalactic to zodiacal light contributions are a precondition for an eventual separation of the extragalactic background light. A separation based on measurements at one wavelength only appears not to be feasible.

The COBE/DIRBE experiment so far has the best opportunity to approach or achieve this goal. But published results will only be really meaningful if **all** assumptions on the handling of solar system contributions are given in quantitative detail.

ACKNOWLEDGEMENTS

I thank Eberhard Grün for clarifying discussions and guidance in some of the topics addressed, as well as Martina Weckauf and Karin Dorn for producing the Figures and the Universities Space Research Association for a travel grant.

REFERENCES

Ceplecha, Z. 1988, Bull.Astron.Inst.Czechosl., 39, 221
Ceplecha, Z. 1992, A&A, 263, 361
Cochran, A., Levinson, H., Stern, A., & Duncan, M. 1995, IAU Circular No 6163
Divine, N. 1993, JGR, 98, 17029
Doynanyi, J.S. 1978, in Cosmic Dust, ed. J.A.M. McDonnel (Chichester: Wiley), 527
Dumont, R., & Levasseur-Regourd, A.-Ch. 1988, A&A, 191, 154
Giese, R.H., & Kneissel, B. 1989, Icarus, 81, 369
Grün, E. and 22 co-authors 1993, Nature, 362, 428
Grün, E., Gustafson, B., Mann, I., Baguhl, M., Morfill, G.E., Staubach, P., Taylor, A., & Zook, H.A. 1994, A&A, 286, 915
Hammel, H.B., Telesco, C.M., Campins, H., Decker, R., Storrs, A.D., & Cruikshank,D.P. 1987, A&A, 187, 668
Kessler, D.J., 1971 in Physical Studies of Minor Planets, ed. T. Gehrels NASA SP-267, 595

Levasseur-Regourd, A.-Ch., Renard, J.B., & Dumont, R. 1991, in Origin and Evolution of Interplanetary Dust, eds. A.-Ch. Levasseur-Regourd & H.Hasegawa (Dordrecht: Kluwer), 131

Leinert, Ch., & Grün, E. 1990, in Physics of the Inner Heliosphere, eds. R. Schwenn & E. Marsch (Berlin/Heidelberg/New York: Springer), 207

Leinert, Ch., Röser, S., & Buitrago, J. 1983, A&A, 118, 345

Luu, J. 1994, in Asteroids, Comets, Meteors 1993, eds. A. Milani, M. DiMartino, & A. Cellino, (Dordrecht: Kluwer), 31

Marochnik, L.S., Mukhin, L.M., & Sagdeev, R.Z. 1988, Science, 242, 547

Southworth, R.B. 1967, in The Zodiacal Light and the Interplanetary Medium, ed. J. L. Weinberg NASA SP-150, 179

Southworth, R.B, & Sekanina, Z. 1973, NASA CR-2316

Staubach, P., Divine, N. & Grün, E. 1993, Planet.Space Sci., 41, 1099

Tedesco, F. 1989, in Asteroids II, eds. R. Binzel, T. Gehrels, & M.Matthews (Tucson: University of Arizona Press), 1090

Weissman, P.R. 1990, Nature, 344, 825

Weissman, P.R. 1991, Icarus, 89, 190

DIFFUSE INFRARED EMISSION FROM THE GALAXY

DEEP STAR COUNTS AND COSMOLOGICAL BACKGROUNDS: A POWERFUL NEW MODEL FOR THE POINT SOURCE SKY

Martin Cohen
Jamieson Science and Engineering, Inc., 5321 Scotts Valley Drive,Suite 204, Scotts Valley, CA 95066 and Radio Astronomy Laboratory, 601 Campbell Hall, University of California, Berkeley, CA 94720

ABSTRACT

I describe the capabilities of "SKY", the model for the point source sky originally developed by Wainscoat *et al.* (1992: ApJS, 83, 111) and considerably enhanced by Cohen (1994: AJ,107,582; also ApJ,427,848), emphasizing those aspects most relevant to interpretation of DIRBE data. SKY predicts cumulative and differential source counts, and integrated surface brightness of the sky due to smeared point sources, in any direction, both for eleven "hardwired" far-UV, optical, and infrared filters and, in a customized mode, for any filter lying entirely within the 2.0-35.0 μm range. This enables it to predict specifically for the DIRBE bands. SKY realistically represents the Galaxy using disk, spiral arms, local spurs, Gould's Belt, molecular ring, bulge, and halo. It also offers a good representation of the mid-IR extragalactic sky. One can use SKY to generate all-sky maps of the surface brightness of the Galactic point source foreground. These brightnesses fulfil one of the prerequisite elements for seeking cosmological background radiation from DIRBE. The contributions of the Galactic foreground at the South Ecliptic Pole are compared with DIRBE brightnesses.

1. Introduction

Many models have been proposed to describe the infrared sky. Some treat the surface brightness only (by effectively smearing all the stars); others deal directly with the point source sky. A selective sampling of models is as follows: Eaton *et al.* (1984), Garwood & Jones (1987), Ishida & Mikami (1982), Kawara *et al.* (1982), Kent *et al.* (1991), Maihara *et al.* (1978), Mikami & Ishida (1981), Mikami *et al.* (1982), Ortiz & Lepine (1993), Ruelas-Mayorga (1991), Wainscoat *et al.* (1992). Some models include solely an exponential disk, others add bulges, molecular rings, etc. None yet contains all the detail of the SKY model, namely: disk, halo, bulge, ring, arms, two local spurs, Gould's Belt and the local molecular clouds (for mid-infrared fidelity). No model other than SKY incorporates non-stellar point sources (reflection nebulae, planetary nebulae, HII regions, and galaxies). Several models represent only normal stars and omit the AGB objects that dominate the mid-infrared sky. The major factor contributing to SKY's relevance to the problem of the Galactic foreground emission is its provenance in the mid-infrared (IRAS range) rather than the visible. Even the extension to 2 μm from V is fraught with difficulties (Mamon & Soneira 1982) because many of the IR-dominant types of source are not included in the V luminosity function. What further distinguishes SKY from all other detailed models is its capacity to make predictions specific to passbands of interest because it contains an embedded complete spectral library from 2–35 μm, over which one can integrate an arbitrary (contained) passband.

The development of this model can be traced from Wainscoat *et al.* (1992), Cohen (1993), Cohen (1994), Cohen, Sasseen & Bowyer (1994), with the recent

© 1996 American Institute of Physics

addition of a displacement of the sun to a location 15 pc north of the Galactic plane and a rescaled halo:disk ratio (Cohen 1995).

2. Surface Brightness Products

To show the relevance of SKY in the present context of defining the cosmic background, I describe its surface brightness capabilities here. SKY fundamentally creates cumulative and differential source counts in any hardwired or customized filter with abscissae either magnitude or in-band flux. It can render this information in the form of the total integrated surface brightness by smearing all the point sources expected in the patch of sky requested (either a 1 square degree patch centered on a single Galactic direction, or a finite area computed as a grid of directions with chosen spacings in longitude and latitude). In addition to total surface brightness, SKY also yields the error in this quantity induced by the flux–weighted Poisson count statistics. Surface brightness can be further broken down into the differential contributions of any of the individual geometrical elements (disk, halo, etc.). SKY can be arranged to yield all–sky simulations in arbitrary bandpasses; an example is shown here in Fig. 1, an all–sky surface brightness prediction customized for COBE/DIRBE's band 3 (3.5 μm). Optional, very detailed, log files can be generated that tabulate the separate contributions of each of the 87 categories of celestial source over an arbitrary range of apparent magnitudes, for each sky patch, and for each geometrical component. These provide extremely valuable diagnostics on the breakdown of sources along any line–of–sight.

3. High Latitude Performance

In other publications I have described near– and mid–IR fidelity in the Galactic Plane. However, the search for cosmic backgrounds is dominantly a high–latitude task because here the Galactic point–source and diffuse foregrounds will be minimized, and one can select directions away from the maximum of zodiacal contamination. The ability to match North Galactic Pole (NGP) deep V counts (Wainscoat *et al.* 1992) validates SKY's representation of G–K disk dwarfs and deep FAUST (1660A) counts (Cohen *et al.* 1994) of B–F disk dwarfs. Matching high latitude counts at K validates G–K giants; at 12/25 μm, M giants. What effects limit SKY's performance at high latitudes?

At high latitudes one sees essentially only disk and halo so there is maximal sensitivity to the normalizing ratio of halo:disk counts. An incorrect value for this ratio can contribute to erroneous predictions of sky surface brightness. Similarly, at high latitudes one has the greatest leverage on the displacement of the sun from the Galactic plane. Again, an inaccurate value for $z_\odot$ leads to artificial north–south polar asymmetries in model predictions which have a strong bearing on the quest for near–infrared cosmic background radiation. Cohen (1995) has recently determined an optimal value of 15 pc for $z_\odot$ from FAUST (far–ultraviolet) counts at 1660A, and IRAS at 12 and 25 μm, and has redefined halo:disk in SKY as 1:1250 for SKY, at the lower bound of Schmidt's (1975) suggested approximate range, but within that determined by Bahcall *et al.* (1983). Both halo:disk and $z_\odot$ affect counts and surface brightnesses at about the 10% level. It would be highly desirable to redetermine these parameters in a multi–wavelength context, preferably from source counts across the near–infrared, for comparison with the statistically identical values found at 1660A and 12/25

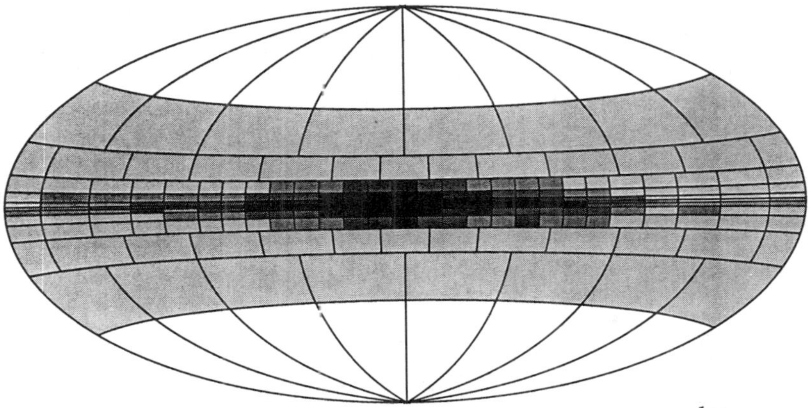

FIGURE 1. All-sky simulation of surface brightness of the DIRBE band 3 sky. Grey levels are logarithmic: no shading, $\leq$0.3% of peak brightness; 0.3$<$ lightest shading $\leq$3.2; 3.2$<$ pale-grey shading $\leq$5.6; 5.6$<$ mid–grey shading $\leq$12.5; 12.5$<$ dark–grey shading $\leq$20.0; 20.0 $<$ black shading. The blackest shading corresponds to a maximum surface brightness of 12.56 MJy/sr between longitudes -5° and 5° and between latitudes -0.5° and 0.5°.

μm. A general agreement across such a broad wavelength range adds credence to the physical realism of the model because such vastly different types of source dominate the FUV and mid–infrared skies.

Only the advent of new reasonably (or very) deep infrared source counts (and perhaps examination of DIRBE data themselves) can help improve SKY's high latitude performance. Such star counts are becoming available through hemispheric surveys like DENIS, and even deeper surveys over several hundred (arcmin)^2 such as that by Meadows (1994). These offer J, K, and/or H data sets to J$\sim$16, K$\sim$14 (DENIS), and K$\sim$18 (Meadows). SKY is already capable of reproducing these observed counts satisfactorily but there is clearly the potential for improved leverage on halo:disk and $z_\odot$.

4. Uncertainties in SKY that bear on the search for cosmic backgrounds

Fig. 2 illustrates SKY's predictions for the stellar foreground at the South Ecliptic Pole (SEP). At the SEP the balance between the "zodi" and Galactic point source foregrounds switches around, compared with the NGP shown by Reach at this workshop. The SEP zodi contribution is lower. It is pertinent to consider two sets of uncertainties: those in the DIRBE observations and those in SKY. DIRBE's exoatmospheric nature obviates the need for removal of the complex environmental emissions that attend balloon and rocket background experiments. The input maps must be "dezodied". Beyond any residuals that arise from imperfect removal of the rather complex and bright zodiacal foreground, what other sources of uncertainty are implicit in DIRBE data? To create the transient absolute magnitude tables with which SKY operates in its "customized filter" mode, we must first know the complete system transmission

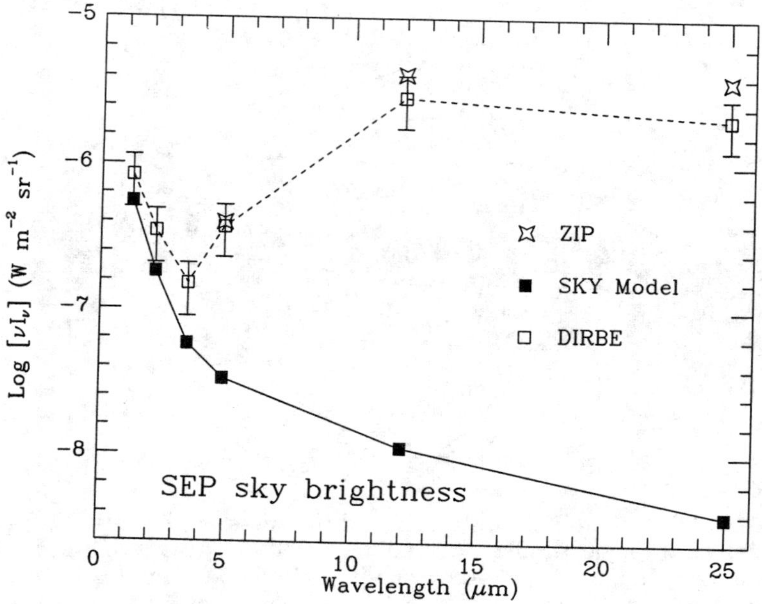

FIGURE 2. SKY's predictions for the stellar foreground (filled squares) at the South Ecliptic Pole (SEP). Longward of 3.5 μm is the powerful zodiacal reradiated emission (observed by both ZIP and DIRBE); shortward, sunlight scattered from zodiacal particles.

profiles (detectors plus all optics), for the actual flight instrument in DIRBE under the on–orbit conditions, with the greatest precision because these are integrated over SKY's spectral library (they are essential inputs to any model).

We must further know precisely, as a function of ambient (dominantly zodiacal) background levels, what DIRBE threshold magnitudes characterize the recognition of individual point sources. SKY provides a total integrated surface brightness for each region in which it predicts counts through the cumulative flux density (or in–band flux) present in those counts. It requires two additional inputs, namely the thresholds for the brightest and faintest sources to be considered in the patch of sky. The faint limit is relatively unimportant compared with the bright threshold, because bright sources clearly dominate the integrated sky brightness. Furthermore, the manner in which SKY has been implemented by the DIRBE team (Arendt & Dwek 1995, this workshop), attempts to blend the distribution of recognizable "bright" point sources in a region with the "faint" ones, as represented by SKY. Variations across the sky in the threshold magnitude for discrimination of bright sources can create problems for this blending process, and can definitely introduce uncertainties as to the relevant surface brightness to extract from SKY (*cf.* Cohen *et al.* 1994) where this issue is addressed in a far–ultraviolet context). Such errors are not systematic, nor readily assessed, and surely exceed simple Poisson–induced brightness errors essentially because one may "overcount" the brightest sources in a sky patch.

As a demonstration of the great sensitivity to this bright limit, consider Fig. 3 which replicates the purely celestial component of emission detected by Noda *et al.* (1992) at points "A","B". On this I have plotted the total surface brightness predicted by SKY for this direction as a function of the faintest source whose passage through the rocket–borne telescope's aperture might have been recognized. Results for two passbands are shown: K and the Japanese narrowband [2.4]. If the background had been bright enough, one could in principle account for almost the entire 2-μm flux by starlight, even without allowance for zodiacal scattered sunlight. The points vertically below one another indicate the predicted surface brightness for four values of bright threshold, namely: none (stars of all magnitudes are accepted); 0^m; 3^m; and 5^m. Clearly, any inference drawn about cosmic backgrounds is critically dependent on the precise definition of this threshold. Noda and colleagues attempted to remove those stars in the beam that were listed in the SAO Catalog. But, in the infrared, this approach is not sufficient because many of the IR–brightest and reddest sources will not have been optically catalogued at all. The same difficulty attends the interpretation of the cosmic K background described by Matsumoto, Akiba & Murakami (1991). Assuming that the stressing terrestrial and zodiacal components of emission were correctly assessed and removed, their estimated cosmic component is critically dependent on the faintest point source that they could have identified during this difficult rocket experiment. With SKY I estimate 2.2× their estimated starlight contribution near 2 μm, for their declared threshold. In part this difference can be anticipated by examination of their figure showing the fit of their own model to Elias's (1978) NGP K counts. Their model systematically overpredicts the bright counts and underpredicts the faint, naturally leading to a severe underestimate of total integrated starlight. (These comments are not meant as an attack on a difficult experiment, nor as criticism of alternative models, but are intended as a cautionary note on the problems that attend the removal of predicted starlight.)

Returning to SKY, the library of spectra may itself contain irregularities, imperfections, and errors. However, for those categories of source that dominate the near–infrared sky, its elements were assembled as far as possible (Cohen 1993) from existing Kuiper Airborne Observatory spectra (thereby, gaps between terrestrial windows are largely avoided) and recalibrated (Cohen *et al.* 1992b) IRAS LRS spectra. This empirical approach bypasses the shortcomings of grids of state–of–the–art model atmospheres; *e.g.*, bands of water vapor and of SiO (notably the fundamental from 7.5–11.5 μm) are currently omitted from stellar atmospheric grids, but are prevalent in K/M giants (Cohen *et al.* 1992c). The library was constructed from many spectral fragments in the published literature and other archives. These fragments were not always calibrated with respect to the hot standards, Sirius and Vega; the most popular calibrators were IR-bright K–giants, typically treated as cool blackbodies, which leads to strictly incorrect spectral slopes. Errors arising from the end–to–end splicing of such fragments could contribute errors of order 10% in DIRBE bands 3 and 4 in the spectral shapes of individual sources, through the change in spectral shape relative to the normalizing wavelength, 12 μm.

The absolute magnitudes and dispersions in SKY may provide additional sources of uncertainty although they are certainly adequate to replicate IRAS counts at both 12 and 25 μm quite faithfully over more than 95% of the sky. The resulting table of absolute DIRBE magnitudes must at some point be converted from magnitudes to in–band fluxes (or equivalently monochromatic flux densities), with a defined zero point. (For a full discussion of a major program designed to unify future infrared absolute calibrations see Cohen *et al.* (1992a,b,c,1995)). There is reason to believe that the zero magnitude scheme detailed in the IRAS Explanatory Supplement comes quite close to current beliefs for the absolute (point source) calibration of IRAS, but may be high by about 2% at 12, and 6% at 25 μm.

5. The absolute calibration of DIRBE

Here I speak solely of DIRBE bands 1–5 from the perspective of the ongoing effort in absolute IR calibration cited above. The conversion of DIRBE's data numbers into physical units is currently based on the absolutely calibrated model of Sirius presented by Cohen *et al.* (1992a). (At longer wavelengths the primary calibrator switches first to NGC 7027, then to Jupiter.) The formal uncertainty in this Sirius calibration is a wavelength independent ±1.47%. Furthermore, a newer model for this star, also created by Kurucz (1994), has virtually identical attributes to the 1991 one described by Cohen *et al.* (1992a). However, there is now a newer Vega model (Castelli & Kurucz 1994). What influence would be exerted on the absolute calibration of DIRBE bands 1–4 were we to switch to the combination of this pair of 1994 models from the current pair of 1991 calibrated models? The strong similarity of the two Sirius models means that any such influence arises solely through the redefinition of "zero magnitude flux densities" from the newest Vega model. The changes in absolute calibration for DIRBE would be -2.8, -2.7, -2.3, and -2.3% for bands 1–4, respectively. These small shifts would also travel in some complex fashion through the process of removing the zodiacal foreground.

Acknowledgements

The development of the SKY model has been primarily supported by Phillips Laboratory through Dr. Stephan D. Price.

References

Arendt, R. G. & Dwek, E. 1995, these proceedings
Bahcall, J. N., Schmidt, M., & Soneira, R. M. 1983, ApJ 265 730
Castelli, F. & Kurucz, R.L. 1994, AA 281 817
Cohen, M. 1993, AJ 105 1860
Cohen, M. 1994, AJ 107 582
Cohen, M. 1995, ApJ May 10
Cohen, M., Sasseen, T., & Bowyer, S. 1994, ApJ 427 848
Cohen, M., Walker, R. G., Barlow, M. J., & Deacon, J. R. 1992a, AJ 104 1650
Cohen, M., Walker, R. G., & Witteborn, F. C. 1992b, AJ, 104, 2030
Cohen, M., Witteborn, F. C., Carbon, D. F., Augason, G. C., Wooden, D., Bregman, J., & Goorvitch, D. 1992c, AJ, 104, 2045
Cohen, M., Witteborn, F. C., Walker, R. G., Bregman, J., & Wooden, D. 1995, AJ, 110, in press

Eaton, N., Adams, D. J., & Giles, A. B. 1984, MNRAS, 208, 241
Elias, J. 1978, AJ 83 791
Garwood, R., & Jones, T.J. 1987, PASP 99 453
Ishida, K., & Mikami, T. 1982, PASJ, 34, 89
Kawara, K., Kosasa, T., & Sato, S. 1982, PASJ,34 389
Kent, S., Dame, T., & Fazio, G. 1991, ApJ 378 131
Kurucz, R.L. 1994, priv. comm.
Maihara, T., Oda, N., Sugiyama, T., & Okuda, H. 1978, PASJ,30 1
Mamon, G. A. & Soneira, R. M. 1982, ApJ, 255, 181
Matsumoto, T., Akiba, M., & Murakami, H. 1988, ApJ 332 575
Noda, M. *et al.* 1992, ApJ 391 456
Meadows, V. S. 1994, *Ph.D. Dissertation*, Univ. of Sydney, Australia
Mikami, T., & Ishida, K. 1981, PASJ,33 135
Mikami, T., Ishida, K., Hamajima, K., & Kawara, K. 1982, PASJ,34 223
Ortiz, R., & Lepine, J.R.D. 1993, A&A, 279, 90
Ruelas–Mayorga, A. 1991 Rev. Mex. Astr. Astrof., 22, 27
Schmidt, M. 1975, ApJ 202 22
Wainscoat, R. J., Cohen, M., Volk, K., Walker, H. J. & , Schwartz, D. E. 1992, ApJS 83 111

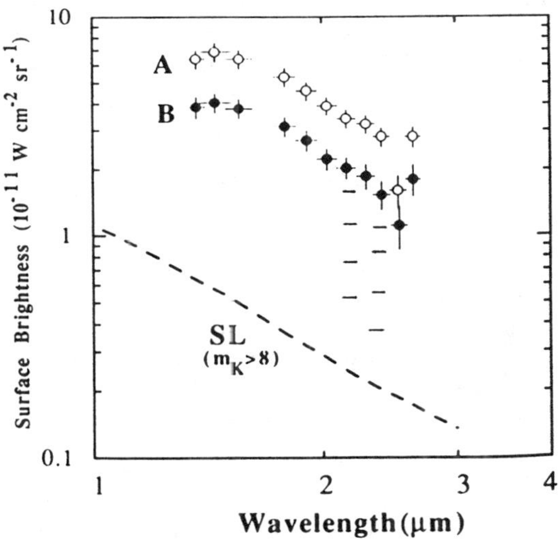

FIGURE 3. The celestial emission detected by Noda *et al.* (1992) at points "A" and "B". The four short bars at the wavelengths of K and [2.4] represent (from bottom to top) the Galactic point source foreground predicted by SKY at point B after removing stars brighter than mag. 5^m, 3^m, 0^m, and finally removing no stars. (The dashed line, "SL", is the stellar foreground predicted by Noda *et al.* , after removing stars brighter than K=8.)

THE INTERSTELLAR DUST CONTRIBUTION TO THE DIFFUSE INFRARED SKY BRIGHTNESS

J. L. Weiland
General Sciences Corp.,
Code 685.3, NASA Goddard Space Flight Center, Greenbelt, MD 20771

R. G. Arendt, T. J. Sodroski
Applied Research Corp.,
Code 685.3, NASA Goddard Space Flight Center, Greenbelt, MD 20771

ABSTRACT

We use *COBE*/DIRBE data from which zodiacal light and Galactic stellar foregrounds have been removed in order to derive an all-sky view of the Galactic cirrus. The dust is seen in emission from 3.5 to 240 μm; the scattered contribution at 1.25 and 2.2 μm is not readily discernible. Similar spatial structure allows the correlation of the dust signal in each DIRBE band with a DIRBE 100 μm cirrus template. The slopes of these correlations are used to derive an average dust spectrum at high Galactic latitudes. The effectiveness of using this average spectrum to remove the contribution of the dust to the diffuse infrared sky brightness is discussed. We also place the DIRBE-derived average dust spectrum in context with observations in the submillimeter (from *COBE*/FIRAS) and optical.

1. INTRODUCTION

Hauser (1995) has outlined the process by which the DIRBE team has chosen to characterize and remove the contributions of solar-system and Galactic foregrounds from the DIRBE data in order to search for the cosmic infrared background. We will describe in greater detail the last step in this 'onion-peeling' process, which addresses the removal of the Galactic foreground arising from interstellar medium (ISM) emission. To aid in this analysis, we will make use of products from previous steps in our foreground removal process: namely, DIRBE maps in which the signal from interplanetary dust (IPD) has been subtracted (Reach et al. 1995a, Franz et al. 1995), and subsequent map products in which the contributions from bright sources and integrated Galactic starlight ('faint sources') have also been removed (Arendt et al. 1995).

In order to construct maps of the ISM emission at DIRBE wavelengths, we adopt a simple methodology. We assume that, for any wavelength, the ISM emission is separable into independent spatial and spectral parts. We adopt a spatial template to represent the angular structure of the ISM at any wavelength, and assume that the spectral energy distribution of the emission has no dependence on sky location. Thus, the ISM contribution at each DIRBE wavelength may be computed as:

$$I_\lambda^{ISM}(l,b) = R_\lambda \times G(l,b) \qquad (1)$$

where R_λ is a scaling factor, constant over the whole sky, and $G(l,b)$ represents the spatial template.

© 1996 American Institute of Physics

We devote section 2 to the description of the spatial template, and section 3 to the derivation of the color factors, R_λ. A discussion of our findings is presented in section 4.

2. CHOOSING A SPATIAL TEMPLATE

Ideally, a spatial template of the ISM angular structure should meet the following requirements: 1) the template would represent only the ISM component: *i.e.*, no signal from solar system, stellar, or extragalactic components should be present. 2) the template would have a similar spatial resolution to that of DIRBE, so that it would reproduce observed structures on comparable angular scales, 3) the template would have good signal-to-noise ratio, so as to not introduce extraneous noise, and should be relatively free of artifacts or calibration errors, 4) the template would represent spatial components seen by all chosen wavelengths, and 5) the template would provide all-sky coverage.

An obvious choice for the template would be an all-sky H I map at roughly 1/3 degree resolution. The best publicly available H I survey with good side-lobe rejection is currently the Bell Labs Survey (Stark et al. 1992), which has an angular resolution of 2° and is missing sky coverage at southern declinations. As this is not entirely sufficient for our purpose, a hybrid solution has been chosen. As a compromise, we use the DIRBE 100 μm residual map after zodiacal emission removal as the spatial template for all DIRBE wavelengths other than 100 μm. This resolves the sky-coverage difficulties and resolution problems presented by use of the Bell Labs survey. However, we still need an independent template for the 100 μm data, and so we use the Bell Labs map as the template for this wavelength.

Before the DIRBE 100 μm residual map may be used as a spatial template for the ISM, its zero-point must be adjusted so that it matches our best determination of the minimum background intensity of the ISM component. To accomplish this, we choose a region of sky for which H I, H_2 and H^+ contributions are well-characterized at DIRBE resolution, and correlate with the DIRBE observations. We are presently using the H I observations of an 8° × 9° region centered on the north ecliptic pole (NEP) obtained by Elvis, Lockman & Fassnacht (1994). The region has been observed at a resolution of 21 arcmin, and the observations have been calibrated to match the Bell Labs H I survey. Firm upper limits on the H_2 contribution are available, although the contribution from H^+ is less certain.

Figure 1 shows the DIRBE 100 μm residual data plotted vs. the H I data of Elvis, Lockman & Fassnacht (1994) for this NEP region. The H I data have been resampled to DIRBE resolution. The two datasets are very well correlated: the straight line is a linear fit through the data. The intercept of this fit is used to define the zero-point for the DIRBE 100 μm template.

With the determination of the intercept above, the spatial template $G(l,b)$ is now fully defined:

$$G(l,b) = \begin{cases} \text{Bell Labs H I map}, & \text{for } \lambda = 100\ \mu\text{m}; \\ I_{100\mu m}^{DIRBE} - ZL - intercept, & \text{for } \lambda \neq 100\ \mu\text{m} \end{cases} \quad (2)$$

We now need to derive the scaling factors, R_λ, for each wavelength.

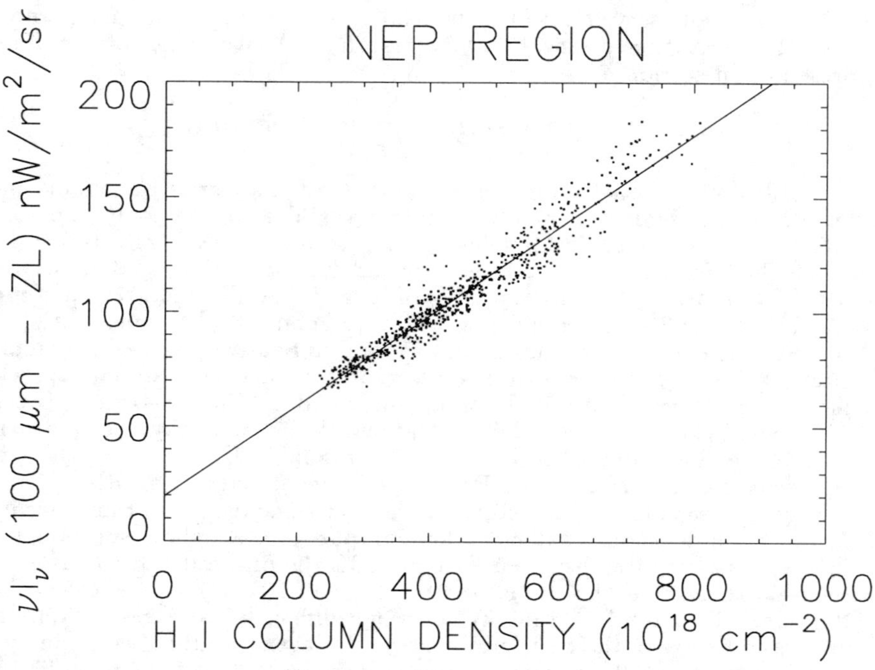

Figure 1: Correlation of DIRBE 100 μm data after zodiacal emission removal with H I column density for the NEP region of Elvis, Lockman & Fassnacht (1994).

3. DERIVATION OF SCALING FACTORS, R_λ; ISM REMOVAL RESULTS

The method used to derive R_λ is wavelength dependent, and falls into three broad regimes: 1) 100 μm, 2) mid- and far-infrared wavlengths other than 100 μm, and 3) the near-infrared, defined as $\lambda \leq 4.9\mu$m.

For 100 μm, R_{100} is simply defined as the slope found in Figure 1. As per equation (1), the ISM intensity at 100 μm is found by multiplying R_{100} by the Bell Labs H I column density map. A residual map which has the IPD and ISM components removed still shows some ISM-related structures with angular scales which are less than the 2° resolution of the Bell Labs H I map. In addition, there is a weak residual signal which is concentrated toward the ecliptic plane, indicating that the zodiacal foreground was not removed completely from the 100 μm map. Thus, the ISM spatial template derived from the 100 μm residual map is not completely free of zodiacal light artifacts, and use of this template introduces an artifact into the ISM-removed maps along the ecliptic plane.

In the mid-infrared (12, 25, 60 μm) and far-infrared (140, 240 μm), where emission from the ISM component is clearly seen, we derive R_λ from the slope of the linear least-squares correlation of the zodiacal emission- and stellar-foreground removed maps with the ISM spatial template. For the far-infrared wavelengths, the correlations were fit for the regions where $|b| > 45°$. Using a cut at lower latitudes resulted in a steeper slope, which we attribute to the influence of bright clouds found only at the lower latitudes. A higher latitude cut resulted in a shallower slope, possibly an artifact of the poorer correlation provided by the fainter, very high latitude cirrus. We settled on the 45° cut as a compromise between the weaker correlations at higher latitudes, and the

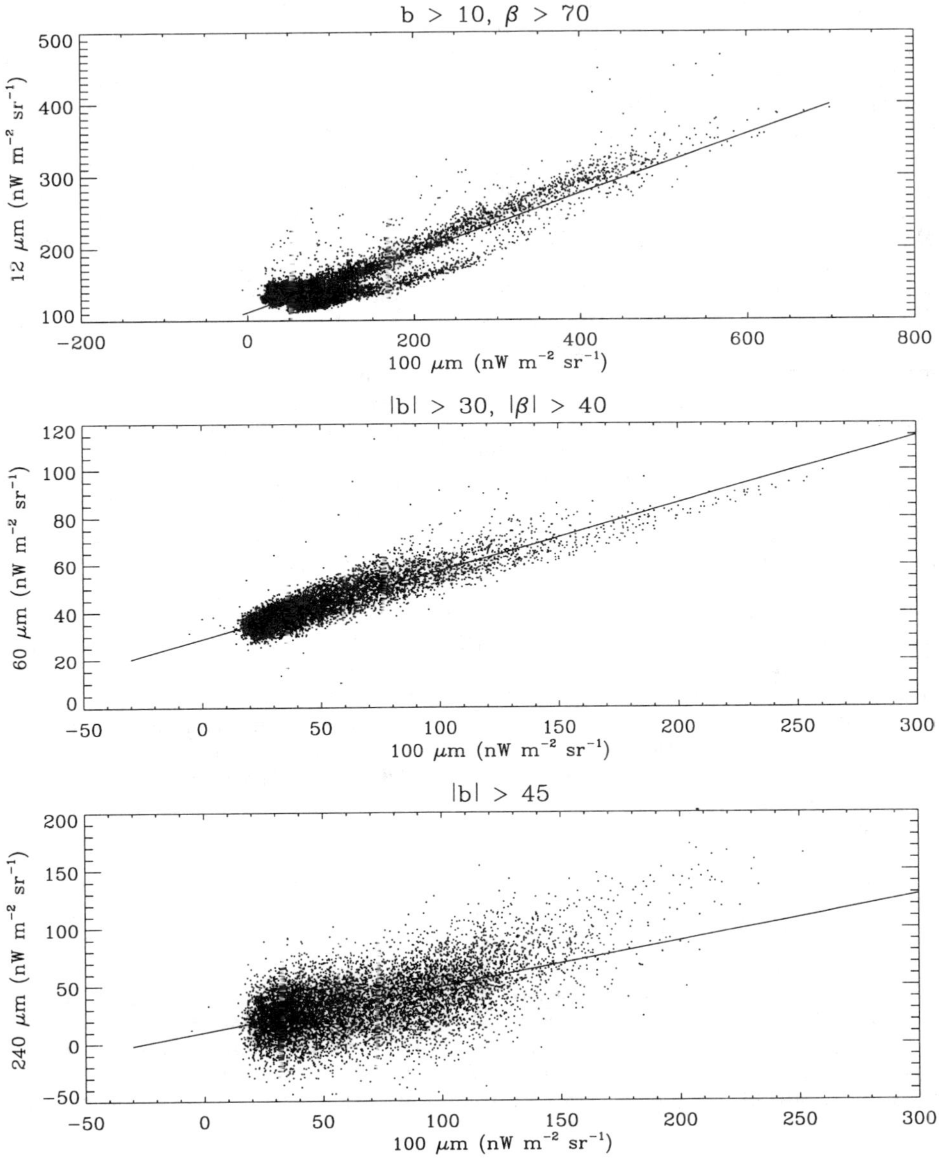

Figure 2: Plots of correlations used to derive R_{12}, R_{60}, and R_{240}. The sky regions used in each correlation are indicated by the limits on Galactic latitude, b, and ecliptic latitude, β

intrinsic temperature variations between the clouds visible at the low latitudes. Residual maps show a great reduction of ISM emission, but still possess clear indications of its presence, especially at the lower Galactic latitudes. At mid-infrared wavelengths, remaining low-level residuals from zodiacal light removal are large enough that we need to restrict the region used for correlation in order to avoid contaminating the result. Smaller regions where there were no visual indications of obvious structures were used: $|b| > 30°$ and $|\beta| > 40°$ at 60 μm, and $\beta > 70°$ at 12 and 25 μm. At these wavelengths, use of a single color factor R_λ provides good apparent removal of the cirrus for most of the sky at $|b| > 20°$. The correlation plots for 12, 60 and 240 μm are shown in Figure 2.

For the near-infrared wavelengths, the stellar foreground removal is not sufficiently accurate to reveal the ISM in the residual intensity maps. This is because the stellar foreground is computed using a statistical model (Arendt et al. 1995, Wainscoat et al. 1992) which does not follow the exact placement of stars on the sky; the high density of residual low-intensity bumps and wiggles serve to hide the cirrus. Therefore, direct correlations between the near-infrared data and $G(l,b)$ are not useful for deriving R_λ. Instead, we backtrack to using the near-infrared maps which have the zodiacal light, but not the stellar, foreground removed, and make use of the color differences between starlight and the ISM emission as a discriminant. Colors formed from the simple ratio of two near-infrared intensities reveal extinction effects as well as intrinsic source variations. To remove the influence of extinction, we construct maps of a reddening-free parameter, Q (Johnson & Morgan 1953). Q is essentially the intercept on a color-color plot, and only varies if the intrinsic source of emission varies. A map of $Q_{1.25,2.2,3.5}$ (generated from the DIRBE 1.25, 2.2 and 3.5 μm maps after zodiacal light removal) clearly shows cirrus structures which match those seen at 100 μm.

The problem of finding R_λ for the near-infrared wavelengths now reduces to one of relating Q to R_λ. We can do this by making some simplifying assumptions. The first of these is that cirrus emission in the 1.25 and 2.2 μm maps is negligible. This assumption is based upon the observation that the 1.25/2.2 μm colors show little or no evidence of diffuse ISM features, whereas 2.2/3.5 μm colors show such features clearly. The second assumption is that the bulk of Galactic stellar emission can be characterized by a single value of Q, which we designate as $Q(stars)$. After algebraic manipulation, we arrive at the following equation relating R_λ and Q:

$$R_\lambda = \frac{I_\lambda}{G(l,b)} \times \left(1 - \frac{Q_{1.25,2.2,\lambda}(stars)}{Q_{1.25,2.2,\lambda}}\right) \tag{3}$$

where λ is either 3.5 μm or 4.9 μm, and I_λ is the intensity at wavelength λ after removal of the IPD component. We then simultaneously solve for R_λ and $Q_{1.25,2.2,\lambda}(stars)$ using linear least squares over a restricted region of the sky. Using this method, we derive values of R_λ at 3.5 and 4.9 μm, but not at 1.25 and 2.2 μm. Residual maps at 3.5 and 4.9 μm show little or no change in appearance after removal of this faint cirrus component.

4. DISCUSSION

The methods by which we derive the ISM emission contribution to the Galactic foreground observed by DIRBE have been described. Our values for the ISM contribution to the North Galactic Pole emission for wavelengths from 3.5 to 240 μm have been presented by Hauser (1995). In his Figure 1, we see that ISM emission has a small contribution to the total observed intensity for wavelengths shortward of 60 μm, and is most important at 100 μm and longward.

It is also interesting to compare the DIRBE results with those derived from other experiments. Figure 3 presents a composite high-Galactic latitude spectral energy distribution, normalized to 1.0 at 100 μm. The DIRBE results are the R_λ values derived in section 3, and are plotted as open diamonds connected by a solid line. The broken line, which extends the distribution into the submillimeter, is the spectrum reported by Reach et al. (1995b) using FIRAS observations for $|b| > 45°$. Optical and IRAS observations of selected high-latitude clouds reported by Guhathakurta & Tyson (1987) are plotted as triangles connected by dashed lines. The data from Guhathakurta & Tyson indicate increasingly large scatter from cloud to cloud as one goes to shorter wavelengths. Although we presently have no DIRBE information about the spectrum at 1.25 and 2.2 μm, where the emission is expected to arise from scattering of the interstellar radiation field, we can see from a crude model (solid line extending from .45 to 2.2 μm, as calculated using the methodology of Guhathakurta & Tyson) that it is not expected to be a large contributor.

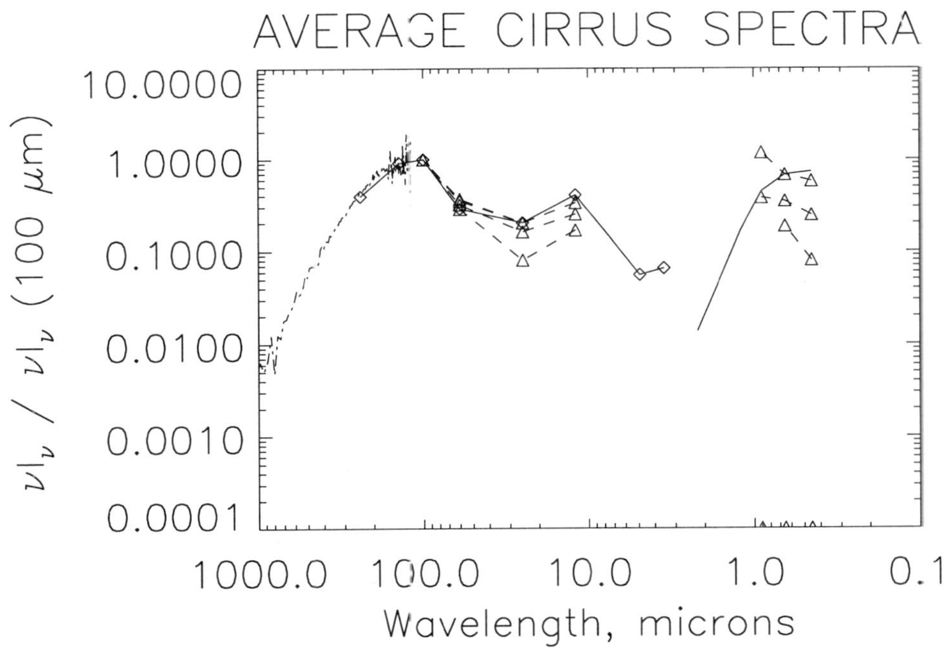

Figure 3: Average DIRBE cirrus spectrum (normalized to 100 μm), in context with other high Galactic latitude cirrus observations. See text for description of data sources and symbols.

ACKNOWLEDGEMENTS

COBE is supported by NASA's Astrophysics Division. Goddard Space Flight Center (GSFC), under the scientific guidance of the *COBE* Science Working Group, is responsible for the development and operation of *COBE*. JLW thanks N.P. Odegard and G.L. Rawley for their assistance in the preparation of this talk.

REFERENCES

Arendt, R.G., et al. 1995, this workshop
Elvis, M., Lockman, F.J. & Fassnacht, C. 1994, ApJS, 95, 413
Franz, B.A., et al. 1995, this workshop
Guhathakurta, P. & Tyson, J.A. 1989, ApJ, 436, 773
Hauser, M.G. 1995, this workshop
Johnson, H.L. & Morgan, W.W. 1953, ApJ, 117, 313
Reach, W.T., et al. 1995a, this workshop
Reach, W.T., et al. 1995b, ApJ, in press
Stark, A.A., et al. 1992 ApJS, 79, 77
Wainscoat, R.J., et al. 1992 ApJS, 83, 111

EXPLORING THE INFRARED STELLAR POPULATIONS OF THE GALAXY WITH DENIS

First results of star counts

Stéphanie Ruphy and Nicolas Epchtein
DESPA, Observatoire de Paris, 5 place J.Janssen, 92195 Meudon Principal cedex, France
ruphy@megasx.obspm.fr

ABSTRACT

The DENIS project (Deep Near Infrared Survey of the Southern Sky) will provide, for the first time, deep star counts on a large scale, in the I, J, K_s photometric bands. A first analysis of star counts in the J and K_s bands in various directions of our Galaxy, using preliminary data obtained with DENIS are presented. The counts are compared to the results of currently available models of stellar population synthesis for the Galaxy, in order to derive some constraints on the parameters of these models.

1. INTRODUCTION

COBE/DIRBE has carried out a large scale exploration of the sky at low spatial resolution in the infrared range from 1.25 to 200 microns. The wide beam of DIRBE ($0.7^o \times 0.7^o$) generally hampers the discrimination of the stellar and diffuse radiation contributions. In the–near infrared, the galactic radiation is largely dominated by the stellar component. The separation of the diffuse background from this component can be achieved by comparison with models of stellar population synthesis (Cohen, this conference), or with high spatial resolution deep near–infrared surveys. Two projects for deep near–infrared surveys of the sky are presently under definition or in commissioning phases, namely the 2MASS project (Kleinmann *et al.*, 1994) and DENIS (Epchtein *et al.*, 1994). It is expected that the star counts that will be derived from these surveys in the J and K bands will be extremely useful to improve the models of the Galaxy and for direct comparison with DIRBE measurements.

2. THE DENIS PROJECT

DENIS will provide, before the end of the century, the first digitized survey of the sky in the declination range ($+2^o$ to -88°) in 3 near-infrared bands (I, J, K_s). This survey will be achieved with a specially designed focal instrument set–up at the 1 meter telescope of the European Southern Observatory at La Silla in Chile. This instrument is equipped with 3 cameras, each dedicated to one spectral band. The infrared channels (J and K_s) are equipped with a NICMOS 3 array made by *Rockwell* and the red channel (I) with a 1024× 1024 CCD array from *Tektronix*. The pixel sizes are 3 arcseconds for the infrared channels and 1 arcsecond for the I channel. The 3 σ detection limit in the nominal integration time of $\approx$ 10 seconds are 18, 16 and 14 in the I, J and K_s bands, respectively. The DENIS instrument has been in operations since September 1994, with the

© 1996 American Institute of Physics

J and K_s channels, for a pilot survey.

3. MODELS OF INFRARED STELLAR POPULATIONS

The stellar populations that dominate the infrared radiation of an evolved galaxy such as our Milky Way are, the K and M main sequence stars, the red giants and AGB stars, the red dwarfs, the young stellar objects embedded in molecular clouds, and the old halo population. Depending on the observed direction in the Galaxy and on the completion limiting magnitudes, one or several of these populations is dominating. Their relative contribution can be estimated with models of the point source sky such as :

* The "Besançon" model (Bienaymé *et al.*, 1987) is a self-consistent model of the Galaxy, in the sense that the model parameters are forced to follow physical laws (physical links between the density, velocity, and metallicity distribution are taken into account).

* The "sky" model (see Cohen, 1994 for reference) is based on a realistic representation of the Galaxy, including features such as disc, spiral arms and local spurs, molecular ring, bulge, halo, and delivers star counts in many filters lying within a large spectral range, from far-UV to far-infrared. Its specificity results also from its provenance in the mid-infrared range rather than in the visible range. It should be emphasized that some parameters describing the stellar populations, such as the scale lengths, the scale heights, the existence of the thick disc, are quite different in the two models.

4. DENIS STAR COUNTS COMPARED WITH MODELS

During the pilot survey, hundreds of square degrees have been observed in J and K_s. Among these data, three fields have been selected for star counts at different galactic latitudes.

4.1 At high and intermediate galactic latitudes

The counts in J and K_s are performed in two areas of the sky, one of 2.6 square degrees, at high galactic latitude (l = 240°, b = -61°), containing 1970 stars in J and 620 stars in K_s, and the other one of 0.2 square degree, at intermediate galactic latitude (l = 202°, b = −17.5°), containing 770 stars in J and 260 stars in K_s.

Figures 1 and 2 show predictions given by the two models, in the J and K bands, compared to observations for the high galactic latitude field. The agreement is good for the two models (the discrepancy for the bright sources is not significant, since they are very few). At intermediate galactic latitude, the agreement is also satisfactory.

4.2 At low galactic latitudes near the anticenter

Star counts in J and K_s are performed in several 0.2 square degree fields, in the galactic plane, around the l = 210° direction. The agreement of the two models with the DENIS data is not as good as in high galactic latitude fields, as shown in Figs. 3 and 4 for the "Besançon" model. However, both models present very similar discords with the DENIS data. These discrepancies cannot be entirely due to an inaccurate estimation of the interstellar extinction in the models, since the extinction in the near-infrared is very low in this direction of the galactic plane. They rather reveal some possible weaknesses of the models, such as, for instance, the scale height of the K giants, that still need to be investigated.

4.3 Integrated flux

In order to allow the comparison of the DIRBE and the DENIS data, we have calculated the DENIS integrated fluxes on the three fields studied in the previous sections. The table summarizes the DENIS data.

Table 1: Number of stars and integrated fluxes detected with DENIS, in three different directions.

direction	l=240° b=-61°	l=202° b=-17.5°	l=213° b=3°
Nb d'*/deg^2 (J≤15.5)	620	3650	8600
Nb d'*/deg^2 (K≤13.5)	240	1200	1425
$Flux_J/deg^2$(Jy) (J≤15.5)	14	56	133
$Flux_K/deg^2$(Jy) (K≤13.5)	9.6	30.3	92

We have not yet attempted to compare the integrated fluxes derived from the DENIS data, with the DIRBE measurements. However, we plan to achieve this comparison on a larger sample of galactic directions.

5. CUTOFF OF THE THIN DISC AT 5.5 KPC FROM THE SUN ?

Star counts in the visible range have recently revealed the possible existence of a sharp cutoff of the thin disc at about 5.5 kpc from the Sun, for a scale length of 2.5 kpc (Robin *et al.*, 1992). These values can be measured with DENIS data, in directions close to the anticenter. The cutoff will be clearly visible at 0.6≤J-K≤1, and 12≤K≤13.5. A first attempt to confirm this cutoff with the DENIS data can be seen in Fig. 5, where a J-K histogram, obtained on a 0.2 square degree field at l=213° and b=3°, is displayed along with J-K predicted histograms with and without cutoff : when the Besancon model takes into account the cutoff, the predicted number of stars is largely outnumbered by the predicted number with no cutoff. A qualitative comparison of these three histograms seems to favour the existence of the cutoff. But this very preliminary result needs to be confirmed on larger samples, with a better limit of completion.

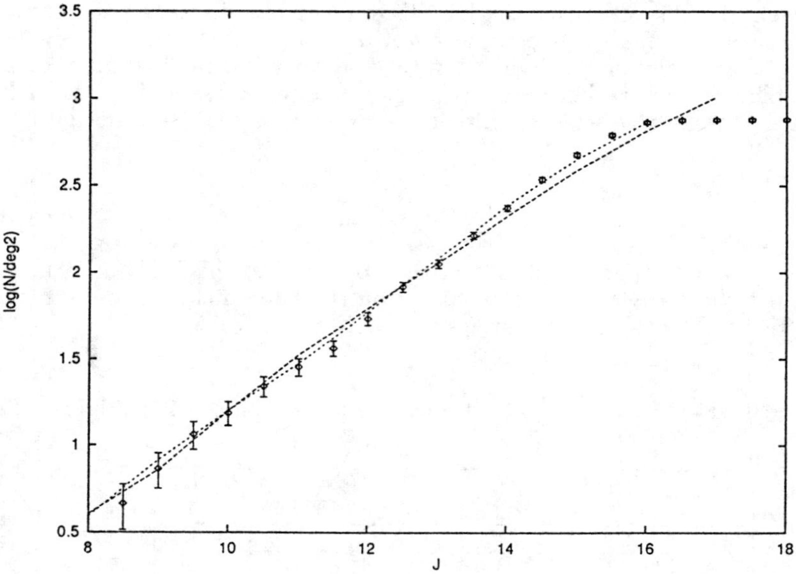

FIGURE 1. Cumulative star count around l = 220° and b = -61°, in the J band, as compared to the model predictions (long dashes : "Besançon" model, short dashes :"Sky" model, diamonds with poissonian error bars : DENIS data)

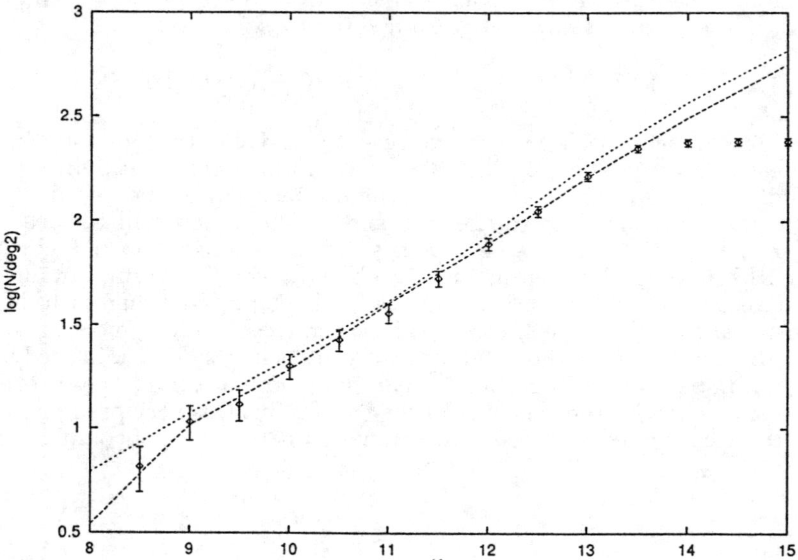

FIGURE 2. Same as fig. 1, in the K_s band.

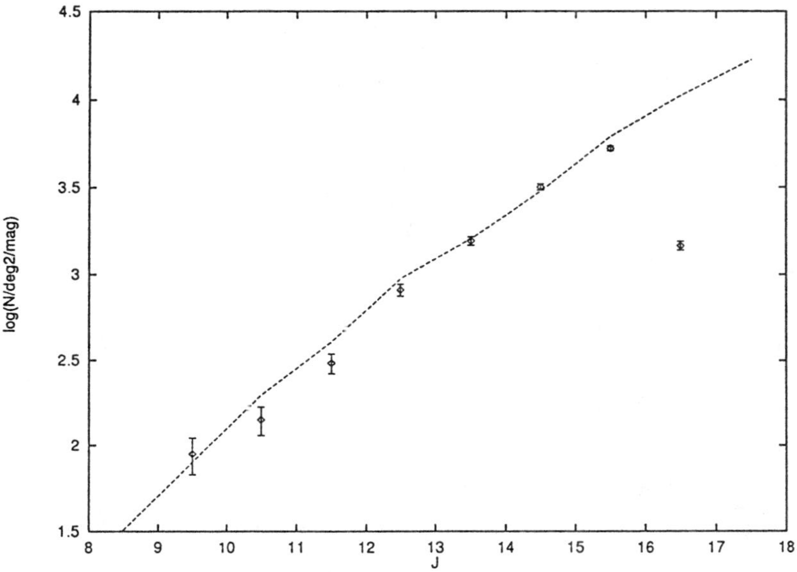

FIGURE 3. Differential star counts around l = 215.8^o and b = 1.5^o, in the J band, as compared to the "Besançon" model prediction (short dashes : "Besançon" model, diamonds with poissonian error bars : DENIS data).

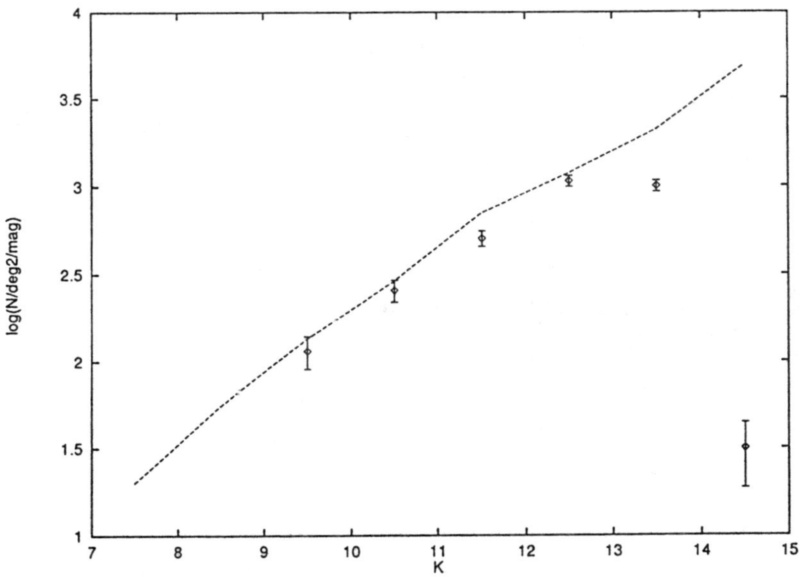

FIGURE 4. Same as Fig. 3, in the K_s band. (In that particular field, due to a poor quality of the K-background, the sample is complete only up to 12.5).

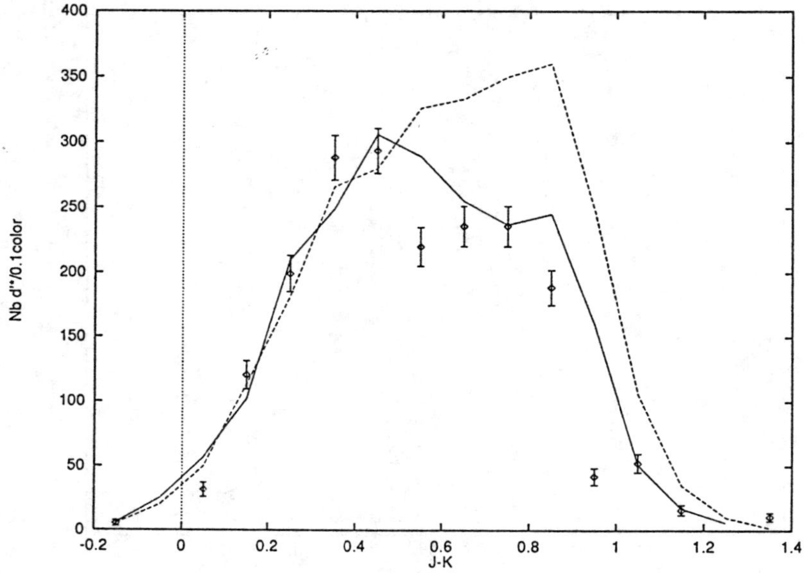

FIGURE 5. J-K predicted distribution by the "Besançon"model, toward ($l = 212.8^o$, $b = 3^o$), in the $12 \leq K \leq 13.5$ interval : solid line, prediction with a cutoff of the thin disc at 5.5 kpc from the Sun ; dashes, prediction without cutoff ; diamonds, DENIS data with poissonian error bars.

ACKNOWLEDGEMENTS

The DENIS team is warmly thanked, specially E. Bertin for providing us with his source extraction software. We also thank A. C. Robin and M. Cohen who kindly made their models of the Galaxy available for us.

REFERENCES

Bienaymé O., Robin A. C., Crézé M., 1987, A&A, 180, 94
Cohen M., 1994, Astrophys. and Sp. Sc., 217, 181
Epchtein N. + 25 authors, 1994, Astrophys. and Sp. Sc., 217, 3
Kleinmann S. G., + 17 authors, 1994, Astrophys. and Sp. Sc., 217, 11
Robin A. C., Crézé M., Mohan V., 1992, A&A, 265, 32

Far-Infrared Emission from the Galaxy in Extra-Galactic Windows

F. Boulanger, A. Abergel, J.P. Bernard, F.X. Désert, G. Lagache, J.L. Puget
Institut d'Astrophysique Spatiale
Université Paris XII, Batiment 121, 91405 Orsay, France

W.B. Burton, D. Hartmann
Sterrewacht Leiden
Postbox 9503, 2300 RA Leiden, The Netherlands

ABSTRACT

A correlation analysis between DIRBE and FIRAS sky maps and the new Leiden- Dwingeloo H I survey enables us to establish the large scale spectral and spatial properties of the far-infrared dust emission over the high Galactic latitude sky. This work provides the information on the Galactic emission necessary to support the search for the far-infrared Cosmic Background.

1. INTRODUCTION

Study of the infrared emission at high Galactic latitude with the IRAS data has provided valuable information about the distribution of nearby interstellar matter and emission properties of dust. A decade later the data collected by the COBE satellite allow to extend this work to the whole emission spectrum of dust. This new perspective holds promises of discoveries which have just started to emerge from the data analysis. But understanding the emission from Galactic dust is also a major hitch in the search for the far-IR cosmic background. Since this was the topic of the workshop, we approach the subject from this last view point. To refer to the announcement poster of the workshop, we show how we can peel off the Galactic skin of the sky.

A natural way to do so is to search for the contribution of each gas component to the IR emission. In terms of column density, the three most important gas components in the nearby interstellar medium are the neutral atomic gas, the diffuse ionized gas and the molecular gas. We start this paper by summarizing information available on these three gas components. We then present results from a spatial correlation between the DIRBE and FIRAS data and the new Leiden-Dwingeloo H I survey. From this we derive the far-IR emission spectrum from dust in the H I gas and a map of IR residuals which is used to look for the contribution of ionized and molecular gas. The readers are referred to other papers in these proceedings for the description of the COBE instruments and to Boulanger et al. (1995) for technical details about the data analysis behind the results presented here.

2. Nearby Interstellar Matter

2.1 Neutral Atomic Gas

The neutral atomic gas is the dominant component of the nearby interstellar medium. It is also the best known since it is traced by the well studied

© 1996 American Institute of Physics

21cm line of atomic hydrogen. The IR contribution from this component may be accurately measured by correlating the spatial distribution of the IR and 21 cm emission. The results presented here make use of the Leiden-Dwingeloo survey which has been recently completed with the 25m antenna at Dwingeloo (Hartmann 1994, Hartmann and Burton 1995). It covers all the at $\delta > -30°$ with an angular resolution of 36' which matches that of DIRBE. The kinematic coverage, $-450 < v_{LSR} < 400 km/s$ at a spectral resolution of $1.03 km/s$ covers the whole range of Galactic velocities. The stray radiation has been fully modeled and subtracted from the raw data. In the regions of lowest emission the stray radiation was observed to account for up to half of the measured emission. The data of the Leiden/Dwingeloo survey represent an improvement over that of earlier large-scale surveys by an order of magnitude or more in at least one of the principal parameters of sensitivity, spatial coverage, or spectral resolution.

2.2 Molecular Gas

Distribution of molecular gas in the nearby interstellar medium has been studied from two different view points. (1) Statistical information about the distribution of molecular gas is available from measurements of UV absorption lines of H_2 in the direction of nearby stars. These observations were done with the Copernicus satellite. Based on a survey of 109 stars Savage et al. (1977) found that the fraction of molecular gas, $f = 2N(H2)/(2N(H2) + N(HI))$ is very low ($f < 0.01$) where $E(B - V) < 0.08$ (N(H) smaller than $5 10^{20} cm^{-2}$ for the normal extinction curve). Around this threshold value the fraction of molecular hydrogen shows an abrupt transition to values ~ 0.1; above the threshold values of f for a given total gas column density have a one order of magnitude scatter but there is a clear trend for the median value to increase steadily with N(H). These results have been interpreted as the effect of the self-shielding against photo-dissociation onto the abundance of H2 (Federman et al. 1979). (2) Since the initial survey by Magnani et al. (1985) numerous high latitude clouds have been observed in CO. But all together only a very small fraction of the sky has been observed. We are thus far from having an overall view at the spatial distribution of molecular gas at high latitude. Also, the question: how much of the molecular hydrogen in clouds of low column density is traced by detectable CO emission at the sensitivity of present mapping programs, is still opened.

2.3 Warm Diffuse Ionized Gas

Observations of H_α emission at high Galactic latitude and dispersion measurements in the direction of pulsars indicate that low density ionized gas (the Warm Ionized Medium, WIM) accounts for about 25% of the gas in the Solar Neighborhood (Reynolds 1989). The column density from the mid-plane is estimated to be in the range 0.8 to $1.4 10^{20} cm^{-2}$. Little is known about the spatial distribution of this gas. Only one region at high latitude has been completely mapped. Comparison between the H_α and H I emission in this region shows a significant but not complete correlation between the two gas components (Reynolds et al. 1995). This is important because IR emission from spatially correlated ionized and neutral gas cannot be separated. From the dispersion measures towards a few pulsar in the Galactic halo, the scale height of the WIM has been estimated to 1.5 kpc (Reynolds 1989). Half of the WIM is at a z distance from the mid-plane larger than $1 kpc$ while this fraction is less than 15% for the H I gas (Lockman, et al. 1986). At $|z| > 1kpc$, the WIM is thus the dominant component of the interstellar medium. The volume filling factor of the WIM has been estimated to be at least $\sim 20\%$ (Reynolds 1991). The WIM which is widespread throughout a thick disk must be rather smoothly distributed over the high latitude sky.

3. Emission from Dust in H I gas

3.1 IR-H I Spatial Correlation

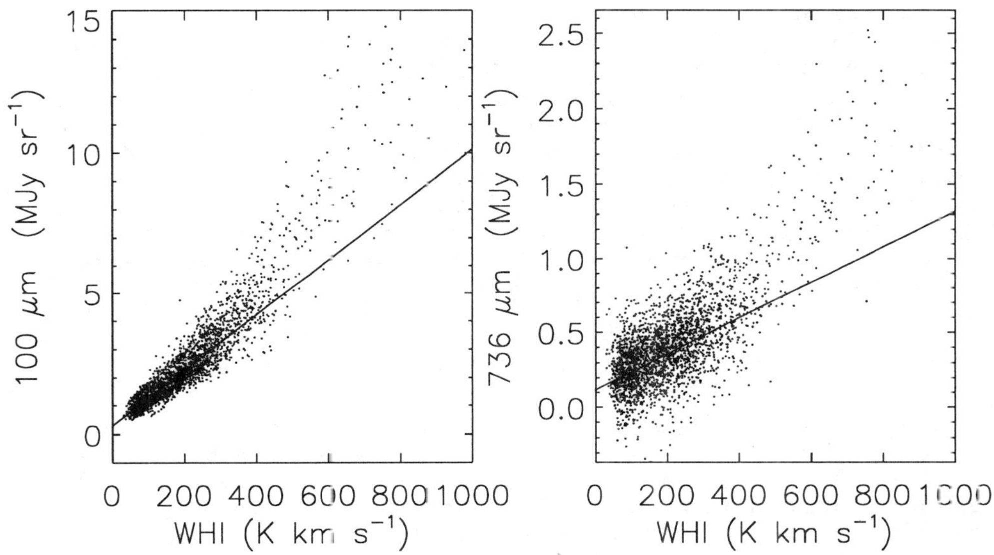

Figure 1: Correlation between IR emission and H I emission at $100\mu m$ (DIRBE data smoothed to the 7° resolution of FIRAS) and at $736\mu m$ (FIRAS LLSS data averaged between 600 and $900\mu m$). All data points at $|b| > 20°$ with measured H I emission are shown. The solid line represents the linear fit to data points with $WHI < 250Kkm/s$. WHI is the HI emission integrated from $V_{LSR} = -100$ to $100km/s$.

Several studies based on the IRAS data have shown that the $100\mu m$ emission outside of the Galactic plane and the main nearby molecular complexes is well correlated with the distribution of the neutral atomic gas as inferred from the H I emission (e.g. Boulanger and Pérault 1988, Deul and Burton 1993). Regions of local excess IR emission exist but they globally represent a minor fraction of the total emission (Désert et al. 1988). We have repeated this analysis exploiting the wavelength coverage and sensitivity of the DIRBE and FIRAS data on the one hand, and the sensitivity and velocity coverage of the Leiden/Dwingeloo H I material on the other.

The initial step in this correlation study is to prepare maps of the Galactic Emission. To do this one must subtract from the data the zodiacal light and the Cosmic Microwave Background. The Cosmic Microwave Background and the dipole component are easy to remove because their parameters are accurately known. It is not so easy to subtract the zodiacal light but at long wavelengths ($\lambda > 100\mu m$ a very crude modeling provides a sufficiently accurate subtraction. The large difference between Galactic and Zodiacal emission spectra allow to use the $25\mu m$ DIRBE map as a template for the zodiacal light. This template is scaled to the longer wavelengths with one single emission ratio. Outside warm sources the $25\mu m$ Galactic emission present in the $100\mu m$ zodiacal map is ~ 0.008 of the $100\mu m$ Galactic emission. This fraction decreases with increasing wavelengths; it is 5 times smaller at $140\mu m$ than at $100\mu m$. It is a clear over simplification to use a single emission ratio for the zodiacal emission, independent of ecliptic coordinates. But it does not produce an obvious distortion on the Galactic maps even at $100\mu m$.

IR-H I comparison diagrams are presented in Figure 1. Where the H I emission integrated over velocities is low , $WHI < 250Kkm/s$ ($1Kkm/s = 1.8210^{18} Hcm^{-2}$ for optically thin emission), a good correlation is observed at all far-IR wavelengths from $100\mu m$ to 1mm. The IR/HI emission ratio derived from a linear fit to these data points are presented in Figure 2. For $WHI > 250Kkm/s$ the correlation persist but with a higher IR/HI emission ratio. This is further discussed in Sect. 4. We checked that the results of the correlation are not biased by an additional Galactic component such as that expected from the diffuse ionized gas by correlating the IR and HI maps after subtraction of a mean cosecant law removing in both data the large scale Galactic latitude gradients. The two procedures led to the same results.

3.2 Emission Spectrum

We have computed the spectrum of the emission associated with the H I gas in the following way. We have selected in the H I map at the FIRAS resolution all sky pixels which have an emission integrated in velocity, WHI, lower than 250 Kkm/s (30% of the whole sky). We have divided these pixels in two sets of roughly equal number of data points with the two criteria: $WHI < 140Kkm/s$ and $140 < WHI < 250Kkm/s$. We have averaged all data points within each of these two sets to produce two mean emission spectra. This data averaging is necessary to gain in signal to noise. The fact that the signal to noise ratio on the H I emission is large compared to the scatter in the IR-HI diagrams ensures that this binning does not bias the measured IR to H I emission ratio. The emission spectrum presented in Figure 2 is the difference between the two mean spectra. The difference removes any additional emission not correlated with the H I emission. The difference spectrum has been normalized per unit HI column density by dividing it by the difference in mean H I emission for the two sets of points.

The FIRAS spectrum is fit well by a single modified Planck curve with an emissivity $\tau = 6.5\ 10^{-5} \times \nu^2$ and a temperature of 17.5 K. Wright et al. (1991) and more recently Reach et al. (1995) have emphasized he fact that the FIRAS spectra for the Galaxy may not be fitted with a single temperature and proposed the existence of a very cold dust component with a temperature of 5-7K. In their spectral decomposition of FIRAS spectra Reach et al. found a correlation between warm and very cold component with a mean optical depth ratio $\tau(\text{cold})/\tau(\text{warm}) \sim 7$. For the spectrum of dust in H I gas, the residuals to the single temperature fit show no sign of an additional very cold component at the level found by Reach et al. (1995). Assuming a temperature of 7K for this component we can set an upper limit on the optical depth ratio on $\tau(7K)/\tau(17.5K) \sim 1$. The cold component found at high Galactic latitude by Reach et al. is unlikely to be Galactic because it shows no correlation with the distribution of interstellar matter nor Galactic coordinates (Puget et al. 1995). Of course, the results presented here do not exclude that localized features may have flatter emission spectra (like that of the Galactic plane, Wright et al. 1991) because of localized changes in dust properties and/or because of variations in the radiation environment.

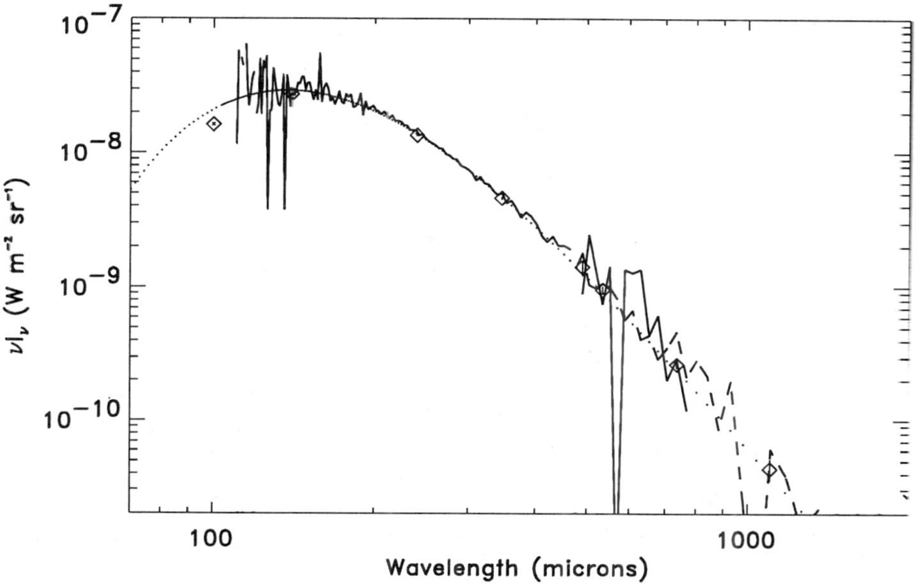

Figure 2: Far-IR spectrum of matter associated with the H I emitting gas for a column density of $10^{20} Hcm^{-2}$. The solid and dashed lines represent two independent FIRAS data sets known as RHSS for the high frequency part of the spectrum and LLSS for the low frequency segment starting at $500\mu m$. The noise increases towards high frequency at $\lambda < 200\mu m$. For $\lambda > 500\mu m$ a few spectral points are very noisy (See Fixsen et al. 1994). The dust emission is well fitted by a Planck curve with an emissivity index of 2 and a temperature of 17.5 K (dashed line). The diamonds represent the emission ratios derived from a linear fit to the IR-HI correlation diagrams where $WHI < 250Kkm/s$ (see Sect. 3.1).

The fact that the far- IR and sub-mm emission spectrum from dust is so well fitted by a single dust temperature is remarkable. The far-infrared ν^2 emissivity law corresponds to the expectation for classical (non-fluffy) dust grains, amorphous or crystalline. We have compared the spectrum in Fig. 2 with the emission spectrum computed by Draine and Anderson (1985) for the Draine and Lee (1984) dust model. The large graphite grains in this model have an equilibrium temperature of 18.5 K sufficiently different from the temperature derived from the data to make a clear shift between data and model. However the presence of silicates with a cooler temperature ($T \sim 14K$) produces only a very small widening of the spectrum compared to a single temperature emission spectrum. The existence of separate carbon and silicate grains with different equilibrium temperature can thus not be excluded. The dust model of Désert et al. (1990) which includes a single type of large grain fits well the FIRAS spectrum if we reduce the intensity of the local interstellar radiation field by 30% to adjust for the difference in calibration between DIRBE and IRAS.

The spectrum in Fig. 2 also shows the C II emission line at $158\mu m$. We derive from this detection a C II emission per H atom of $3.0 \pm 0.4\ 10^{-26} erg/s/H$. This value agrees with recent estimates of this emission from absorption line studies (Morris et al. 1991, Gry et al. 1992) and the IR measurements from a rocket by Bock et al.(1993). The line to continuum emission ratio, $I(C\ II)/\nu I_\nu(158\mu m) = 0.009 \pm 0.0015$, is very close to the mean Galactic plane value 0.0082 reported by Wright et al. (1991). This line provides an additional mean to differentiate the extragalactic background from Galactic emission.

3.3 Variations in the IR/H I Emission Ratio

The large velocity coverage of the Leiden/Dwingeloo H I survey allows to discuss the variations in the IR/H I emission ratio with gas velocity (see Deul and Burton 1990 for an early work on this issue). The all-sky map of the $100\mu m$ residuals after subtraction of the emission correlated with HI emission at low velocities ($|V_{LSR}| < 45 km/s$) is presented in Fig. 3. We chose to only subtract the contribution of low velocity gas to highlight the variations in IR emission for the higher velocity gas. To focus the attention to high Galactic latitudes we have centered the projection on the Galactic poles. At low Galactic latitudes the subtraction which does not include all of the Galactic velocities and does not take into account the variation in the IR/HI emission ratio with Galactocentric distance is meaningless. At $|b| > 20°$ the brightest and most extended structures are associated with molecular gas. They are discussed in the next section. Here we point at structures related to intermediate ($45 < |V_{LSR}| < 90 km/s$) and high velocity gas ($|V_{LSR}| > 90 km/s$). The bright spot at $l = 100°$ and $b = 37°$ and the faintest emission extending towards $l = 135°$, $b = 60°$ coincides spatially with intermediate velocity gas. In the H I maps (Hartmann 1994) this spectacular structure of intermediate velocity gas extends beyond what is seen in the $100\mu m$ image with several bright spots around $l = 225°$, $b = 70°$ which have no IR counterparts. The structure at $b = 40°$ from $l = 135°$ to $165°$ also coincides with intermediate velocity gas which in part overlaps with the Ursa Major high latitude molecular clouds. The brightest HI structures at intermediate velocities $b < -20°$ have no IR counterpart. We also see no IR counterpart for the high velocity complex M located around $l = 175°$, $b = 60°$. At the resolution of the HI survey the peak H I column density of high velocity clouds in this complex is around $10^2 0 H cm^{-2}$ which would produce a detctable IR emission if the IR/HI emission ratio was the same for these clouds than for the low velocity gas. IR emission from this complex was also unsuccessfully searched in the IRAS data (Wakker and Boulanger 1991). We thus observe significant variations of the IR/HI emission ratio with gas velocity with only some of the intermediate velocity gas having an IR emissivity comparable to that of low velocity H I gas. A more thorough study is necessary to be more quantitative.

Comparison of both emission and absorption H I spectra show that the 21 cm emission comes from both the cold and warm neutral medium. Calculations of dust destruction by fast shocks produced by supernovae explosions show that most of the grain destruction occurs in the low density components of the interstellar medium (Mc Kee 1989). The abundance of dust and therefore the strength of the associated IR emission could thus differ between the warm and cold phases. Such a difference could go unoticed if the two components share the same spatial distribution. It is natural to speculate that the warm component is more smoothly distributed than the cold component. It is thus worth pointing out that linear fits to the DIRBE data give the same IR to H I emission ratio for the full 40' resolution than for the 7° FIRAS beam. The C II 158μm line is potentially the best discriminator since negligible line emission is expected from the low density warm component. It will thus be interesting to compare the large scale line intensity derived from the FIRAS data with values on small angular scales within H I clouds which should soon be provided by the Infrared Space Observatory.

4. Residuals to the IR-HI Correlation

4.1 Contribution from Molecular Gas

Most of the features in the residual map of Fig. 3 are not readily identifiable but we speculate that most of them may be associated with molecular gas for the following two reasons. Firstly, as pointed out in Sect. 3. there is a break in the IR-HI correlation at $N_H = 510^{20} H cm^{-2}$, the column density for which the Copernicus data show that the fraction of molecular gas becomes significant (see Sect. 2). Note that Deul and Burton (1993) interpreted the changing slope in terms of a non-negligible optical depth of the HI line at some directions through the Galactic gas layer. This alternative explanation deserves to be investigated further. Secondly, the brightest and most extended structures in the residuals map are found within and around well known nearby molecular complexes such as Polaris ($l \sim 125°$), Taurus ($l \sim 180°$), Orion ($l \sim 210°$) and Ophiucus ($l \sim 350°$),). In these last three directions, bright features are seen up to $|b| = 50°$. We think that they correspond to high latitude extensions of the molecular complexes. The network of bright filaments at negative latitudes around $l = 90°$ is in the direction of the nearby OB association I Lacerta. The presence of an OB association makes it very likely that some molecular gas would be found in a CO survey of this area.

4.2 Warm Ionized Medium

In Figure 3, the spot at $l = 316°$, $b = 51°$ with an extended halo coincide with the nearby B1 star Spica(αVir). This star is known to lie at 85 pc from the sun in a low density ($n \sim 0.5 H cm^{-3}$) region of the interstellar medium. In th IRAS Sky Images one can see emission from dust heated by Spica up to 30′ from the star . This region is the entry DBB463 in the catalog of IR excesses of Désert et al. (1998). In the DIRBE data, the diameter of the halo, 15°, corresponds to the extent over which the stellar radiation dominates the average interstellar radiation. There is thus little doubt that the IR halo represents emission from dust in the low density gas heated by the star. The steady decrease of the brightness with distance to the star also supports this interpretation. The IR halo coincides in extent with the faint H II region observed by Reynolds (1985). This object is in the caltalog of IR excesses of Désert et al. (1998) as DBB463. is the best evidence we have for IR emission from the WIM. The brightness of the halo is consistent with a normal dust abundance in this piece of the WIM.

Boulanger et al. (1995) have attempted to estimate the overall contribution of the WIM by looking for a Galactic latitude gradient in the IR residual maps. What they found is consistent with an emission spectrum like that of the H I gas for $N_H \sim 3 - 5 10^{19} H cm^{-2}$. This column density is slightly less than half of the estimated column density of H II gas (see Sect. 2). Such a measurement is difficult because it relies on an accurately determined IR/HI emission ratio, with no variation with Galactic latitude, and on the absence of a contamination of the latitude profiles by emission associated with molecular gas. The interpretation of the result also depends on the degree of correlation between the H I and H II gas. What is reported by Boulanger et al. (1995) is consistent with a normal dust abundance in the WIM if about half of the H II gas is spatially correlated with H I gas.

ACKNOWLEDGEMENTS

We are grateful to the Goddard Space Flight Center team for introducing us with the COBE data. This work was supported by a CNES-INSU grant.

REFERENCES

Bock, J., Hristov, V.V., Kawada, M., Matsughara, H., Matsumoto, T., Matsuura, S., Mauskopf, D., Richards, P.L., Tanaka, M. and Lange, A.E. 1993, Ap. J. 410, L115

Boulanger, F. and Pérault, M. 1988, Ap. J. 330, 964

Deul, E.R., Burton, W.B., 1990, AA 230, 153

Deul, E.R. Burton, W.B. 1993, in the Galactic Interstellar Medium (D. Pfenniger and P. Bartholdi, eds.), Springer Verlag, Heidelberg, p. 79

Désert, F.X., Bazell, D. and Boulanger, F. 1988, ApJ 334, 815

Désert, F.X., Boulanger, F., and Puget, J.L. 1990, A.A. 327, 215

Draine, B.T., and Lee, H.M. 1984, Ap. J. 285, 89

Draine, B.T., and Anderson, N. 1985, Ap. J. 292, 494

Federman, S.R., Glassgold, A.E., and Kwan, J. 1979, Ap. J. 227, 466

Fiksen, D.J. et al. 1994, Ap. J. 420, 457

Fischer, M.L. et al. 1995, Ap. J. 444, 226

Gry, C., Lequeux, J. and Boulanger, F. 1992, AA 266, 457

Hartmann, D., 1994, Ph. D. Thesis, University in Leiden

Hartmann, D., Burton, W.B. 1995, An Atlas of Galactic Neutral Hydrogen Emission, Cambridge University Press, under contract

Lockman,F.J., Hobbs, L.M. and Shull, J.M. 1986, Ap. J. 301, 380

Magnani, L., Blitz, L. and Mundy, L. 1985, Ap. J. 295, 402

Mc Kee, C.F. 1989, in Interstellar Dust, eds. L.J. Allamandola and A.G.G.M. Tielens, p. 431, Kluwer

Morris, S.L., Weymann, R.J., Savage B.D., Gilliland R.L. 1991, Ap. J. 377, L21

Reach, W.T. et al. 1995, Ap. J. in press

Reynolds, R.J. 1985, A.J. 90, 92

Reynolds, R.J. 1989, Ap. J. 339, L29

Reynolds, R.J. 1991, Ap. J. 372, L13

Reynolds, R.J., Tufte, S.L., Kung, D.T., Mc Cullough, P.R., Heiles, C. 1995, ApJ 448, 715

Wakker, B.P. and Boulanger, F. 1991, A.A. 170, 84

Wright, E. L. et al. 1991, Ap. J. 381, 200

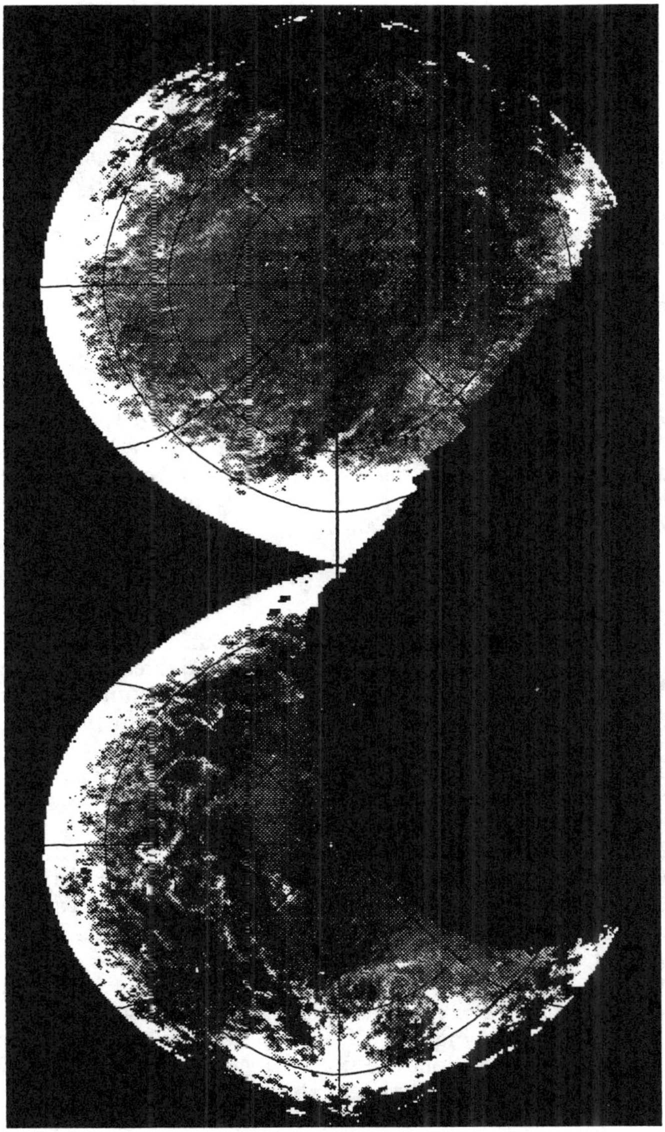

Figure 3: Sky Image at $100\mu m$ emission after subtraction of the emission spatially correlated with the distribution of the low velocity H I gas $|V_{LSR}| < 45 km/s$. The maps are in an Aitoff projection centered on the Northern (top) and Southern (bottom) Galactic poles. The Galactic center is at the center of image. Lines of constant Galactic Latitude are drawn at $|b| = 20°$, $40°$, $60°$ and $80°$. Lines of constant Galactic Longitude correspond to multiples of $45°$.

LIMITS ON THE FAR INFRARED CIBR FROM DIRBE, FIRAS AND HI SURVEYS

F.–X. Désert, A. Abergel, J.–P. Bernard, F. Boulanger & J.–L. Puget
Institut d'Astrophysique Spatial, Bat 121, Université Paris XI
F–91405 Orsay Cedex France

W. B. Burton & D. Hartmann
Sterrewacht Leiden, Postbox 9503, 2300 RA Leiden, The Netherlands

ABSTRACT

After having removed the various foregrounds and the cosmic microwave background from DIRBE and FIRAS whole–sky maps, we find a positive residue at high galactic latitude in the whole submillimetre domain. If interpreted as an extragalactic background (CIBR), it could be the important trace of an intense star formation at redshifts larger than 3.

1. INTRODUCTION

The detection of a Cosmic Infrared Background Radiation (CIBR) would give an important clue about the formation of galaxies, and more generally about the so–called dark age of the Universe (see all the conference proceedings), between the matter recombination ($z = 1000$) and the near past ($z \leq 1$). But so far, only upper limits on the existence of the CIBR have been observed (Hauser 1995a). The main difficulty of the task of observing the CIBR, once satellites can be sent in order to avoid atmospheric absorption and emission, can be pinpointed to the presence of much stronger foregrounds which are spectrally and spatially very complex. This is unlike the situation for the Cosmic Microwave Background Radiation (CMBR) and the X–ray extragalactic background radiation which are the main contributors of the sky brightness at their respective wavelengths. False detections of "submillimetre excesses" in the last twenty years have lead to very cautious approaches in the search for a CIBR.

Nevertheless, the launch of COBE with its absolute and sensitive photometric capabilities and the improvements in observations of the interstellar medium via the Leiden/Dwingeloo HI survey have prompted us to renew the search for the CIBR. In the following, we present the adopted decomposition method, its implementation and the analysis of the residues that we obtain. The results are then discussed.

© 1996 American Institute of Physics

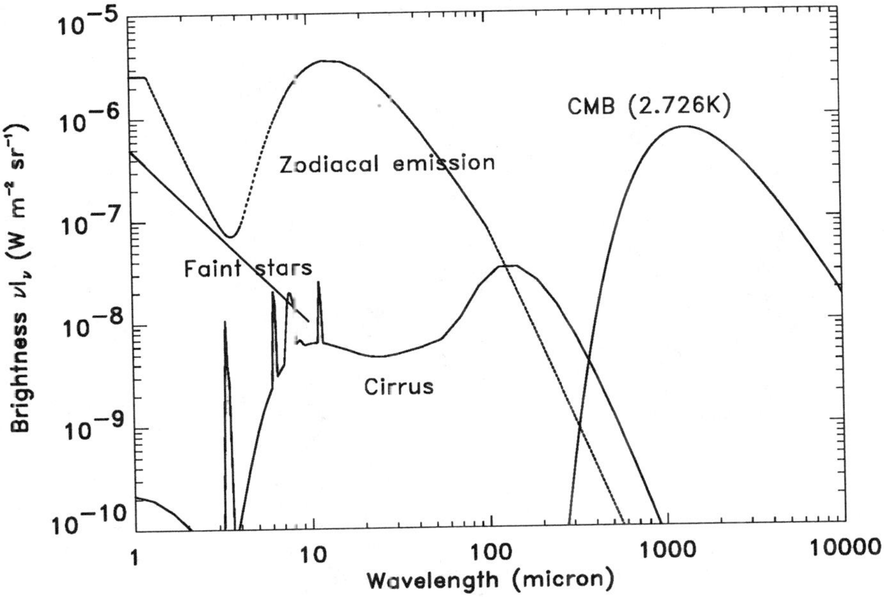

Figure 1: Brightness of the infrared sky per logarithm of wavelength, outside the terrestrial atmosphere and at high galactic and ecliptic latitude. The zodiacal emission and scattering are in dashed line. The stellar component comes from the unresolved stars in DIRBE beam. The cirrus component (HI diffuse interstellar medium) is taken from the model by Désert *et al.* (1990) fitted to IRAS photometry for a column density of $10^{20}\,\mathrm{H\,cm^{-2}}$. The CMB dominates the millimetre spectrum.

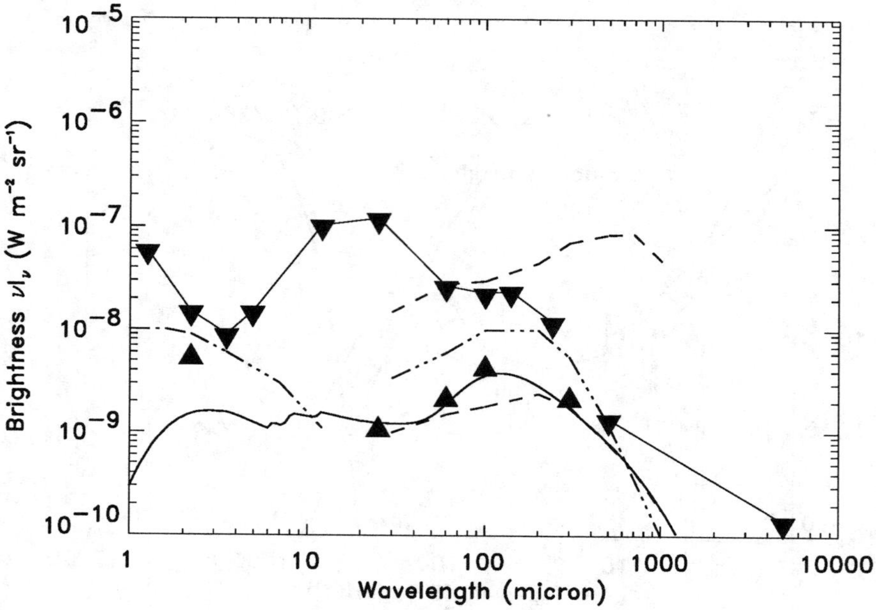

Figure 2: Best observational limits on the CIBR brightness in the same scales as Fig. 1. The down arrows are upper limits on the CIBR from DIRBE and FIRAS (Hauser 1995a). The up arrows are lower limits on the CIBR from (nearby) galaxy number counts in the K and IRAS bands (see references in Hauser 1995a). The full line model is the CIBR expected with a mild evolution in the past (Maffei 1994). The dot-dot-dot-dash line is a strong evolution model from Franceschini *et al.* (1994). The two dashed curves are from Beichman & Helou (1991) extreme cases of CIBR models.

2. THE DECOMPOSITION METHOD

To illustrate the complexity (or richness) of the infrared sky, once above the terrestrial atmosphere, Fig. 1 gives the typical spectrum of the different components that contribute to the IR brightness. The level of these components has been chosen at their near-minimum in any line–of–sight. Actually, all these components are spectrally and spatially variable.

Fig. 2 shows the expected possible levels of the CIBR along with the observational limits obtained with COBE data, K-band and IRAS galaxy number counts. The near infrared CIBR, which would come from unprocessed stellar light in all galaxies, has so far eluded detection mainly because of the number of foreground components to subtract in the near IR range. We decide thereafter to concentrate on the submm domain where the foregrounds and the CMB leave the so–called cosmological spectral window. Indeed, many models have predicted (*e.g.* Stecker et al. 1977, Franceschini *et al.* 1994) that dust inside galaxies could reprocess the stellar light into a bright far IR and submm component. The method to set constraints on or detect the submm CIBR is directly inspired from the method Boulanger & Pérault (1988) used with IRAS data. It is identical to what Boulanger *et al.* (1995) describe in this conference. We subtract from COBE brightness measurements a model of the zodiacal light and the CMBR, then the main difficulty is in the contribution from the interstellar medium (ISM). Obviously, the work is to be done at hight galactic latitudes. Then, from seeing Fig. 2, due to the redshift effect, one could expect *the extragalactic background to be colder than interstellar dust emission* and thus to exhibit a better ratio of CIBR to foregrounds in the submm domain.

3. IMPLEMENTATION OF THE METHOD

3.1 The COBE data and Zodiacal emission subtraction

We used both DIRBE and FIRAS data (see *e.g.* Silverberg *et al.* 1993, and Mather *et al.* 1993). DIRBE is a photometer that covers 1 to 240 μm at an angular resolution of 0.7 °. FIRAS is a spectrometer that covers the submm and mm domain at a resolution of 20 GHz and with a 7 ° beam. We used in particular the elongation 90 ° DIRBE sets for wavelengths smaller or equal to 100 μm in order to be able to remove reasonably well the zodiacal emission. The complete DIRBE data sets were used at 140 and 240 μm to improve the sensitivity. The RHSS particular set of data was used for FIRAS because of its sensitivity and wavelength range. Data are delivered in units of $\mathrm{MJy\,sr^{-1}}$ which are easily transformed in the conference official units with

$$\nu B_\nu = 3\times10^{-8}\,\mathrm{W\,m^{-2}\,sr^{-1}}\left(\frac{B_\nu}{1\,\mathrm{MJy\,sr^{-1}}}\right)/\left(\frac{\lambda}{100\,\mu\mathrm{m}}\right).$$

Zodiacal light is removed in the far infrared and submm by first making an ecliptic latitude profile of the minimum 25 μm brightness for all longitudes and scaling that profile for the other wavelengths with $B_{\nu\,z}/B_{25\,\mu\mathrm{m}\,z} = 0.17\,(\lambda/100\,\mu\mathrm{m})^{-3}$. This last zodiacal submm emission dependence with wavelength was found by Reach *et al.* (1995). Uncertainties in this subtraction method cannot lead to significant errors at wavelengths longer than 200 μm.

The CMBR and its dipole were also removed from FIRAS spectra with the parameters found by Mather *et al.* (1994).

3.2 The HI data and the ISM emission subtraction

The Leiden/Dwingeloo galactic neutral atomic hydrogen survey (Hartmann & Burton 1995) provides very accurate maps of HI, the main phase of the ISM at high galactic latitudes in terms of IR emission tracers. North of -30 ° declination, the maps are sampled every 30 arcmin with a 36 arcmin beam and a velocity resolution of 1 km/s. Corrections for stray light were thoroughly done. Here we used a map of the HI spectrally integrated between -100 and 100 km/s (see Boulanger *et al.* 1995), and degraded its resolution to the 7 ° FIRAS beam.

Once the subtraction in Section 3.1 is done, the FIRAS spectra and DIRBE photometric measurements correlate remarkably well with the HI column density. In order to avoid over–heated parts of the ISM (close to stars) as well as denser regions for which hydrogen could be molecular, we limit the correlation to low HI column densities (less than 4.5×10^{20} H cm^{-2}) and high galactic latitudes ($|b| > 20°$). The HI correlated dust emission spectrum is shown by Boulanger *et al.* (1995). Because it is obtained with a correlation method, any possible residual zero level does not influence this HI spectrum.

4. RESULTS

4.1 The residue

For each line of sight with low HI column density, we now can subtract the HI–related submm emission from the previous FIRAS and DIRBE measurements. The residue is then averaged over all allowed lines of sight and is shown in Fig. 3 and given in Table 1, which summarizes the HI dust emission as well. The upper curve and DIRBE points are the direct results of the previous subtraction and as such can be considered as the best estimate of an upper limit to the far IR CIBR. DIRBE values are consistent with the results presented by Hauser (1995b) and are not dealt further here. Once the H^+ estimated component (0.4×10^{20} H cm^{-2} Reynolds 1989, 1992) is subtracted one gets the middle curve which is the best estimate of the residue. The main conclusion is that the residue is definitely positive over the whole submm range and is different from (not as steep as) the ISM spectrum. The lower curve indicates that the residue cannot be nulled by removing an additional HI-like dust spectrum (here by removing the equivalent of 0.6×10^{20} H cm^{-2} from the upper curve). This curve shows already many points with unphysical negative brightness for wavelengths lower than 200 μm.

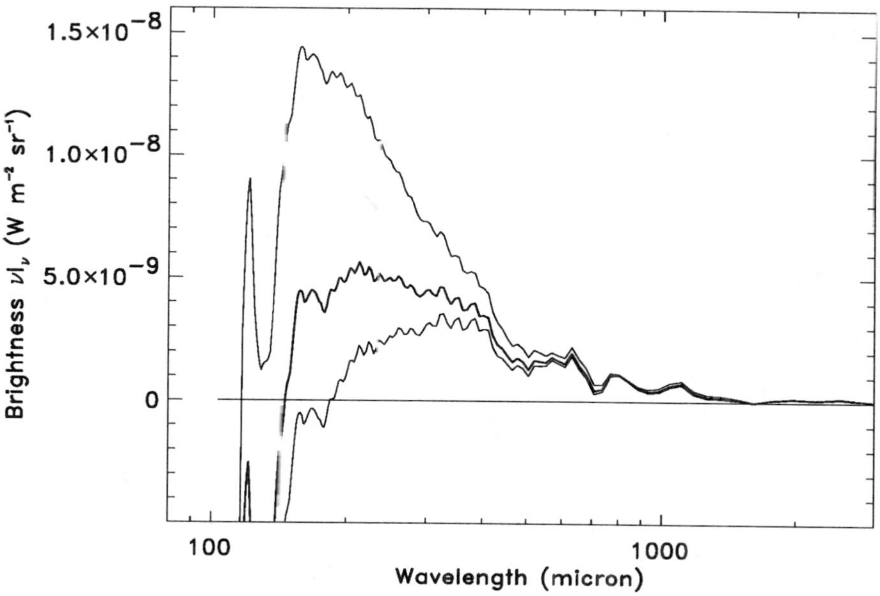

Figure 3: Residual brightness in DIRBE and FIRAS measurements after the zodiacal, ISM and the CMBR components have been removed. Smoothing with a constant $\Delta\nu/\nu = 0.05$ has been applied. Some wavelength–correlated noise is visible but does not influence our conclusions. The FIRAS noise increases exponentially at short wavelengths and the spectra are unreliable below 130 μm. (see the text for the different curves)

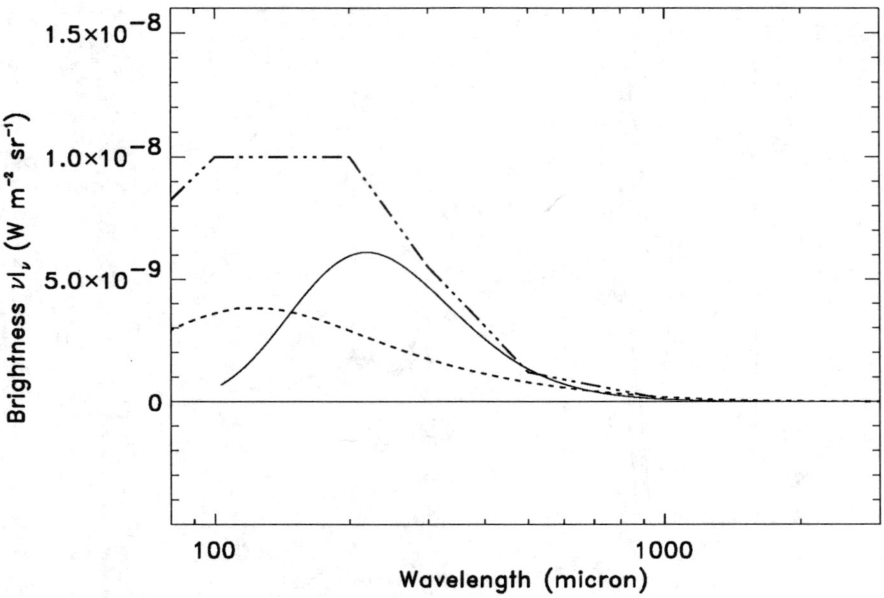

Figure 4: Models for the CIBR in the same scales as Fig. 3. The lower curve corresponds to mild evolution of present galaxies (Maffei 1994) and represents the lower limit to the expected CIBR in the submm. The upper curve is taken from Franceschini *et al.* 1994. The dash curve shows a blackbody curve at 11 Kelvin with a ν^2 emissivity law and optical depth of 2×10^{-5} at $333\,\mu$m, for illustration.

νB_ν HI-spectrum $\mathrm{nW\,m^{-2}\,sr^{-1}\,/10^{20}\,H\,cm^{-2}}$	100	140	240 μm
DIRBE	21 ± 1.5	19 ± 1	16 ± 1
FIRAS	$T = 17.5\,\mathrm{K}$	ν^2 emiss.	$\tau_{333\,\mu\mathrm{m}} = 5.7 \times 10^{-6}$

Residues νB_ν $\mathrm{nW\,m^{-2}\,sr^{-1}}$	100	140	240 μm
upper limits	24	32	15
upper limits FIRAS	$14(\lambda/200\,\mu\mathrm{m})^{1.5}$		
unexplained	15	21	10
unexplained FIRAS	$T = 11\,\mathrm{K}$	ν^2 emiss.	$\tau_{333\,\mu\mathrm{m}} = 2 \times 10^{-5}$

Table 1: HI and residual component characterisation

4.2 Discussion

The FIRAS positive residue is rather unsensitive to the choice of galactic latitude or HI column density cuts. It is well above the expected systematic offset that could be present in the observations (see Hauser 1995b). It was computed with observations from about 150 FIRAS beams distributed over a large fraction of the sky (about 15%). It does not vary by more than 15% from one face of the sky cube projection to another (see FIRAS explanatory supplement). It is slightly above the upper limits given by Mather *et al.* (1994) but their bootstrapping method assumed a negligible background at 240 μm which is in contradiction with what is seen in Fig. 3. Leinert (1995) argues that it is difficult to mimick any submm background with some dust in the solar system. Reach *et al.* (1995) have found a cold dust component in the ISM that could produce the observed positive residue. Boulanger *et al.* (1995) show that this component is not apparent in the neutral atomic medium, when averaged over large areas. Although the existence of cold dust is in principle possible (*e.g.* Wright *et al.* 1991), it is difficult to envisage this dust to be completely decoupled from the gas. We cannot exclude some dust in the galactic halo that could produce the observed background (this is the same type of uncertainty as the gamma-ray bursts origin). Synchrotron radiation is excluded by its steep spectral dependence. Free–free emission is excluded by its expected level (Bennet *et al.* 1992). Distortions of the CMBR with respect to the assumed one cannot produce a background in the submm range which decreases with wavelength. We are left with the most plausible explanation that the positive residue is indeed the integrated ligth from present and past galaxies. Fig. 4 shows the expected level of this CIBR in different models. The color temperature of the residue directly indicates that a large fraction of the galaxy formation occured at large redshifts (above 2–3). It is consistent with the scenario advocated by *e.g.* Franceschini (1995) and Silk (1995), where radiation coming from the formation of spheroids has been absorbed and reemitted by early dust. This radiation is more important than the one coming from spirals which evolved at a slower pace. The large scale isotropy of this positive residue is thorougly tested in the forthcoming publication by Puget *et al.* (1995). Small scale anisotropies are expected if the residue presented here is indeed the submm CIBR. Angular fluctuations are predicted to increase at small angular scales (Guiderdoni 1995) and could be the ultimate limit to the detection of the CMBR anisotropies. The CIBR small–scale anisotropies could be measured with existing and planned ground-based and balloon experiments in the near future. This would yield the critical test for the origin and existence of this submm CIBR.

ACKNOWLEDGEMENTS

The COBE datasets were developed by the NASA Goddard Space Flight Center under the guidance of the COBE Science Working Group and were provided by the NSSDC. Part of this work was supported by a CNES-INSU grant.

REFERENCES

Beichman, C. & Helou, G., 1991, ApJL, 370, L1

Bennett, C. L., 1992 ApJL 396, L7

Boulanger, F., & Pérault, M., 1988, ApJ, 330, 964

Boulanger, F. *et al.* 1995, This conference and A & A, in preparation

Désert, F.–X., Boulanger, F., & Puget, J.–L., 1990 A & A, 237, 215

Franceschini, A., et al., 1994, ApJ, 427, 140

Franceschini, A., 1995, This conference

Guiderdoni, B., 1995, This conference

Hartmann, D., & Burton, B. W. 1995, in *An Atlas of Galactic Neutral Hydrogen Emission*, Cambridge University Press, in press

Hauser, M. G., 1995a, in Proceed. of IAU Symp. 168, *Examining the Big Bang and Diffuse Background Radiations* The Hague, Aug 94

Hauser, M. G., 1995b, this conference

Leinert, Ch. 1995, this conference

Maffei, B., 1994, PhD thesis, Univ. of Paris

Mather, J. C., et al., 1994, ApJ, 420, 439

Mather, J. C., et al., 1993, in SPIE Conference on *Infrared Spaceborne Remote Sensing*, San Diego, July 93.

Puget, J.–L., *et al.* 1995, A & A, in preparation

Reach, W. T., *et al.* , 1995, ApJ, in press

Reynolds, R. J., 1989 ApJL, 339, L29

Reynolds, R. J., 1992 ApJL, 392, L35

Silk, J., 1995, This conference

Silverberg, R. F. , *et al.* 1993, in SPIE Conference on *Infrared Spaceborne Remote Sensing*, San Diego, July 93.

Stecker, F. W., Puget, J.–L., & Fazio, G. G., 1977, ApJL, 214, L51

Wright, E. L., *et al.* 1991, ApJ, 381, 200

Dust in the diffuse ISM as revealed by DIRBE observations

J.P. Bernard, F. Boulanger, F.X. Désert, J.L. Puget
IAS, bât 121, Campus d'Orsay, 91405 Orsay Cedex, France.

G. Helou
IPAC/JPL, MS100-22, Pasadena CA, 91107, USA.

W. Langer
JPL, MS169-506, 4800 Oak Grove Drive CA, Pasadena CA 91109, USA.

ABSTRACT

The weekly averaged DIRBE full sky images have been processed to separate the various components contributing to the total brightness in the various bands. The zodiacal emission, which dominates at 12 and 25 μm and the zodiacal dust scattering at $\lambda < 5\,\mu$m are both accounted for using an empirical fit to the data. The diffuse stellar emission which dominates at $\lambda < 5\,\mu$m is determined using the shortest DIRBE photometric bands at 1.25 and 2.2 μm and a standard NIR extinction law. Preliminary results based on the first release of the DIRBE data have been presented in Bernard et al. 1994. When the zodiacal light and stellar emission are subtracted, significant emission remains above 2.2 μm, which follows the general distribution of the dust emission as seen in the IRAS bands. The DIRBE images therefore allow to extend our knowledge of the dust emission spectrum below 12 μm and above 100 μm. In the L(3.5 μm) and M(4.9 μm) bands, the dust emission can be seen not only toward the galactic plane but also in diffuse regions above the plane as well as toward closeby molecular complexes (ρ-Ophiuchi, Orion, Taurus,...). The existence of NIR dust emission in cold and diffuse regions strongly suggests transiently heated small dust particles as the carrier. The dust NIR spectrum is generally consistent with the dust model of Désert et al. 1990. In particular, the dust emission increases from 4.9 to 3.5 μm, which can be attributed to the contribution of the 3.3 μm emission feature of Polycyclic Aromatic Hydrocarbons (PAH). Significant continuum emission, or other feature emission, is also required to explain the observed brightness in the L band and the AROME ballon experiment results at low galactic latitude.

1. ZL subtraction

The thermal emission and scattering by dust associated to the solar system (Zodiacal Light: ZL) is visible in most all-sky DIRBE maps and often dominates the emission from the Galaxy. The thermal emission is particularily strong in the 12 and 25 μm bands but is also seen at 4.9 μm and above 60 μm, while at $\lambda \leq 5\,\mu$m, the ZL is dominated by scattered solar light. The ZL must therefore be precisely removed, in order to evidence weaker emission associated to the Galaxy.

In this study, we adopt an empirical approach to remove the ZL contribution from the DIRBE weekly averaged maps. Unlike more comprehensive methods based on physical models of the zodiacal cloud (see Reach et al. in this volume), this method arbitrarily removes any isotropic component present in the map. Therefore, the resulting subtracted maps cannot be directly used for the purpose of evidencing the cosmological IR background. However, it provides

© 1996 American Institute of Physics

maps with very little residus, which is very usefull in investigating the origin of the dust emission.

The data recorded by DIRBE during 1 week of observation cover about half the sky. Each weekly averaged map was binned into fixed intervals of solar elongation $E_\odot$ and the data in each bin, plotted as a function of eccliptic lattitude β, is fitted using the function proposed by Hauser 1993:

$$ZL(\beta) = A\,\beta_c/|\sin(\beta - \beta_0)|(1 - \exp(-\xi - \alpha\xi^2)) + B(\beta - \beta_0) \qquad (1)$$

$$\xi = |(\beta - \beta_0)/\beta_c|$$

where the amplitude A, the ecliptic latitude of the maximum β_0, the width β_c, the parameter α describing the behavior near $\beta = \beta_0$ and the slope of an additive baseline B are free parameters. In adjusting these parameters, the data points corresponding to the zodiacal bands at $\beta \simeq 9°$ and/or affected by strong galactic emission are ignored. At a given solar elongation, the fit is peformed independently into 2 zones of ecliptic longitude and for $\beta \leq \beta_0$ and $\beta \geq \beta_0$, leading to 4 sets of parameter values for each solar elongation bin and week. The accuracy of the fit is generally better than a few percent of the ZL intensity at 12 μm for one profile.

The above procedure is repeated for each of the 41 weekly averaged maps and the fitted values for each free parameter are gathered into 4 maps on a regular grid of week number and solar elongation. The individual parameters show systematic variations in these coordinates, which reflect properties of the zodiacal cloud as seen from varying location along the earth orbit. The amplitude A decreases with angular distance from the sun roughly as $1/E_\odot$. The peak position β_0 varies with time as a 1 year period sin function, due to the inclination of the zodiacal cloud on the ecliptic plane. The width β_c steadily increases toward high solar elongation, probably reflecting actual radial variations of the shape of the zodiacal cloud with distance from the sun (see Reach et al.). Each parameter map is then fitted using a smooth 2D function matching the behavior described above, or a low order polynomial if no systematic trend is observed. The smooth parameter surfaces are then interpolated to construct the ZL estimate for each pixel of each weekly average map, using equation 1. After subtraction of the ZL, all 41 week maps are averaged into an annual average map. As ZL subtraction absolute uncertainties are smaller at high solar elongation, we restricted the annual average to $E_\odot \geq 90°$ and used a weighting fuction proportional to $1/E_\odot$. The procedure was repeated independently for all DIRBE bands from 1.25 to 100 μm, leading to a set of subtracted annual maps. Due to the weakness of the ZL contribution at $\lambda > 100\,\mu$m, the above procedure cannot be applied and unsubtracted maps were used in the following at 140 and 240 μm.

2. Stellar light subtraction

The stellar light from the Galaxy dominates the total sky brightness below $\lambda = 5\,\mu$m in the DIRBE data. It includes emission from individual bright stars, as well as a diffuse stellar component produced by late type stars. This emission as to be removed, in order to evidence dust emission below 5 μm.

The subtraction of the diffuse component can be done efficiently assuming that (1) the sky brightness in the DIRBE 1.25 and 2.2 μm bands is dominated by star light (i.e. the dust emission in these bands can be neglected relative to the

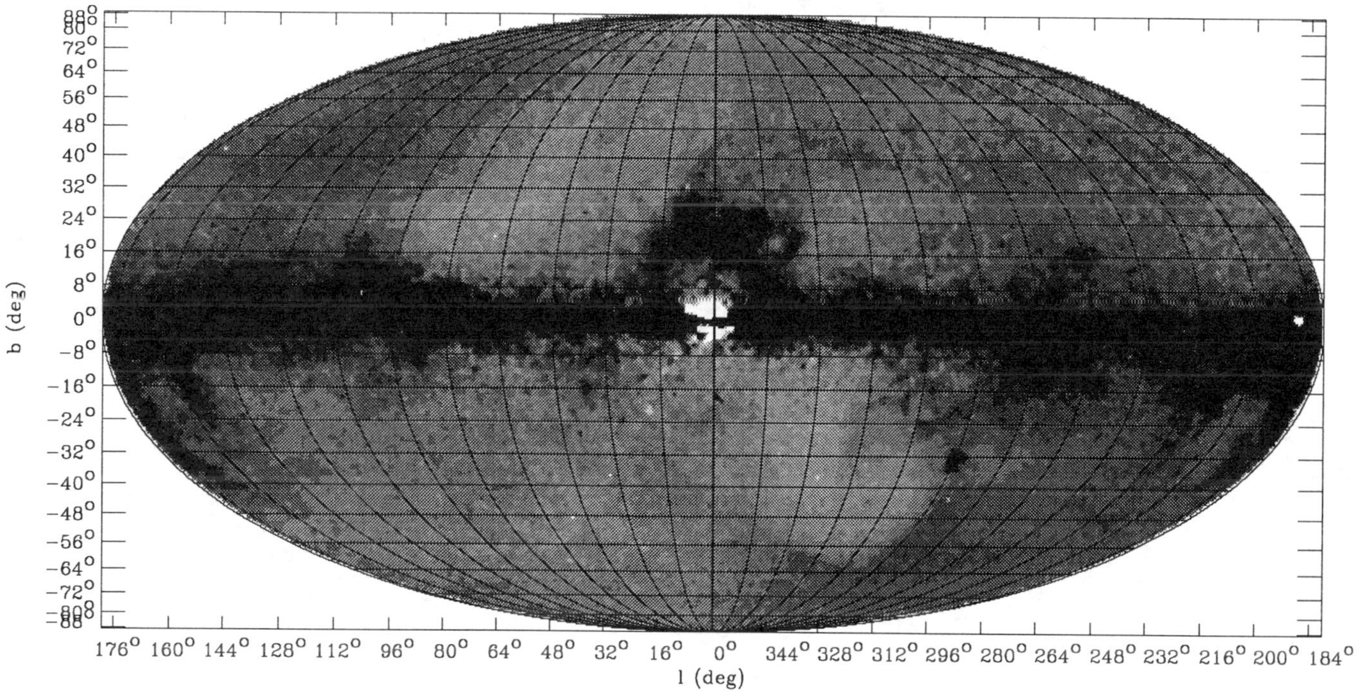

Fig 1: Dust emission in the L(3.5 μm) DIRBE band.

stellar emission) and (2) the dust clouds are mainly located in front of most of the stellar diffuse emission. The last assumption is generally verified along line of sights above a few degrees above the plane because the scale height of dust is smaller than that of stars. It may however fail toward the plane. Under these assumptions, the intrinsic stellar colors are related to the observed intensities through

$$(\mathrm{I}^*_\lambda/\mathrm{I}^*_{1.25})_{\mathrm{int}} = \mathrm{I}^*_\lambda e^{\tau_\lambda}/\mathrm{I}^*_{1.25}e^{\tau_{1.25}}. \tag{2}$$

The diffuse stellar contribution in any DIRBE band therefore writes

$$I^*_\lambda = I^*_{1.25}\,(I^*_\lambda/I^*_{1.25})_{int}\,e^{(-a_\lambda \tau_{1.25})}$$

where $a_\lambda = \tau_\lambda/\tau_{1.25}$ is taken from the average IR extinction law derived by Rieke and Lebofsky 1985. $\tau_{1.25}$ can be derived from eq 2:

$$\tau_{1.25} = (\mathrm{a}_{2.2} - 1)^{-1}\ln[\mathrm{I}^*_{1.25}/\mathrm{I}^*_{2.2}(\mathrm{I}^*_{2.2}/\mathrm{I}^*_{1.25})_{\mathrm{int}}]$$

The intrinsic colors are derived by averaging into the DIRBE maps in regions of low dust emission and low $\tau_{1.25}$. The colors derived ($(I^*_{1.25}/I^*_{2.2})_{int} = 1.4$, $(I^*_{1.25}/I^*_{3.5})_{int} = 2.8$, $(I^*_{1.25}/I^*_{3.5})_{int} = 6$) are in general agreement with those measured by Arendt et al. 1994 extrapolated to low extinction. The diffuse stellar contribution derived above was subtracted from DIRBE maps in the L(3.5) and M(4.9) bands and the remaining contribution from individual stars was subtracted using a simple median filter. At IRAS wavelengths, the IRAS PSC convolved to the DIRBE beam pattern was used to subtract the contribution of known IRAS point sources. The dust map obtained in the L band after subtraction of the ZL and stellar emission is shown in figure 1. Low frequency residuals of the ZL subtraction can be seen near the ecliptic poles. Toward the Galactic center, the stellar diffuse emission is over subtracted because of the peculiar intrinsic colors of the stars in the bulge.

3. Discussion

The maps subtracted from ZL and stellar emission show extended emission down to the L band. The spatial distribution correlates closely to the dust thermal emission at $\lambda > 100\,\mu$m and the observed brightness levels are 1 to 2 order of magnitude above what is expected from stellar scattered light at the surface of clouds (Cutri, private comunication). We therefore attribute the L and M emission to dust emission. Figure 1 shows that the L band emission is mainly located along the galactic plane ($|b| < 10°$) and in well known high latitude regions such as ρ-ophiuchi ($-20° < \mathrm{l} < 20°$, $10° < \mathrm{b} < 32°$), Taurus ($155° < \mathrm{l} < 180°$, $-20° < \mathrm{b} < -10°$), Orion ($200° < \mathrm{l} < 220°$, $-20° < \mathrm{b} < -10°$), ... The short wavelength emission can often be followed up to $|b| \simeq 40°$. A similar behavior is observed in the M band, although stronger residuals of the ZL subtraction are present in the map and the dust emission is generally weaker.

3.1 *Comparison with the AROME data*

Along the Galactic plane, the emission in the $3.3\,\mu$m PAH feature and the adjacent continuum has been observed by the AROME balloon born experiment (see Giard et al. 1994). The agreement between the AROME data and the DIRBE total L band emission map was found to be $\simeq 5\%$, well within the calibration uncertainty of the AROME data (see Bernard et al. 1994). The comparison also reveals that the prominent regions emitting the 3.3 μm feature along the Galactic plane are also maxima of the L DIRBE map in figure 1. However, the 3.3 μm feature detected by AROME contributes to only about 20% of the DIRBE L band dust emission. This ratio seems constant with Galactic latitude in the range $-50° < l < 30°$. This indicates that the regions emitting the PAH feature along the galactic plane also contribute to a strong 3 μm continuum. It is worth noticing that, in order to explain both the L and M DIRBE dust emission with 20% of the L band produced by the feature, the underlying continuum should have colors comparable to the continuum observed by Sellgren et al. 1990 toward reflection nebulae ($T \simeq 800\,K$). Wether or not the continuum is also present at high galactic latitude cannot be answered with the present data only, as no 3.3 μm feature observation of high or intermediate latitude cirrus clouds is yet available. Note that significant continuum dust emission at high latitude could seriously affect the transparency of the cosmological window near 3 μm. The IRTS and ISO satellites will soon bring data relevant to this important issue.

3.2 *Average colors and comparison to the PAH dust model*

In order to derive average colors for the dust emission, the DIRBE maps described above were averaged for various individual clouds. The template for clouds ($\tilde{I}_{100}$) is the full sky 100 μm DIRBE map filtered using a lowest percentile filter in elongated boxes of galactic longitude in order to remove low spatial frequency variations along galactic latitude. Individual clouds are defined as contiguous regions above a given threshold (Thr) in $\tilde{I}_{100}$. We then define a background region (1.2° wide with $\tilde{I}_{100} < Thr$) surrounding each cloud. At each wavelength, a low degree polynomial surface is fitted in the background region, in order to subtract any remaining background and the data in the cloud region with $Thr < \tilde{I}_{100} < Thr + \Delta Thr$ is used to produce the average spectrum. The threshold is then increased by ΔThr and the above steps are repeated. This procedure allows to separate the often closely associated diffuse extended regions from the bright localized ones (e.g. HII regions) that can easily dominate the total emission of a given cloud, without counting any individual region twice. The average spectra of individual regions with various average 100 μm brightness are shown in fig 2, normalized to 240 μm. The spectra with high average 100 μm brightness ($\overline{I}_{100}$) correspond to hot and bright clouds, generally located close to the galactic plane. It can be seen that the lowest spectra, corresponding to the most diffuse regions, agree well with the PAH model by Désert et al. 1990. Going from low to high $\overline{I}_{100}$, the NIR and IR spectrum increases as a whole, so that bright clouds generally have higher NIR/FIR ratio. Such a behavior is expected under the PAH model if hot clouds are illuminated by higher energy UV photons (e.g. due to the presence of closeby young stars) which are preferentially absorbed by small particles, or if hot clouds have higher small particles abundances. Note that a relative decrease of the 12 μm emission is generally observed in the IRAS data around highly irradiated regions such as HII regions (e.g. Ryter et al. 1987), which is attributed to the destruction of the smallest particles in high UV density regions. This trend is also observed in the average colors at $\overline{I}_{100} > 100$ MJy/sr (not shown in figure 2). The NIR increase with $\overline{I}_{100}$ is specially strong at wavelengths corresponding to continuum in the dust model (e.g. 4.9, 25, 60 μm) and only moderate in the PAH feature dominated bands (3.5 and 12 μm). This trend will have to be explored in more details using the model, and may require a larger feature/continuum ratio of the PAH emission, in particular in the 10 μm region.

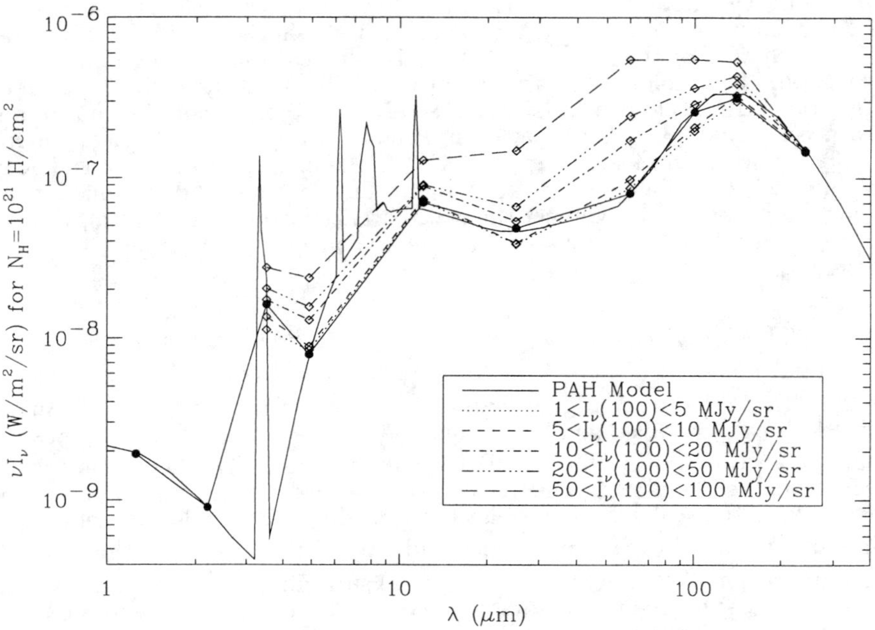

Fig 2: Prediction of the PAH model (dots) and DIRBE average colors.

4. Conclusions

The zodiacal light and stellar emission has been subtracted from the DIRBE all-sky maps from 3.5 to 100 μm, in order to evidence galactic dust emission. Significant emission is found down to the 3.5 μm DIRBE band and up to galactic latitudes as high as 40°. The spatial distribution of the DIRBE M(4.9 μm) and L(3.5 μm) band emission are generally very similar. The NIR emission follows the IR and FIR distribution and cannot be attributed to starlight scattering at the clouds surface but must be interpreted as dust emission. Average dust emission spectra of individal regions over a range of excitation conditions has been built. The spectral energy distribution of low brightness, high latitude regions is in good agreement with the predictions of the PAH model in the solar neighborhood. The average NIR/FIR ratio increases with increasing excitation, which also seems in qualitative agreement with the model, although a more detailed analysis is required. The spatial distribution of the L and M dust emission closely follows the 3.3 μm PAH feature emission observed by the AROME experiment along the galactic plane, but a strong NIR continuum accounting for as much as 80% of the DIRBE L band dust emission is also required to explain the DIRBE L band emission levels at low latitude.

Acknowledgements

We are grateful to the Goddard Space Flight center team for their help in reducing the DIRBE data. This work was conducted in part at the Jet Propulsion Laboratory, California Institute of Technology under contract to NASA and supported in part under the LTSA program and a post-doctoral fellowship from the European Space Agency.

References

Arendt R.G. et al., 1994 ApJ Let 425, 85.
Bernard J.P., Boulanger F., Puget J.L. 1993, A&A 277, 609.
Bernard J.P., Boulanger F., Désert F.X., Giard M., Helou G., Puget J.L., 1994, A&A Let 291, L5.
Désert F.X., Boulanger F., Puget J.L., 1990, A&A 273, 215.
Giard M. et al. 1994, A&A 286, 203.
Hauser M.G. 1993 in AIP Conf. Proc. 278, "Back to the Galaxy", eds S.S. Holt, F. Verter.
Rieke G.H., Lebofsky M.J., 1985, ApJ 288, 618.
Ryter C., Puget J.L., Pérault M., 1987, A&A 186, 312.
Sellgren K., Tokunaga A.T., Nakada Y., 1990, ApJ 349, 120.

OBSERVATIONAL LIMITS ON THE EXTRAGALACTIC INFRARED BACKGROUND

CLUSTERING OF DIRBE LIGHT AND IR BACKGROUND

A. Kashlinsky[1], J.C. Mather[2], S. Odenwald[2,3], M.G. Hauser[4]
[1]NORDITA, Blegdamsvej 17, Copenhagen, DK-2100 Denmark
[2]Code 685, Goddard Space Flight Center, Greenbelt, MD 20771 USA
[3]Applied Research Corporation
[4]Code 680, Goddard Space Flight Center, Greenbelt, MD 20771 USA

ABSTRACT

We outline a new method for estimating the cosmic infrared background using the spatial and spectral correlation properties of infrared maps. The cosmic infrared background from galaxies should have a minimum fluctuation of the order of 10% on angular scales of the order of 1°. We show that a linear combination of maps at different wavelengths can greatly reduce the fluctuations produced by foreground stars, while not eliminating the fluctuations of the background from high redshift galaxies. The method is potentially very powerful, especially at wavelengths where the foreground is bright but smooth.

1. INTRODUCTION

Diffuse background radiation fields contain information about the entire history of the universe, including those periods in which no discrete objects exist or can be detected by telescopic study. The cosmic infrared background (CIB) should contain much of the luminosity of the early generations of stars and galaxies, since the cosmic expansion redshifts the wavelength of peak luminosity from these sources from the UV and visible bands into the infrared. In addition, dust absorbs much of the original UV luminosity and re-emits it in the IR. The CIB therefore contains important information about the evolution of the Universe between the redshift of last scattering (probed by the microwave background) and today (probed by optical surveys).

The CIB has still not been discerned in the sky brightness data measured by the COBE Diffuse Infrared Background Experiment (DIRBE), which was designed for the purpose, because it is hard to distinguish the CIB from foreground sources such as interplanetary and interstellar dust, and stars. Extrapolations of galaxy counts clearly predict a near IR CIB well above the DIRBE noise level. The most direct analysis of the DIRBE data gives the limits on the CIB derived from the darkest parts of the sky (Hauser 1995). At the DIRBE J, K, and L bands they are respectively $\nu I_\nu = 393 \pm 13$, 150 ± 5, and 63 ± 3 nW m^{-2}sr^{-1}. The latest attempt by the DIRBE team to account for foreground emission is reported by Hauser (1995), giving residuals for these bands ranging from 50 to 104, 15 to 26, and 15 to 24 nW m^{-2}sr^{-1} respectively.

An alternative detection method was suggested by Gunn (1965) based on the correlation properties of the diffuse backgrounds. It was successfully applied by Shectman (1973, 1974) in the V band and Martin and Bowyer (1989) in the UV. Indeed, if it comes from the redshifted light emitted by galaxies, the diffuse background must show fluctuations to reflect the correlation properties of galaxies.

We present here an overview of this study: section 2 gives a brief theo-

© 1996 American Institute of Physics

retical background for the method, data analysis is presented in section 3 and we conclude in section 4. More details and final numbers will be found in our forthcoming paper (Kashlinsky, Mather, Odenwald, and Hauser 1995; hereafter Paper I).

2. THEORETICAL PRELIMINARIES

If the primordial density fluctuations were Gaussian as most theories assume, then the three-dimensional spatial correlation function $\xi(r)$ determines all properties of the primordial density field. $\xi(r)$ has now been measured to large distances ($< 100h^{-1}$Mpc) in the B_J (Maddox *et al.* 1990; hereafter the 'APM' survey) and r bands (Picard 1991; hereafter the 'Palomar' survey). These estimates are consistent with other measurements of the mass distribution (cf. Kashlinsky 1994 and references therein), although there is no proof that all types of dark matter follow luminous matter. While deeper than before, the APM and Palomar surveys map the galaxy distribution at rather low redshifts (the APM survey limit is $b_J = 20.5$, and the Palomar survey limit is $r = 19$).

The three-dimensional correlation function is not directly observed in two dimensional maps. We therefore consider the projected two dimensional correlation function as a directly observable quantity. For diffuse light, the projected 2-D correlation function between fluxes in two different wavebands λ_1, λ_2 is defined as $C_{\lambda_1,\lambda_2}(\theta) \equiv \langle \lambda_1 \delta I_{\lambda_1}(\mathbf{x}) \cdot \lambda_2 \delta I_{\lambda_2}(\mathbf{x}+\theta)\rangle$, where $\delta I = I - \langle I \rangle$. It is related to the underlying 2-point correlation function via the analog of the Limber equation, which in the limit of small angles $\theta < 1$ reads:

$$C_{\lambda_1,\lambda_2}(\theta) = \int_0^\infty dz[\partial\lambda_1 I_{\lambda_1}/\partial z][\partial\lambda_2 I_{\lambda_2}/\partial z] \int_{-\infty}^{\infty} d\Delta\xi(r_{12}; z), \tag{1}$$

where $\Delta \equiv (z_2 - z_1)$, $r_{12}^2 = (\Delta cdt/dz)^2 + x^2(z)/(1+z)^2\theta^2$, and $x(z) = c\int_0^z (1+z)dt$ is the comoving distance to z.

The angular scale of the DIRBE beam probes comoving linear scales of $< 20-30h^{-1}$Mpc for the range of redshifts contributing to the bulk of the (near) IR extragalactic flux. On these scales, the correlation function implied by the APM survey can be approximated as a power law of slope $\gamma + 1 \simeq 1.7$. Thus as $\theta \to 0$ we can model $\xi(r;z)$ as $\xi(r;z) = (r/r_*)^{-1-\gamma}\Psi^2(z)$, where $r_* = 5.5h^{-1}$Mpc and the function $\Psi(z)$ accounts for the evolution of the clustering pattern, and is defined to be unity for $z = 0$. This leads to a predicted zero-lag correlation signal (mean square deviation) seen by DIRBE with the top-hat beam of:

$$C(0) = 1.6 \times 10^{-19} D Q_\lambda^2 \int \frac{[E(z)S(\lambda;z)\Psi(z)]^2}{[(1+z)^6\sqrt{1+\Omega z}]}[\frac{R_H(1+z)}{x(z)}]^{0.7} dz \tag{2}$$

where $C(0)$ is in $\mathrm{W^2m^{-4}sr^{-2}}$, $Q_\lambda = \lambda f_\lambda(\lambda;0)/\Delta\lambda_B f_\lambda(\lambda_B;0)$. Here λ_B and $\Delta\lambda_B$ refer to the B band filter wavelength and bandwidth used to define the galaxy luminosity functions. For galactic spectra and galaxy type mixes adopted from Yoshii and Takahara (1988) we get $Q_J = 8.9$, $Q_K = 3.8$, $Q_L = 1.7$ for the J, K, L bands. D is the beam dilution factor which for a circular top-hat beam and $\gamma = 0.7$ is $D(\gamma) \simeq 1.4$. We evaluated the effective radius of a circular beam corresponding to the DIRBE instrument to have radius $\simeq$0.45°, after accounting for the effects of the beam shape, the beam scanning during the observation time, the pixelization process, and the attitude errors. $E(z)$ and $S(\lambda, z)$ are the evolution parameters and are defined in Paper I. Thus we expect the zero-lag signal to be near $C_\vartheta(0) \sim 10^{-2}(\lambda I_\lambda)^2$.

Even if the data do not allow a detection, this method allows translation of an upper limit on $C(0)$ into an upper limit on the total flux in that band. To see how the measurements of the correlation signals constrain the levels of the background it is illustrative to rewrite the above equation as:

$$C(0) = 3.6 \times 10^{-3} \int (d\lambda I_\lambda/dt)^2 \Psi^2(t)(R_H/r)^\gamma H_0^{-1} dt \tag{3}$$

where t is the cosmic time and $r = x(z)/(1+z)$. In the Friedman-Robertson-Walker Universe the ratio of $[r(z)/R_H]^\gamma$ reaches a maximum of 0.45 at $z \sim 1.5$. The value at the maximum is almost independent of Ω being 0.43 for $\Omega = 0.1$ and 0.47 for $\Omega = 1$. This translates into $[R_H/r(z)]^\gamma \geq 2.2$. Thus any detection of the cosmological $C(0)$ or an upper limit on it would set an upper limit on the rms value of the rate of growth of the total flux of

$$(\frac{d\lambda I_\lambda}{dz})_{\rm rms} \equiv \sqrt{\int (d\lambda I_\lambda/dz)^2 dz} \leq 11\sqrt{C(0)}. \tag{4}$$

To relate this rms rate of growth of the flux to the mean rate of growth and the total flux requires a model for the z dependence of the rate of growth $d\lambda I_\lambda/dz$. Reasonable classes of models based on recent galaxy formation (z of a few) give the ratio of λI_λ to its rms rate of growth of the order of unity, and any specific model can be evaluated easily. This method allows for a simple and efficient way of constraining the CIB from material clustered like galaxies, based on the measurements of the mean square deviation in the DIRBE maps.

In the conventional DIRBE analysis one derives upper limits from the DC component of the signal, using various methods for modeling and subtracting the smooth and discrete foreground sources. As eq. (4) shows the correlations method could be quite efficient when the foreground emission is very bright but homogeneous. This is typically the case in wavebands longward of the ones considered here, where zodiacal light dominates stellar emission. The two methods measure different quantities, one depends on foreground models and one on background models, and both provide valuable constraints on the processes of galaxy formation, evolution, and spatial clustering. The analysis at these longer wavelengths will be presented in a future paper. Nevertheless, applying the techniques in the near-IR already gives interesting results.

3. DATA REDUCTION

We analyzed data from the three shortest wavelength bands observed by DIRBE, which approximate the ground-based photometric bands J, K, L, at 1.25, 2.2, and 3.5 microns. We selected four fields of 128×128 pixels, or 38.5 × 38.5°, located outside the Galactic plane, $|b^{II}| > 20°$. Two fields are located in the Ecliptic plane and two near the north and south Ecliptic poles. We also studied the same five fields that are used in setting the strongest current limits on the CIB (Hauser 1995): four fields are 32 by 32 pixels in size (10° by 10°)

and are centered on the two ecliptic (hereafter referred to as SEP and NEP) and two galactic (SGP and NGP) poles. The fifth field is 5° by 5° (16 by 16 pixels) and is centered on the Lockman hole (hereafter LH), the region of the minimum HI column density at $(l, b) = (148^{\circ}, +53^{\circ})$ (Lockman *et al.* 1986). We analyzed two versions of these maps. Map A was constructed by interpolating the data from 4 consecutive weeks of observation to a constant solar elongation of 90°, thus reducing the effects of the variation of the zodiacal light with time. Map B was the grand average of all 42 weeks of observations, as delivered to the National Space Science Data Center (Leisawitz 1995), with the DIRBE model of the zodiacal light subtracted directly (Reach *et al.* 1995; Franz *et al.* 1995). The model uses the time dependence of the measured sky brightness as the unique signature of the zodiacal light.

A measurement of $C(0)$ must satisfy three criteria to be considered a firm detection. First, it must be isotropic to a high degree. Second, it must have a level significantly above the random noise levels from the instrument and data processing. The spatial power spectrum of the fluctuations must be consistent with an origin outside the instrument and data reduction process. Third, it must be clear that it can not be produced by foreground sources. It must not be spatially correlated with foreground sources like the Galactic or ecliptic plane, or the local supercluster. Even without passing all these tests, the measured limits on $C(0)$ and their implications for the clustered part of the CIB are still very interesting as an example of this promising method.

We now describe two steps that were applied to reduce the effects of the zodiacal and stellar foregrounds from the DIRBE maps: discrete source masking, and color subtraction.

Except near bright objects like the Large Magellanic Cloud or the Galactic Plane, the appearance of each map in the selected J, K, L bands is characterized by a large number of bright point sources, mostly known stars, situated on a smooth distribution. For purposes of estimating the properties of the CIB we must eliminate the contribution from the discrete foreground as well as possible. Each field was processed by an iterative background modelling and source removal program developed by the DIRBE team. The smooth component was modeled as a 4th-order polynomial surface, which was fit to the entire field after application of a 5×5-pixel spatial filter. Pixels with fluxes greater than N_{cut} standard deviations above the fitted background model were blanked, as were the surrounding 8 pixels. In other words, each star was assumed to occupy a 3×3-pixel region 0.97° square, a factor of ~ 2 larger in area than the nominal DIRBE beam size of 0.7° square.

The procedure was repeated for the remaining non-blanked pixels in the map until no further points were masked by the algorithm. The background model was recomputed for the non-blanked pixels in the field, and the blanking process performed again at the same clipping threshold factor, N_{cut}. Since the clipping threshold is defined by a factor relative to the standard deviation of the residuals of the remaining pixels, it changes with each iteration and is slightly different for different values of N_{cut}. This process was iterated until no new peaks were identified in the current map iteration. Because this process does not actually subtract a background from the data at each step in the iteration, the final product is a sky map containing all of the original low spatial frequency background information, but in which bright points sources have been deleted by blanking their pixels. We tested cutoff thresholds of $N_{\mathrm{cut}} = 3.5$, 5, and 7; at lower clipping thresholds we have too few pixels remaining to allow for a meaningful analysis.

After the masks were created individually for all three bands, a combined mask was made so that only those pixels which were considered valid for all three bands would be used for further analysis. This step guarantees that the data are consistent for all the bands and for the color-subtracted analysis to be described below.

We examined the histograms of the fluctuations for each field and map and value of $N_{\rm cut}$. We found the following expected effects. The histograms are quite skew, with a high-intensity tail falling off roughly as $dN/d\delta I \propto \delta I^{-5/2}$, in accord with the prediction for stars filling infinite space uniformly. The low-intensity side of each histogram drops abruptly, since it is very improbable for no stars to be found in the large DIRBE beamwidth. Histograms for Map A, for which the zodiacal light model has not been subtracted, show double peaks for regions near the ecliptic plane. In Map B, all the fields give histograms with roughly comparable widths, as expected if the stellar foreground fluctuations are roughly the same for regions all at high galactic latitude. The measured width of the central part of the distribution is moderately sensitive to the clipping threshold used. We find that $N_{\rm cut} = 3.5$ is the lowest value where enough pixels still remain for proper analysis, because each deleted pixel also removes its 8 neighbors. The measured dispersion or $C(0)$ is still far too anisotropic to be interpreted as cosmological, but can be used to give upper limits. To obtain much lower values would require very careful analysis of the observational noise, the stellar foreground fluctuation contributions, and the detailed response properties of our nonlinear analysis algorithms.

The signal from the foreground radiation may also be disentangled from that of galaxies by considering the properties of the foreground stellar emission that are different from the CIB. The extragalactic objects have somewhat different colors than the Galactic stars, and they are redshifted as well. Furthermore, galaxies of different morphological types have different stellar populations and this too should help distinguish the colors of the foreground emission from our Galaxy (type Sc) from those of more representative galaxies (E/S0, Sa, etc.) which are expected to dominate the CIB at J, K and L-bands.

On the other hand the near-IR DIRBE maps exhibit a clear color with a relatively small dispersion. We define the color between any two bands, 1 and 2, as $\alpha_{12} = I_1/I_2$. Within each field the dispersion of colors is small, between 5% and 10%. The colors confirm that most of the light of the Galaxy in these bands comes from K and M giants (Arendt *et al.* 1994).

This empirical property of the constancy of color found in the DIRBE near-IR maps can be used to reduce the foreground contribution to $C(0)$ by taking the scaled difference of two maps. On the other hand, as discussed above most of the CIB in the near-IR is expected to come from galaxies whose spectra are significantly redshifted, and thus the CIB is not expected to be washed out by this procedure. These observations are the basis for the color subtraction method we present below.

There are two *independent* contributions to the measured fluctuations in the linear combinations of the DIRBE maps: the one that comes from the *total* foreground (stars, zodiacal light, and cirrus), which we denote as $C_{*,\Delta}(0)$, and the one that comes from the extragalactic background which we label $C_{g,\Delta}(0)$. We use the Δ notation to indicate that these variances refer to the differences of maps at two different wavelengths. The total $C_\Delta(0)$ is the sum of these two components. Let us assume that the Galactic and zodiacal foreground component has the mean color between any two wavelengths 1 and 2 of $S_{12} = \langle\alpha\rangle$ and the dispersion in the color is $s_{12}^2 = \langle(\delta\alpha)^2\rangle$, where $\delta\alpha$ is the deviation of the pixel color from the mean of the pixels in the field. Assuming that the color fluctuations are uncorrelated with the intensities, the zero-lag signal due to the foreground in Band 1, $C_{*,1}(0)$, can be written as

$$C_{*,1}(0) \simeq (S_{12}^2 + s_{12}^2)C_{*,2}(0) + s_{12}^2\langle I_{*,2}\rangle^2. \tag{5}$$

Note that the approximations used in this equation are not required for the validity of the analysis of the maps, but are only to indicate that major improvements are achievable with color subtraction. Let us construct the quantity which depends on the parameters in two bands (we define $\delta_i \equiv I_i - \langle I_i \rangle$), $\Delta_{*,12} \equiv \delta_{*,1} - \beta\delta_{*,2}$, where the quantity β is for now an arbitrary number which will be quantified later. The zero-lag signal in the "color-subtracted" quantity defined above would be given by

$$C_{*,\Delta}(0) \equiv \langle \Delta_{*,12}^2 \rangle = [(S_{12} - \beta)^2 + s_{12}^2]C_{*,2}(0) + s_{12}^2\langle I_{*,2}\rangle^2. \tag{6}$$

It is clear from the above that if we were to choose β close to the mean color for the foreground, $\beta \simeq S_{12}$, then the largest term in Eq.(6) could be made nearly zero and the contribution of the Galactic foreground confusion noise to the zero-lag signal $C_{*,\Delta}(0)$ in the difference map would be minimized. Furthermore, if $s \ll S$, that contribution can be decreased over the one specified in Eq.(5) by a fairly substantial amount.

Measurement shows that indeed we have $s \ll S$. Thus the variance in the color-subtracted fluctuation Δ due to the confusion noise can be reduced by a substantial factor, with a minimum when $\beta = S_{12}$. The plots of $C_\Delta(0)$ for the data from all five fields as a function of β for $J-K$, $J-L$ and $K-L$ subtraction, for map B and $N_{cut} = 3.5$ show that the minimum of $C_\Delta(0)$ is indeed quite deep. Therefore, in constructing the color-subtracted quantities we use the values of β for each field that correspond to the minimum of $C_\Delta(0)$. Unlike the single band zero-lag signal, which would correspond to $\beta = 0$, the minimum of $C_\Delta(0)$ is distributed significantly less anisotropically. The major anisotropy in the color subtracted signal comes from the SEP field. This field, however, contains the LMC galaxy and it is fair to exclude it from further discussion of the color subtracted signal.

The color subtraction works well to suppress the fluctuations due to the Galactic foreground. On the other hand the predicted extragalactic background fluctuations are unlikely to be removed by the color subtraction method because the redshifts give the background a different color than the foreground. The extragalactic signal would change by a different amount, and may even increase because of the combination of redshift and morphological effects. This can be shown explicitly using eqs.(2)-(4) and is discussed in more detail in our Paper I. Since the Galaxy colors are typical of general stellar populations our method effectively reduces the contribution from the nearby galaxies, but because of the redshift effects on galaxies the overall zero-lag signal remains at the levels of the single band cases. In other words, the color subtracted signal probes galaxies in a window of redshifts centered on an early epoch. The distant galaxies do not have a single color, and can not be removed by the color subtraction.

4. CONCLUSIONS

We have tested a promising method for recognizing the contribution of high-redshift galaxies to the cosmic infrared background, using their spatial and spectral correlation properties. These are different from the foreground stars, which have a single dominant color with small dispersion. Bright foreground stars can also be eliminated by simple masking techniques. We prepared linear combinations of maps at different wavelengths and adjusted the coefficient to minimize the resulting variance. The residuals are much smaller than those of the single band maps, and are much more isotropic. Theoretically, the fluctuations of the CIB should have different colors from the foreground stars and should still be detectable in these linear combination maps.

Even the single band fluctuation limits can be interpreted as (model-dependent) limits on the CIB originating in material clustered like galaxies. In the J, K, L bands, with reasonable galaxy evolution models, they are significantly lower than direct estimates of the upper limits on the CIB derived from the DIRBE maps and are comparable to the residuals in the direct sky brightness analysis (Hauser 1995).

There is still work to do in identifying the sources for all of the measured fluctuations. We have demonstrated that the instrument noise is quite small in the single band maps, but not negligible in the analysis of the color subtracted maps. The possible sources for such fluctuations could be the long-sought CIB, the residuals of the foreground star fluctuations, or effects of patchy dust absorption. A spatial power spectrum analysis is needed to show which of the above contributes to the detected signal.

REFERENCES

Arendt, R. G., *et al.*, 1994, Ap.J., 425, L85
Franz, B. A. *et al.*, 1995, this Workshop
Gunn, J. 1965, Ph.D.Thesis, Caltech. (unpublished)
Hauser, M. G, 1995, this Workshop
Kashlinsky, A. 1994, in "Evolution of the Universe and its Observational Quest", ed. K. Sato, p.181, Universal Academy Press, Inc. - Tokyo
Kashlinsky, A., Mather, J., Odenwald, S., and Hauser, M. 1995, Ap.J., to be submitted
Leisawitz, D. T., 1995, this Workshop
Lockman, F. J. *et al.*, 1986, Ap.J., 302, 432
Maddox, S. *et al.*, 1990, MNRAS, 242, 43p
Martin, C. and Bowyer, S. 1989, Ap.J, 338, 677
Picard, A. 1991, Ap.J., 368, L7
Reach, W. T. *et al.*, 1995, this Workshop
Shectman, S. 1973, Ap.J., 179, 681
Shectman, S. 1974, Ap.J., 188, 233
Yoshii, Y. and Takahara, F. 1988, Ap.J., 326, 1

NEAR INFRARED BACKGROUND CORRELATIONS

Shoba Veeraraghavan
Code 685, NASA Goddard Space Flight Center, Greenbelt, MD 20771
shoba@stars.gsfc.nasa.gov

ABSTRACT

The infrared background contribution from galaxies will exhibit angular correlations due to the spatial clustering of galaxies. The amplitude of these correlations and their evolution with redshift are determined by the evolutionary history of galaxies, specifically, the evolution of galaxy spectrophotometric properties and clustering. I estimate the expected correlations in the DIRBE near IR bands for various galaxy evolutionary models.

1. INTRODUCTION

It is believed that the infrared background contains much information on the details of galaxy formation, including the epoch when stars and galaxies were first formed, their luminosity evolution, the evolution of their spectra, and the strength and evolution of their clustering (Bond, Carr, & Hogan 1986, 1991). However, due to foreground contamination from interplanetary dust and our galaxy, it has proved quite difficult to isolate the isotropic background signal from the data, and the cosmic infrared background is yet to be detected (Hauser 1995).

It may prove easier to measure the angular power spectrum of the IR background fluctuations rather than the isotropic mean, since galaxies are known to be strongly clustered. Kashlinsky *et al.* (1995, these proceedings) are attempting to estimate the rms value of the correlated cosmological signal in the DIRBE near IR J, K, and L bands, using an approach suggested by Gunn (1965) and implemented by Schechtman (1973,1974), see also Cole, Treyer & Silk (1992). I discuss below the theoretical expectations for the strength of such a signal from in the near IR background light.

2. DIFFUSE IR BACKGROUND AND ANGULAR CORRELATIONS

The diffuse IR background from galaxies is the sum of the flux from galaxies at all redshifts integrated over the galaxy luminosity function at those redshifts. Its mean value in a band n is,

$$(\lambda I_\lambda)_n = \frac{1}{4\pi}\int_0^{z_{\max}} dz \frac{dx_c/dz}{(1+z)^2}\frac{\mathcal{F}_n}{\mathcal{F}_{n'}}\mathcal{L}_{n'}(z)10^{0.4\Delta_{nn'}(z)}, \qquad (1)$$

$$\Delta_{nn'}(z) = K_{n'}(z) + E_{n'}(z) - K_n(z) - E_n(z)$$

Here $\mathcal{F}_n/\mathcal{F}_{n'}$ is the ratio of the *present epoch* fluxes in the chosen band and a reference band n', $\mathcal{L}_{n'}(z)$ is the luminosity density in the reference band as a function of redshift, and E_n and K_n are the evolutionary and K-corrections respectively in the band n. Finally $x_c(z)$ is the comoving distance to redshift z. Since the galaxy distribution, and therefore the luminosity density, is clustered,

© 1996 American Institute of Physics

this IR background from galaxies will display angular correlations. It can be shown (see, for instance, Peebles 1980, §§50,58) that the IR correlation function is an integral over the galaxy spatial 2-point correlation function $\xi(r;z)$:

$$C_n(\theta) = \frac{\theta}{16\pi^2}\left(\frac{\mathcal{F}_n}{\mathcal{F}_{n'}}\right)^2 \int_0^{x_c^{\max}} \frac{x_c dx_c}{(1+z)^4} 10^{0.8\Delta_{nn}(z)} \mathcal{L}_{n'}(z)^2 \int_{-\infty}^{\infty} dy \xi(x_c\theta\sqrt{1+y^2};z) \tag{2}$$

The luminosity function at high redshifts in any band is poorly known. Locally, it is best characterised in the optical (Efstathiou, Ellis, & Peterson 1988, Loveday *et al.*1992), and is not well determined or not known at all in most other bands, including the near IR. A reasonable assumption is that the galaxies represented by the blue luminosity function are the same ones that provide the near IR background, thus the luminosity densities in the two bands can be related via a spectrophotometric model for galaxies. With a prescription for the evolution of this spectrophotometric model, the luminosity density in the near IR bands at any redshift can be estimated. Additionally, with some ansatz for the nature and evolution of galaxy clustering, we can calculate the expected correlation function and its zero-lag amplitude C_0. Where the clustering is still linear, $\xi(r;z)$ can be expressed in terms of the power spectrum $P(k;z)$ of primordial density fluctuations, with the result,

$$C_n(\theta) = \frac{1}{4}\left(\frac{\mathcal{F}_n}{\mathcal{F}_{n'}}\right)^2 \int_0^{z_{\max}} dz \frac{dx_c/dz}{(1+z)^4} 10^{0.8\Delta_{nn'}(z)} \mathcal{L}_{n'}(z)^2 \int_0^{\infty} kdkP(k;z)J_0(kx_c\theta) \tag{3}$$

I will defer consideration of the linear theory to a later paper, and here explore the background expected from nonlinear and quasi-linear clustering. In this regime, the power law form of the correlation function can be used. Letting r be a physical (not comoving) distance, the evolution with redshift is often parametrised as (Efstathiou *et al.* 1991),

$$\xi(r;z) = \xi(r;0)(1+z)^{-(3+\epsilon)}, \qquad \xi(r;0) = (r/r_0)^{-\gamma} \tag{4}$$

Two representative values for ϵ are, $\epsilon = 0$, which is equivalent to the clustering being stable in physical coordinates (galaxies are unbiassed), and $\epsilon = \gamma - 3$ which follows if the clustering is stable in comoving coordinates (the galaxy distribution is biassed with galaxies at the density peaks of an underlying mass field). Numerical simulations have shown that in both hierarchical and pancake clustering models, $\epsilon > 0$ is obtained. Adopting such a parametrisation leads to,

$$C_n(\theta) = \frac{H_\gamma}{16\pi^2}\frac{r_0^\gamma}{\theta^{\gamma-1}}\left(\frac{\mathcal{F}_n}{\mathcal{F}_{n'}}\right)^2 \int_0^{z_{\max}} dz \frac{dx_c/dz}{x_c^{\gamma-1}(1+z)^{4-\gamma+\epsilon}} 10^{0.8\Delta_{nn'}(z)} \mathcal{L}_{n'}(z)^2 \tag{5}$$

which, not surprisingly, is a power law in θ, $C(\theta) = C_0\theta^{1-\gamma}$. Here H_γ is a numerical constant.

Here I consider the predictions for C_0 from various models for the evolution of the galaxy spectral, photometric, and clustering parameters. Generically, we expect an enhanced luminosity density in the past. For example, the Bruzual spectrophotometric models predict an enhanced luminosity in the past when there was more fuel available for star formation. In merger models, the number densities are higher prior to the merger, whilst the luminosities are enhanced post-merger. On the other hand, spatial correlations are expected to be generally weaker in the past, although the data are consistent with no evolution of the correlation function (Cole *et al.* 1994).

3. RESULTS

The reference model consists of prototypical "average" galaxy SED which is a linear combination of the SEDs of the different Hubble types, heavily weighted towards early-type galaxies (Tinsley 1980). In this model neither the galaxy SED nor the luminosity function evolve in redshift, and the spatial clustering is stable in physical coordinates. A no evolution model predicts a lower bound to the near IR background compared to many models. The rms fluctuations are approximately (1.5, 0.5, 0.5) $nW/m^2/sr$ in J, K, and L bands respectively. The contribution to the integrated amplitude at zero lag is maximum at redshift $\approx 0.1 - 0.3$. Figure 1 shows the redshift dependence of the contribution to $\sqrt{C_0}$ in the reference model.

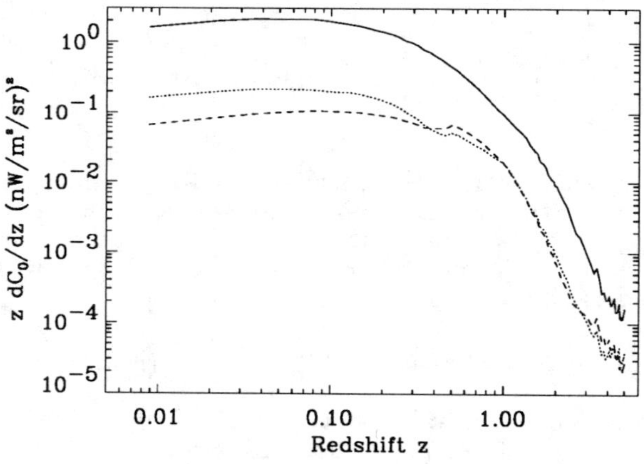

Figure 1: Evolution of the differential zero-lag correlation amplitude dC_0/dz for the reference no-evolution model described in the text. Galaxy formation is initiated at $z = 5$. The solid line represents J, the dotted line is K, and the dashed line is L. Cosmological parameters are $H_0 = 100$ km/s/Mpc, $\Omega_0 = 1$, $\Omega_\Lambda = 0$.

Reducing the rate of clustering growth by assuming a clustering model in which the galaxy distribution is biassed increases the contribution at high redshift, but since the bulk of the contribution to the intensities comes from low redshift the enhanced clustering does not change $\sqrt{C_0}$ significantly. The evolution of the luminosity function, in particular density evolution, enhances both dC_0/dz and $d/dz(\lambda I_\lambda)$, and hence C_0 and λI_λ, and hence C_0 considerably. Luminosity evolution produces a more modest enhancement. These models are shown as the dotted, dashed and dot-dashed lines in Fig. 2 on the left. More realistically, the SED and luminosity density will evolve as star formation takes place in galaxies. Such a model is shown as the solid line in Fig. 2 on the right. The fluxes are somewhat lower than those in the no-evolution model, because the galaxy luminosities initially increase and then decrease with redshift. Modifying the blend of galaxies to allow for more late-type and irregular galaxies (Sandage, Binggeli, & Tammann 1985) slightly lowers the fluxes and C_0 since the latter galaxies are on average less luminous. Finally the effect of a low density model or a vacuum-dominated cosmology are similar, both show enhanced contribution to the flux at redshifts ≈ 1 so that the amplitude of C_0 is boosted in all bands.

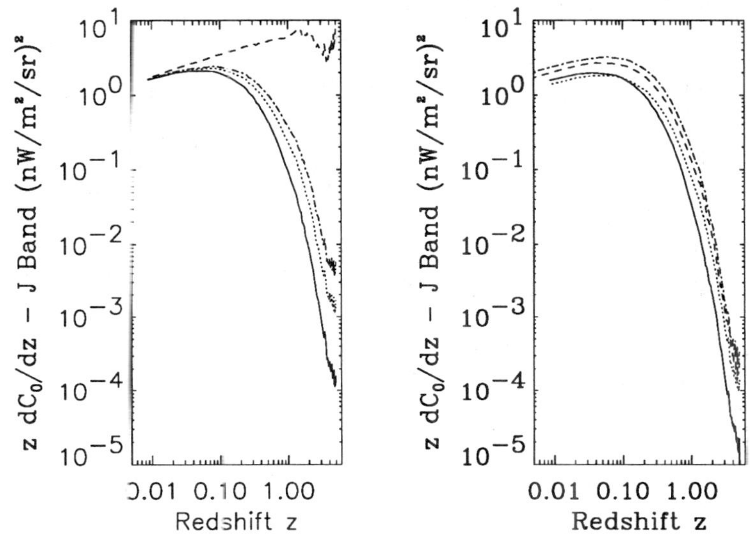

Figure 2: Evolution of J-band dC_0/dz for the models described in the text. On the left, the solid, dotted, dashed and dot-dashed lines show the reference no-evolution model with clustering stable in physical coordinates, no-evolution model with clustering stable in comoving coordinates, comoving density evolution model, and luminosity evolution model. On the right, the solid, dotted, dashed and dot-dashed lines show the same model with SED evolution, a no-evolution model heavily weighted towards late Hubble types, and no-evolution $\Omega_0 = 0.1$ and $(\Omega_0 = 0.2, \Omega_\Lambda = 0.8)$ models.

The values of C_0 in the near IR bands do not change significantly unless rather drastic density or luminosity evolution is assumed. In a wide range of models, $\sqrt{C_0}$ is in the range $[2.5 - 4, 1 - 2, 0.5 - 1.5]$ nW/m^2/sr. The best publicly available DIRBE results are the dark sky fluxes with amplitudes $[393 \pm 13, 150 \pm 5, 63 \pm 3]$ nW/m^2/sr (Hauser 1994). To compare the models above with DIRBE, the intrinsic sky fluctuations have to be folded in with the DIRBE beam of $\theta_b = 0.7°$. The beam convolved mean square intensity equals $\mathcal{A}C_0/\theta_b^{\gamma-1}$, where A is a small numerical constant. In order to be confident of a detection of the diffuse near IR background correlations even using a colour subtraction technique (Kashlinsky *et al.* 1995), one would have to achieve accurate elimination of foregrounds at the few percent level.

ACKNOWLEDGEMENTS

I thank Gustavo Bruzual for providing me with his GISSEL package, and John Mather for useful discussions.

REFERENCES

Bond, J.R., Carr, B.J., & Hogan, C.J. 1986, ApJ, 306, 428
Bond, J.R., Carr, B.J., & Hogan, C.J. 1991, ApJ, 367, 420
Cole, S., Treyer, M-A., & Silk, J. 1992, ApJ, 385, 9
Cole, S., Ellis, R., Broadhurst, T., & Colless, M. 1994, MNRAS, 267, 541
Efstathiou, G., Ellis, R., & Peterson, B. 1988, MNRAS, 232, 431
Efstathiou, G., *et al.* 1991, ApJ, 380, L47
Gunn, J. 1965, Ph.D. thesis, Caltech (unpublished)
Hauser, M.G. 1995, in Extragalactic Background Radiation, STScI Symp. Ser. 7, ed. D. Calzetti, M. Livio, & P. Madau (Cambridge: Cambridge Univ. Press), 135
Kashlinsky, A. *et al.* 1995, preprint
Peebles, P.J.E. 1980, The Large-Scale Structure of the Universe (Princeton: Princeton University Press)
Sandage, A., Binggeli, & Tammann, S. 1985, AJ, 90, 1759
Schechtman, S. A. 1973, ApJ, 179, 681
Schechtman, S. A. 1974, ApJ, 188, 233
Tinsley, B. 1980, ApJ, 241, 41

NUMBER COUNTS AND COLOURS OF GALAXIES IN DEEP NEAR-INFRARED SURVEYS

J. P. Gardner
University of Durham
Department of Physics
South Road
Durham DH1 3LE
UNITED KINGDOM

ABSTRACT

We present deep galaxy number counts and colours of $K-$band selected galaxy surveys. We argue that primeval galaxies are present within the survey data, but have remained unidentified. There are few objects with the colours of an L^* elliptical galaxy at a redshift of $z \approx 1$, in contradiction to standard luminosity evolution models. We calculate the extragalactic background radiation at 2.2 microns wavelength produced by resolved objects.

1. INTRODUCTION

Galaxy number counts and the colours of the field galaxies making up those counts have proven to be a powerful tool for investigating galaxy evolution and the cosmological geometry. Observations made at a wide range of magnitudes and wavelengths can, in principle, distinguish between different evolutionary and cosmological models. Studies of galaxy evolution are statistical in nature, and field galaxies avoid the short-timescale effects associated with radio galaxies and quasars which make their selection function difficult to determine. Similarly, galaxies in clusters may show environmental effects. Optical galaxy surveys at bright magnitudes have covered very large areas by scanning photographic plates (Maddox et al 1990). Long exposures with CCDs have reached $B = 28$, (Metcalfe et al 1995) while spectroscopic redshift surveys to the current limit of $B < 24$ (Cowie et al 1991) have placed limits on the models of evolution proposed to explain the counts.

The near-infrared, or K band, provides several advantages for faint galaxy studies, including smooth K$-$corrections and well-understood luminosity evolution. The K band represents the reddest we can observe from the ground without the very high backgrounds of the thermal infrared. The flux is dominated by long-lived near-solar mass stars which are comparable in lifetime and timescales of evolution to the age of the galaxy. These stars make up the bulk of the galaxy, and the absolute K magnitude is a measure of the mass of light matter in a galaxy.

2. PRIMEVAL GALAXIES AND THE NUMBER COUNTS

Much early near-infrared survey work was motivated by a theoretical model of dust-free protogalaxies developed by Partridge & Peebles (1967; PP). In this model, galaxies form through the collapse of overdense regions, pass

© 1996 American Institute of Physics

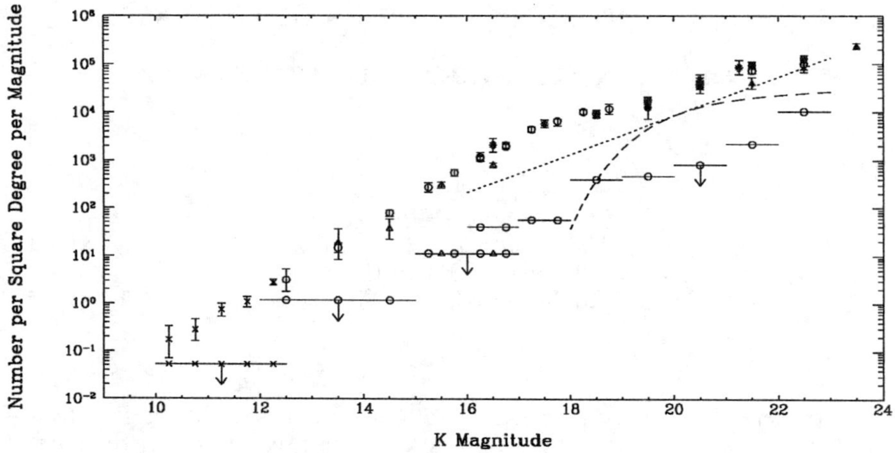

Figure 1: Primeval galaxy non-detection limits; The $K-$band number counts from the compilation in Gardner at al (1995), plotted with a simple model for primeval galaxies. The dotted line provides a locus along which a suitable luminosity function, the dashed line, will slide. Also plotted are the upper limits on the density of objects which could be identified, but are not detected, for each of the surveys. Since the primeval galaxy model is above the upper limits on undetected objects, it is likely that primeval galaxies are hiding within the survey catalogs, but have the same colours as normal galaxies, and remain unidentified.

through a very luminous phase in which the majority of the stars are formed, and appear as large, bright objects at $10 < z < 100$, with Lyman α redshifted beyond the optical.

It is interesting to consider the large red primeval galaxy model of PP in light of the current $K-$band number counts. Early researchers often used a very optimistic set of assumptions to justify conducting surveys using the limited technology available at the time. Lilly & Cowie (1987), however, constructed a model which is fairly independent of assumptions of cosmological geometry and timescale of star formation by assuming that a primeval galaxy will have the flat spectrum ($f_\nu \propto \nu^{-1}$) characteristic of an unreddened young stellar population.

In the initial star-bursting phase of the PP model, the majority of the metals in the universe were produced, and this metal production by stellar nucleosynthesis would have produced a surface brightness on the sky dependent only on the mass fraction of metals, the density of matter processed to produce the metals, and the redshift of galaxy formation. Assuming that the primeval galaxies have a flat spectrum, the $K-$band surface brightness is independent of z_{form} unless $z_{form} > 20$ and Lyman α has passed out of the K band. This surface brightness defines a line on the number count diagram along which an appropriate luminosity function would slide, depending upon the specifics of the model, such as z_{form}, the cosmological geometry, and the duration of the star-bursting phase. Geometry then gives the number density of the progenitors of objects with a given present density. Figure 1 is the $K-$band galaxy number counts, from Gardner et al (1993, 1995), with this model plotted.

In this model, the predicted number of primeval galaxies is lower than the number of normal galaxies for $K > 20$, and so a very important consideration

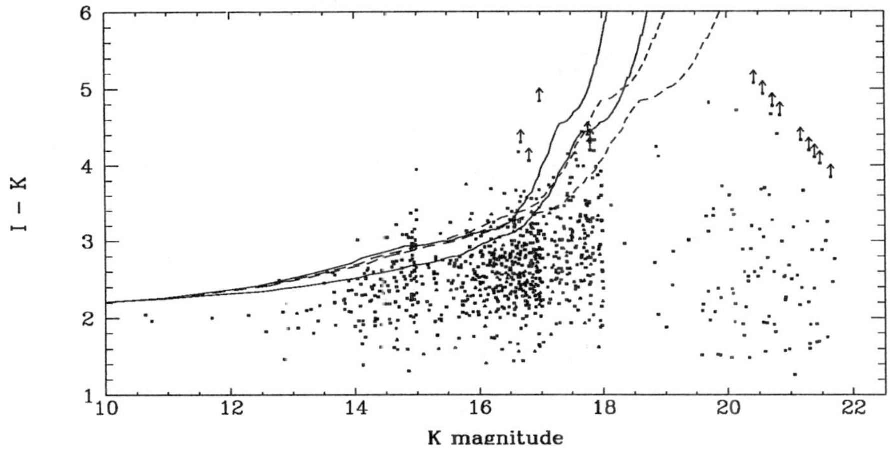

Figure 2: $I - K$ vs K from Gardner (1995). Also plotted is a model for a K^* elliptical galaxy. The dashed lines are the no-evolution model, with the lower line being for $q_0 = 0.5$, and the upper line being for $q_0 = 0.02$. The solid lines include Yoshii & Takahara (1988) evolution. Most galaxies at the bright end are elliptical and Sa galaxies at near K^*. At the faint end, however, the galaxies drop away from the lines, and there are very few galaxies with the colours of L^* ellipticals at $z > 1$.

is whether we can recognize them for what they are. In the surveys discussed in this paper, there are no large ultrared objects of the kind predicted by PP to be primeval galaxies at high redshifts ($z > 7$). If we assume that this failure to find primeval galaxies means that they are not there (i.e., if we assume primeval galaxies would not be confused with normal galaxies or stars), then we have placed a much stronger constraint on the model. Plotted in Figure 1 are the upper limits implied by the failure to identify primeval galaxies in the area covered at each magnitude bin.

Figure 1 shows that a simple model of the surface brightness on the sky of metal formation predicts number counts of primeval galaxies which are below the observations of normal galaxies, but above the limits of detection in the current surveys. While these surveys contain no unambiguous detections of primeval galaxies, this figure says that primeval galaxies are hiding as a minority population within the survey data. There are a wide range of primeval galaxy models, predicting everything from bright extended red objects to faint compact blue ones. However, various observational programs, including this one, are beginning to put strong constraints on some of the models, and it appears likely that the field is close to making an unambiguous discovery of a population of primeval galaxies. This will come either through the spectroscopic identification of a flat spectrum object at high redshift, or, more subtly, through statistical arguments about a moderately high redshift active star forming population.

3. COLOUR DATA

The K−band number counts were presented by Gardner et al (1993, 1995), and were discussed there in the context of the B−band counts and the median

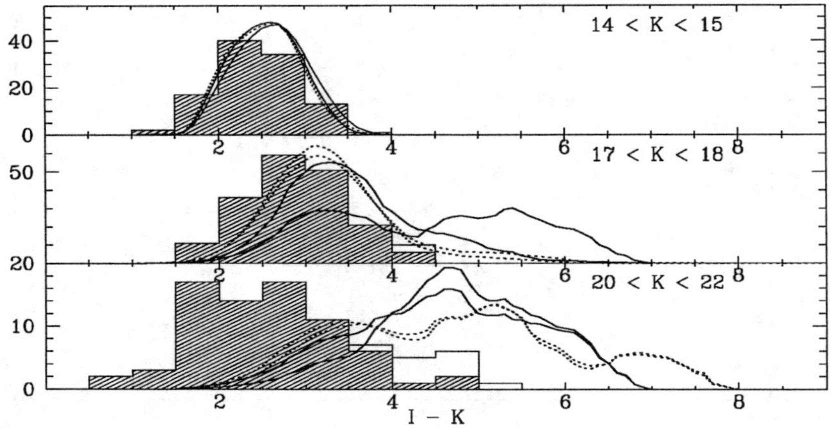

Figure 3: Histograms of the $I-K$ colour of the galaxies in the surveys of Gardner (1995) in 3 different magnitude bins. At the brightest magnitudes, the galaxies are well fit by the models, but in the two fainter bins, the data show a departure from the models. The models predict many more red galaxies, and many fewer blue galaxies than are observed.

$B-K$ colour as a function of K magnitude. Figure 2 is a plot of the $I-K$ colour as a function of K magnitude from Gardner (1995). Also plotted are a no-evolution model for an elliptical galaxy at K^*, for $q_0 = 0.02$ and $q_0 = 0.5$ ($\Lambda = 0$), and a model of this galaxy type with Yoshii & Takahara (1988) (YT) luminosity evolution. The models are the reddest that galaxies can normally be expected to get, although the data does show some scatter around this line due to noise. The general trend of the $I-K$ colours of the galaxies is to follow the models for elliptical galaxies at the bright end. Photometric surveys of this type generally pick out the intrinsically brightest galaxies at each apparent magnitude, since they are the most distant, and therefore sample the largest volume. A $K-$band selected survey will pick out the reddest objects. Thus one expects, in K selected surveys, to be dominated by L^* elliptical and S0 galaxies, and this is evident in the data at the bright end. At the faint end, however, these L^* unevolved elliptical galaxies are no longer seen. Essentially, objects with the colours of non-evolved L^* elliptical galaxies at a redshift of $z \approx 1$ do not appear in the data. YT evolution is not enough to explain this deficiency. With $z_{form} = 5$, the elliptical galaxies at $z \approx 1$ show little difference from the present day SED, and the K−corrections dominate the colours, making the galaxies very red. While a much lower redshift of galaxy formation might save this model, that would create other problems. The data say that there is much more star formation in the galaxy population at $z \approx 1$ than the standard luminosity evolution models predict.

Figure 3 contains histograms of the $I-K$ colours of the galaxies in the surveys of Gardner (1995), along with the predictions of evolution and no-evolution models. Three different magnitude bins are plotted. In the brightest magnitude bin, the galaxies fall close to the model predictions. By K=17, the galaxies begin to show a departure from the models, and at the faintest levels, the models are a very poor fit to the data. The deficit of the reddest class of galaxies is very evident in this figure. A Kolmogorov-Smirnov test gives a probability of 87%

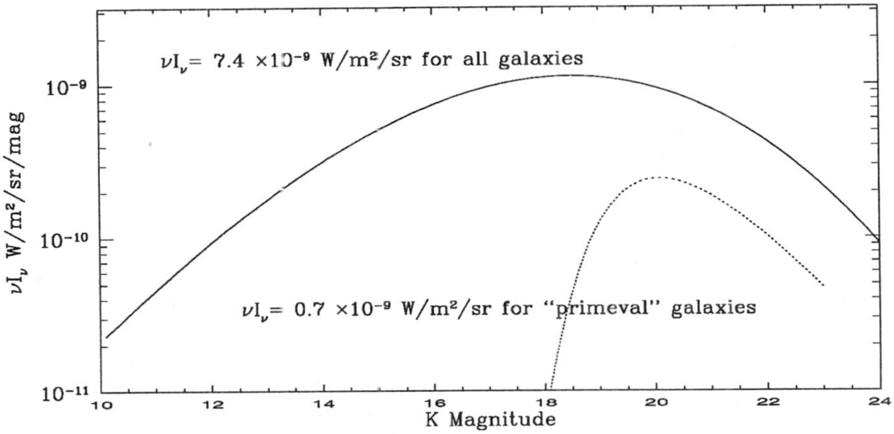

Figure 4: The extragalactic radiation at 2.2 microns coming from resolved objects. Also plotted is the flux from the primeval galaxy model discussed in section 2.

that the galaxies in the HWS are drawn from the best fit model. The other two samples have less than 1% probability of being drawn from any of the model distributions. The few objects which were undetected in I, and therefore have lower limits to their $I - K$ color do not affect this conclusion.

4. THE EXTRAGALACTIC BACKGROUND RADIATION

Figure 4 is a plot of the extragalactic radiation in the K-band from the galaxies making up the number counts in figure 1. From this figure, it is apparent that the majority of the flux comes from galaxies at intermediate magnitudes, $16 < K < 20$, and that observations at fainter magnitudes is unlikely to affect the measurement of the integrated flux significantly. Also plotted on this figure is the primeval galaxy model discussed in section 2. It can be seen that this model does not dominate the extragalactic radiation in the K-band at any magnitude. The integrated flux at 2.2 microns is $\nu I_\nu = 7.4 \times 10^{-9} W/m^2/sr$ for all galaxies, and the primeval galaxies are approximately 10% of this.

REFERENCES

Cowie, L. L., Songaila, A., & Hu, E. M. 1991, Nature, 354, 460

Gardner, J. P., Cowie, L. L., & Wainscoat, R. J. 1993, ApJ, 415, L9

Gardner, J. P. 1995, ApJ, in press

Gardner, J. P., Shanks, T., Campos, A., & Metcalfe, N. 1995, in preparation

Lilly, S. J., & Cowie, L. L. 1987, in Infrared Astronomy with Arrays, ed. C. G. Wynn-Williams & E. E. Becklin, (Honolulu: University of Hawaii), 473

Maddox, S. J., Sutherland, W. J., Efstathiou, G., Loveday, J., & Peterson, B. A. 1990, MNRAS, 247, 1P

Metcalfe, N., Shanks, T. Fong, R., & Roche, N., 1995, MNRAS, in press

Partridge, R. B., Peebles, P. J. E., 1967, *Ap. J.* **147**, 868

Yoshii, Y., & Takahara, F. 1988, ApJ, 326, 1

GALAXY EVOLUTION AND FAINT COUNTS IN THE NEAR-INFRARED

Brigitte Rocca-Volmerange and Michel Fioc
Institut d'Astrophysique de Paris, 98bis Bd Arago, F-75014 Paris
rocca@iap.fr and fioc@iap.fr

ABSTRACT

The contribution of distant galaxies to the diffuse near-infrared and submillimetric extragalactic backgrounds can be predicted with the help of a multispectral modelling of the faint galaxy counts. In particular, star-forming galaxies have to be taken into account as well as old evolved galaxies implying a coherent simulation of the stellar emission from the *blue* to the *near-infrared* with gas and dust contributions. For this reason, an extension of our previous UV and visible models is worked out with a detailed synthesis population model for strong, short timescale starbursts in a dusty medium in the near IR (Lançon and Rocca-Volmerange, 1995, hereafter LRV95)by using a spectral library of stars from 1.4 to 2.5 μm observed with the FTS instrument at the 3.60m/CFHT and fitted on starburst spectra observed with the instrument. Then an extension of our atlas of synthetic galaxies was carried out, the near-IR emission (K band) being carefully normalised to the blue (B, V or J_+ bands) emission. Galaxy counts in the K band are modelled in various cosmologies. The comparison with observations arise questions to be discussed with results from visible counts.

1. INTRODUCTION

The faintest counts of galaxies as a function of the apparent magnitude are clues to determine the geometry of the Universe. But they also are sensitive to others effects such as the intrinsic evolution (pure luminosity or merging) of galaxies. So the main difficulty of interpreting the deepest surveys is to significantly disantangle these various effects. From a modelling in the far-UV and visible, we found that a pure luminosity evolution would concile observations in an open universe while an increase of number density with redshift through a merging effect is in agreement with a flat Universe (Rocca-Volmerange and Guiderdoni, 1990 hereafter RVG90, Guiderdoni and Rocca-Volmerange, 1990, hereafter GRV90) A more complete interpretation of the number counts is given by a multispectral approach, and in particular, in the near-infrared (K band). This critical wavelength range, on the boundary between spectral regions dominated by emission from stars or from dust and non-thermal sources, constrains the relative contributions of these components and reveals recent star-formation through bright supergiant stars. At such wavelengths, stellar evolution is lower and extinction fainter but difficulties arise from a less complete knowledge of stellar properties, in particular of red supergiants, horizontal branch stars and AGB stars. Moreover different luminosity functions were proposed in the K band (Mobasher et al, 1993, Glazebrook et al, 1994), their respective availability being not clear. We improved our model in the near-infrared, with a significant coverage of the Herzsprung-Russell diagram and a consistent set of stellar inputs to simulate galaxies of the Hubble Sequence as well as starbursts (section 2). The predictions of faint galaxy counts in the K band will be then compared to recent observations in section 3.

© 1996 American Institute of Physics

2. IMPROVEMENTS OF MODELLING IN THE NEAR-INFRARED

In view to extend our atlas of synthetic spectra of galaxies (RVG, 1988) to the near infrared, we observed a stellar library and a series of template galaxies (from starbursts to ellipticals) at an intermediate resolution ($\leq 5\AA$) with the Fourier Transform Spectrograph at the CFH Telescope. The stellar energy distributions (and colors) are given by their continua, extended from 1.4 to 2.5μm, corresponding to typical colours H-K, as well as by molecular features, essentially due to the CO respectively shortward of 2.3μm and between 1.56μm and 1.8μm, the wings of the H_2O features and the resulting "H-band bump". All these absorption lines are sensitive to stellar surface gravity and effective temperature.

2.1 The role of M stars

The M stars reach effective temperatures down to 2000K. They are preferentially contributing to the stellar infrared emission, so that their role in modelling the energy distributions of galaxies worths a better understanding. Their spectra fundamentally differ from a black-body distribution. Most of theoretical models, related to the visible as Kurucz's models, 1991 limited by $T_{eff} \geq 3500K$, do not reach the temperatures dominant in the near-infrared and do not include all H_2O absorption lines. These H_2O bands are needed to distinguish between red giants from young supergiants, while CO bands are sensitive to an age effect and to dwarf effect (LRV92). Our spectral library includes several M giants with $3750K \leq T_{eff} \leq 2200K$. They are supposed to be of normal metallicity. But the question of metallicity in relation with the evolutionary stages of these stars is not clearly solved. From a recent analysis in the low-metal abundance Magellanic Clouds (Bessell, 1995), color-magnitude diagrams show more numerous and luminous red supergiants than predicted from recent isochrones (Bertelli et al, 1994).

2.2 Synthetic energy distribution of galaxies and starbursts in the near-infrared

The evolutionary population synthesis model is an updated version of Guiderdoni and Rocca-Volmerange, 1987, based on observational libraries.

A first approach (LRV95) was to simulate through the H+K bands, short bursts of star formation with an integration time lower that 10^5years, roughly corresponding to the lifetime duration of a $100M_\odot$ star. The library of near-IR stellar spectra was obtained with the same instrument than starbursts and elliptical galaxies. The adopted stellar evolutionary tracks include the Asymptotic Giant Branch and the Horizontal Branch phases which are energetically dominant in the near-IR. The evolution of stellar signatures (the H-K colour, the photospheric Brackett line absorption, the equivalent widths of CO and H_2O) is used for a datation and for determining the main properties of current starbursts. The emission lines of H and He (Brγ, Paα, HeIλ2.06μm), as well as [FeII]λ1.64μm and rotation-vibration lines of H_2 detected in most starburts are fitted on the current number of Lyman continuum photons and the number of supernova remnants. The ratio HeI/Brγ is no more considered as a significant indicator of the superior limit of the Initial Mass Function, as proposed by Doyon et al, 1992. Various distributions of dust and stars (dust screen, homogeneous mixing of gas or more complex structures) were considered into a multicomponent extinction modelling.

Another approach uses stellar energy distributions extended from the far-UV and visible to the near-infrared, by respecting the observed colors in the Bessell-Brett, 1988 photometric system and a unique calibration on the whole wavelength range. Moreover all the stellar input data (stellar spectra, bolometric corrections, luminosity classification) have to be coherent from the visible to the near-infrared.The extension to the near-infrared is then consistent with the visible and so available for calculating faint counts at various redshifts. A complete atlas of synthetic spectra is in preparation.

3. THE FAINT GALAXY COUNTS IN THE NEAR-INFRARED

3.1 A status from visible counts

A series of predictions in the visible is computed with the help of our evolutionary model and compared to observations (GRV90, RVG90). The two main parameters constraining the faint galaxy counts as a function of magnitude are the redshift of formation z_{for} and the density Ω_0 of the Universe which modifies the size of the volume element. A straightforward interpretation of observations with a pure luminosity evolution model gives an open Universe ($\Omega_0 \leq 0.05$) and a high redshift of formation ($z_{for} \geq 5$). To reconcile the observations with a flat Universe ($\Omega_0 = 1$), another effect is needed. A possibility is to increase the number of galaxies at high redshift by a factor $(1+z)^{1.8}$ (RVG90), but if the merging process is observed, an observational statistical confirmation was not found. Analyses relating the blue and red counts were proposed by Cowie et al, 1994. From their new Hawaï surveys at various depths, they found that populations of galaxies dominating in the red and in the blue are different but their model is not specifically fitted on these two bands and this difference can be due to an incomplete modelling.

3.2 Modelling the faint K counts

The advent of IR arrays made feasible very faint K counts, even if the high sky background in the near infrared requires extremely long exposure times to reach depths comparable to the visible. From the nearby and intermediate counts by Mobasher et al, 1993, Glazebrook et al, 1994 and Hawaii wide-field surveys (Cowie et al, 1994) to the deepest ones down to K=23-24 (Garner et al, 1993, Soifer et al, 1994), a comparison with models is possible. Outputs of the new version of our evolution model are synthetic spectra of eight different spectra as in RVG88. Cosmological k- and evolutionary e- corrections are calculated in various cosmologies. The adopted luminosity function is given from Efstathiou et al, 1988 corrected of colors consistently with the galaxy spectra (see Fioc and Rocca-Volmerange, 1995 for details). Faint counts in the K band (fig 1) were calculated for various cosmologies ($H_0 = 50kms^1Mpc^{-1}$, q_0=0.05 and q_0=0.5 and a cosmological constant nul). The evolution is based on pure luminosity scenarii as described in RVG88. Clearly the best model corresponds to a pure luminosity evolution in an open universe (q_0=0.05) as in the visible but the counts at magnitude K=23 and 23.5 from Soifer et al, 1995 appear too high by a factor 2. A comparison with the corresponding analysis of the blue faint counts is being in progress.

REFERENCES

Bertelli, G., Bressan, A., Chiosi, C., Fagotto, F., Nasi, E., 1994, Astron. Astrophys., 289, 665
Bessell, 1995, private communication
Bessell, M.S., Brett, J.M., 1988, Publications of the Astron. Soc. of the Pacific, 100, 1134
Cowie, L. L., Gardner, J.P., Hu, E.M., Songalia, A., Hodapp, K.W., Wainscoat, R.J., 1994, Astrophys. J., 434, 114
Doyon, R., Puxley, P.J., Joseph, R.D., 1992, Astrophys. J., 397, 117
Efstathiou, G., Ellis, R.S , Peterson, B.A., 1988, M. N. R. A. S., 232, 431
Fioc, M., Rocca-Volmerange, B., 1995, to be submitted
Garner, J.P., Cowie, L.L., Wainscoat, R.J., 1993, Astrophys. J., 415, L9
Glazebrook, K., Peacock J.A., Miller, L., Collins, C.A., 1994 M. N. R. A. S., 266, 65
Guiderdoni, B., Rocca-Volmerange, B., 1987, Astr. Astrophys., 186, 1
Guiderdoni, B., Rocca-Volmerange, B., 1990, Astr. Astrophys., 227, 362 (GRV90)
Lançon, A., Rocca-Volmerange, B., 1992, Astron. Astrophys. Supp. S., 96, 593 (LRV92)
Lançon, A., Rocca-Volmerange, B., 1995, submitted (LRV95)
Mobasher, B., Ellis, R.S., Sharples, R.M., 1993, M. N. R. A. S. 263, 560
Rocca-Volmerange, B., Guiderdoni, B., 1988, Astr.Astrophys. Supp. S., 75, 93
Rocca-Volmerange, B., Guiderdoni, B., 1990, M. N. R. A. S., 247, 166 (RVG90)
Soifer, B.T., et al, 1994, Astrophys. J. , 420, L1

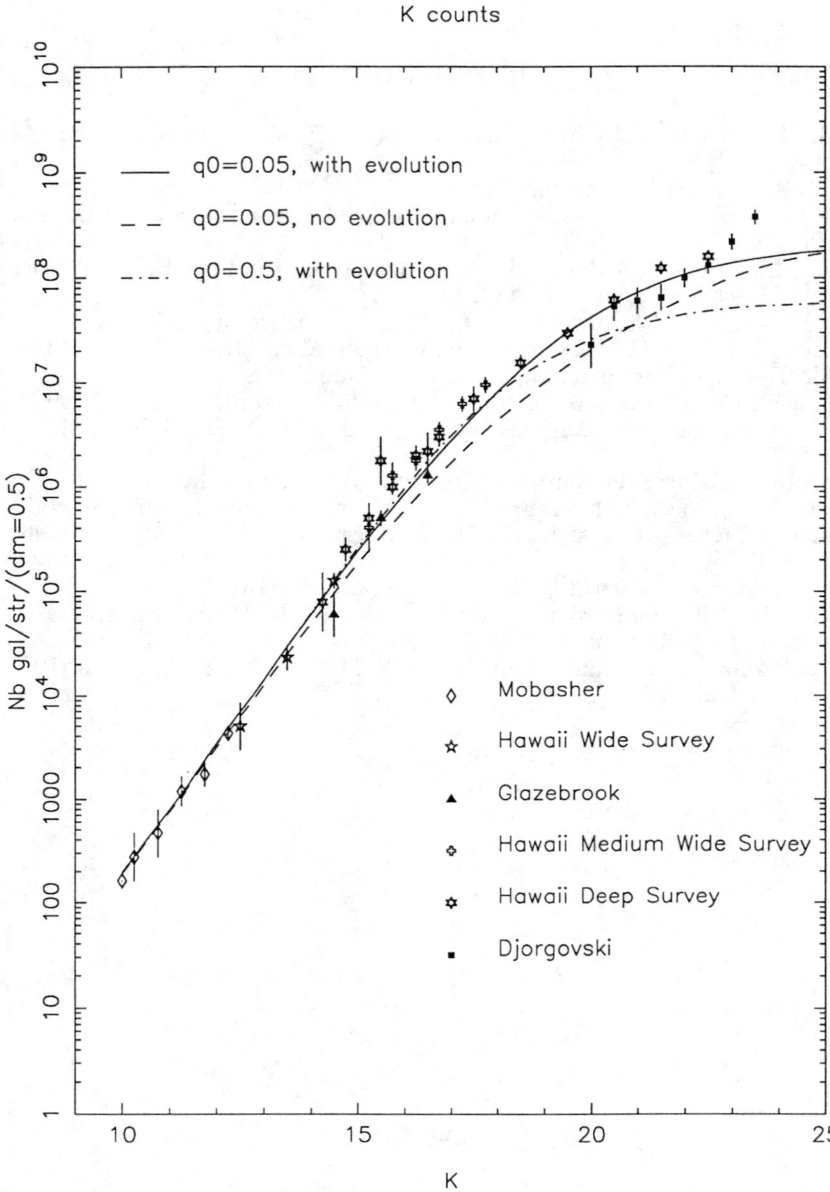

Fig 1. The faint galaxy counts of galaxies in the K band are fitted with a new near-IR model of pure luminosity evolution, calibrated on the visible. Various cosmologies are proposed, as in the visible the best model corresponds to an open Universe.

THEORIES ON THE ORIGIN OF THE COSMIC INFRARED BACKGROUND

ORIGIN OF THE COSMIC FAR INFRARED BACKGROUND

Joseph Silk
Departments of Astronomy and Physics, and Center for Particle Astrophysics,
University of California, Berkeley, CA 94720

silk@pac2.berkeley.edu

ABSTRACT

Increasing evidence indicates that elliptical galaxies formed at high redshift and synthesized considerable amounts of heavy elements. I will summarize arguments that suggest that protoellipticals underwent a luminous starburst phase, and are thereby likely to have been a prominent contributor to the far infrared background.

1. Introduction

The cosmic infrared background (CIB) provides a reservoir for all of the massive star formation that occurred in the early universe. I shall argue that forming elliptical galaxies, S0's and spheroids leave their signature in the diffuse background at far and intermediate wavelengths, whereas forming disks are visible in the near infrared and visible. A close analogy will be drawn between the former systems and starbursts, and the implication for metallicity production in the early universe will be described.

2. Galaxy Formation and the Cosmic Infrared Background

Disk galaxy formation occurs at low redshift, $z \sim 1$. This has become evident from deep galaxy counts and deep redshift surveys. The HST counts reveal that at $I \sim 22$ one is seeing a population dominated by Sd and irregular galaxies in a much greater proportion than found locally (1, 2). The redshift surveys to a similar depth ($I \sim 22$ and $B \sim 24$) show that the blue galaxies are evolving, by about a magnitude in luminosity, while the red galaxy population shows no evidence for evolution (3,4). The MDS results from HST show that these blue galaxies often show evidence of tidal interaction (5). A similar conclusion is drawn for galaxy clusters, where the fraction of blue, star-forming galaxies, and especially of post-starburst galaxies (6), increases dramatically at $z \geq 0.3$. There is little doubt that one is seeing considerable evolution of late-type galaxies at $z \sim 1$.

However, not only do the early type galaxies show no evidence for evolution, but their colors and the lack of dispersion in the colors in nearby clusters (7), and even in distant clusters at $z \sim 0.5$, argue for a high, and for any given cluster well defined, redshift of formation. Evidently elliptical were mature galaxies by $z \sim 1$. This implies a high star formation rate when ellipticals formed. In contrast, spirals have had a nearly constant rate of star formation over the past Hubble time. The star formation history of galactic disks is reasonably well understood, in terms of a gravitationally unstable cold disk that is self-regulated by energy input into the interstellar medium from massive star formation and death. Star formation is inefficient in disks, whereas in the forma-

© 1996 American Institute of Physics

tion of a spheroid, the star formation process was highly efficient. We lack any theory for star formation in spheroids, so any discussion of their evolution is inevitably rather qualitative. One has to resort to indirect evidence to reconstruct their past history. I will describe below how the metallicity of the intracluster medium can be attributed to a vigorous early star forming phase of ellipticals. Both the high iron abundance in clusters and the connection between metallicity and escape velocity in spheroids argue strongly for an early wind, driven by star formation. Indeed one can relate the energy radiated by massive stars to the amount of metals synthesized.

To evaluate the contribution of disks and spheroids to the CIB, one needs to know in addition to the total radiation produced and the redshift at which it was generated, the fraction of starlight absorbed by dust and reradiated in the far infrared. Armed with this information, one can then compute the contribution of disks and spheroids to the CIB. For disks, local studies of IRAS galaxies result in the IRAS luminosity function, comparison of which to the Schechter function demonstrates that about 5 percent of the stellar radiation is reradiated by dust in the far infrared. This should be a good approximation for the past stages of disks, since their star formation rate has undergone modest change.

Observations of IRAS source counts probe to sufficient depth ($z \sim 0.1$) that evolution is seen. The evolution cannot be extrapolated beyond $z \sim 2$ without exceeding IRAS limits on the CIB. However any details as to what types of objects are evolving are ignored in such a backwards evolution approach to the CIB.

A more physical approach begins with the deep optical counts and redshift surveys, and converts optical to infrared fluxes with the IRAS luminosity function that describes star forming disk galaxies. One then finds that the contribution by forming disks to the CIB peaks in the near infrared, with a small ($\sim$ 5 percent) far infrared component. An alternative means of studying star formation in disks, described in the contribution by Fall to these proceedings, identifies damped Lyman alpha clouds with protodisks and uses the evolution in the associated gas content of the universe with redshift to infer a disk formation rate. Allowing for dust in disks (with $\tau_v \sim 0.1$ in damped Lyman alpha clouds at $Z \sim 0.1Z_\odot$) in order to estimate the number of systems that are missed in quasar surveys, Fall finds that disk formation peaks at $z \sim 1$. This independent approach would result in a similar estimate for the contribution of disk star formation to the CIB.

The role of star formation in spheroids in contributing to the CIB is much more difficult to evaluate. In the following section, I will describes how one can tackle this by drawing on an analogy between spheroid formation and starbursts.

3. Spheroid Formation and Starbursts

There are several features in common between the observed properties of a starburst galaxy and those inferred for a forming spheroid. In both cases the star formation rate is high, typically $100\mathrm{M}_\odot\,\mathrm{yr}^{-1}$ per $10^{11}\mathrm{M}_\odot$, as is the efficiency of star formation. The efficiency at which gas is converted into stars within a dynamical time is at least 10 percent, an order of magnitude higher than is observed for our galactic disk. The gas fraction is high, with the molecular mass alone being a significant fraction of the dynamical mass in the starburst region. Starbursts are centrally concentrated, usually occurring within the central kiloparsec of the merged galaxy. Near infrared images reveal a de Vaucouleurs-like profile for the light distribution, similar to the light profile of a bulge. Star bursts are often, but not always, tidally triggered. For the most luminous star bursts, with infrared luminosity in excess of $10^{12}L_\odot$, there almost invariably is evidence of a strong interaction with a companion galaxy. Star bursts are rare today but the IRAS counts indicate that the frequency of luminous star bursts evolves strongly with redshift.

These attributes of the starburst phenomenon, without exception, can be transposed into a description of the presumed properties of a forming spheroid. I note in particular that the bottom-up sequence of structure formation is directly inferred from large-scale structure studies and their numerical simulations. Normalization of the initial density fluctuation spectrum, from which the structure hierarchy develops by gravitational instability in the expanding universe, is set by the COBE DMR detection of temperature fluctuations in the cosmic microwave background. This leads to a picture in which objects of mass as small as $10^6M_\odot$ develop at high redshift ($z \sim 30$) and merge into successively larger structures, culminating in galaxy formation at $z \sim 5$ and cluster formation at $z \sim 1$. Mergers and tidal interactions are inevitable in a hierarchical model of galaxy formation. In particular, mergers play an important role in spheroid formation, since the associated dynamical heating generally forms dynamically hot systems unless gas dissipation plays a major role. This occurs once the merger rate is sufficiently infrequent so as to allow the formation of relatively fragile disks. This scheme fits in qualitatively with the inference that spheroids form early and disks form late, although detailed modelling via numerical simulations has proven to be difficult, because of the large dynamic range that must be followed. In addition, there are further complexities, inevitably arising once massive star formation and feedback of energy are incorporated that frustrate any attempt to compare with observations.

I will describe next an alternative approach to starbursts (and implicitly, spheroid formation) that develops simplistic phenomenological arguments in order to estimate the star formation rate and efficiency, and thereby provide a basis for studying the contribution of spheroid formation to the CIB.

4. A Phenomenological Starburst Model

A tidal interaction or merger with a gas-rich galaxy is the necessary precondition for a star burst. The gas clouds respond by developing non-circular orbits. Collisions between the highly inelastic gas clouds rapidly lead to a central concentration of gas. A central bar plays a similar role, but the external perturbation is necessary for a strong starburst.

Simulations show that a strong interaction will result in a sufficient accumulation of gas within the central few hundred parsecs that the gas is self-gravitating. What happens next is presumably an outbreak of star formation. Local prescriptions of star formation in the Milky Way disk are not applicable in the spheroidal or thick disk geometry of the starburst. Feedback of energy into the gas is a crucial ingredient, given the high rate of star formation that is initially induced in a relatively small volume. The following simple model is a schematic guide.

The key ingredient is an estimate of the gas pressure in the vicinity of the starburst. I assume that the pressure is determined by the collective momentum of the many supernova remnants that are produced within the initial 10^7 years or so of the starburst. I choose a minimum time-scale over which any global outflow or wind could not yet have occurred, but sufficiently long so that an initial generation of massive stars would have formed and died. A two-phase interstellar medium develops, with gas at $\sim 10^6$ K in the interior of supernova remnant-driven bubbles embedded in a network of cold gas in the form of intersecting shells of swept-up matter.

The specific momentum injected by a supernova remnant can be written

$$v_{SN} = \frac{2E_{SN}}{v_{cool}\, M_{SN}}.$$

Here v_{cool} is the velocity ($\sim 300\,\mathrm{km\,s^{-1}}$) at which radiative cooling first becomes important, E_{SN} is the initial kinetic energy ($\sim 10^{51}$ erg), and M_{SN} is the amount of mass formed in stars per supernova explosion (or $\sim 200\mathrm{M}_\odot$ for a normal IMF). Suppose that the central concentration of gas forms stars at a rate $\dot{M}_*$ and is of radius r. The pressure due to the cumulative effect of the supernova remnants is

$$p = \frac{\dot{M}_* v_{SN}}{4\pi\, r^2} = 5\times 10^{-9}\left(\frac{\dot{M}_*}{100M_\odot \mathrm{yr}^{-1}}\right)\left(\frac{v_{SN}}{10^3 \mathrm{km\,s}^{-1}}\right)\left(\frac{1\mathrm{kpc}}{r}\right)^2.$$

The starburst drives a pressure wave, and the over-pressure extends out to a distance given by

$$r_{sb} = \sqrt{\frac{p}{\rho_{gas}}}\cdot t_{sb},$$

where for the starburst duration one may take a dynamical time-scale for the inner galaxy. This results in

$$r_{sb} = 1\,\mathrm{kpc}\left(\frac{t_{sb}}{t_{dyn}}\right)\left(\frac{\dot{M}_*}{100M_\odot \mathrm{yr}^{-1}}\right)^{1/2}\left(\frac{v_{SN}}{10^3 \mathrm{km\,s}^{-1}}\right)^{1/2}\left(\frac{10^3 \mathrm{M}_\odot \mathrm{pc}^{-2}}{\Sigma_{gas}}\right) f_{gas}^{-1/2},$$

where f_{gas} is the gas fraction.

The gas concentration is sufficiently great that a wind does not develop. This may be verified by estimating the porosity of the hot phase:

$$\begin{aligned} P &= (SN\ rate)\times(SNR\,4-volume)\\ &= 0.8\,R_{SN}^{-0.4}\ r_{kpc}^{-1.4}\ \Delta m_{10}^{-0.7}\\ &= E_{51}^{1.3}\ m_{200}^{-1} n^{-0.6}\ \sigma_{200}^{-0.7}\ r_{kpc}^{-1}\ v_{1000}^{-1}. \end{aligned}$$

Here R_{SN} is the SN rate per kpc^3 per yr, given by

$$R_{SN} = \dot{M}_*\,(m_{SN}\,\frac{4\pi}{3} r^3)^{-1} = 0.2\,\dot{M}_{100}\ m_{200}^{-1}\ r_{kpc}^{-3}\,\mathrm{yr}^{-1}\ \mathrm{kpc}^{-3},$$

and Δm_{SN} in units of $10\mathrm{M}_\odot$ is the gas mass heated per supernova.

Star formation can consequently proceed with great efficiency. The high gas pressure helps in two ways. The Jeans mass is reduced, from the value characteristic of a cloud in our local interstellar medium, to that of a massive star;

$$M_{Jeans} = \frac{v_s^4}{G^{3/2}\, p_{gas}^{1/2}} = 30 \frac{T_{50}^2\, r_{kpc}}{\dot{M}_{100}^{1/2}\, V_{1000}^{1/2}} \mathrm{M}_\odot .$$

This means that the Jeans barrier to efficient star formation can be overcome. The pressure is also limited by the self-gravity of the gas, and constrained to satisfy

$$p_{gas} \leq 2\pi G \Sigma_{gas} \Sigma.$$

Suppose that, as more and more massive stars form and produce supernovae, the pressure saturates at this limiting value. It follows that

$$\dot{M}_* \approx \frac{M_g}{t_{dyn}} \cdot \frac{\sigma}{v_{SN}}$$

The star formation efficiency, or fraction of gas converted into stars over a dynamical time, is consequently given by σ/v_{SN}. Numerically, one finds,

$$\dot{M}_* = 500\, \sigma_{200}^4\, f_{gas}\, v_{1000}^{-1}\, \mathrm{M}_\odot \mathrm{yr}^{-1}.$$

The corresponding starburst luminosity is

$$L = 10^{12}\, \dot{M}_{100}\, t_8^{2/3}\, \mathrm{L}_\odot,$$

where t_8 is the starburst duration in units of 10^8 years. Note that the star formation rate is

$$\dot{M}_* \approx M_g \frac{\sigma^2}{r v_{SN}},$$

suggesting that stars naturally form with a steeper density profile. If the gas distribution reflects that of the gas clouds orbiting in an isothermal dark halo, the star formation efficiency alone, without any dynamical rearrangement, could suffice to give a de Vaucouleurs or Hubble-like profile. This is an interesting possibility for accounting for bulge profiles, because dynamical relaxation would act indiscriminately on gas clouds and stars, and hence tend to produce identical profiles. In a disk, one can appeal to viscous relaxation to produce an exponential light profile, but for a spheriod the origin of the de Vaucouleurs-like r^{-3} profile of the stars must lie in dynamical relaxation or in star formation efficiency.

The estimated star formation rate in a starburst allows one to infer the metallicity. I assume that the enrichment is only effective over a dynamical timescale. Once the formation and death rates of massive stars become negligible, the pressure will be dominated by Type I supernovae, and be sufficiently reduced to make a wind inevitable. If the enrichment of the stars is determined by the star formation efficiency, one immediately obtains

$$Z \approx y \frac{\sigma}{v_{SN}} \frac{t_{sf}}{t_{dyn}}, \quad \text{or} \quad Z \propto \sigma.$$

The star formation rate required to produce the observed metals is similar to that inferred with the preceding efficiency estimate:

$$\dot{M}_* = 300 \left(\frac{M}{10^{11}\mathrm{M}_\odot} \right) \left(\frac{Z}{Z_\odot} \right) \left(\frac{10^8\mathrm{yr}}{\mathrm{t_{sb}}} \right) \mathrm{M}_\odot \mathrm{yr}^{-1}.$$

A corollary of this result is that the mass-to-light ratio increases with galactic mass as the star formation efficiency increases. The mass is ZM_*, so that $\frac{M}{L} \propto \sigma$.

The increase in mass is due to compact stellar remnants. The weak rise in M/L with increasing mass is similar to what is measured for spheroids in the fundamental plane, $M/L \propto M^{1/5}$. One must supplement this argument with a relation between σ and spheroid mass in order to obtain the fundamental plane relation. Interestingly, both the (Z, σ) and $(M/L, L)$ relations have a narrow dispersion over a wide range of spheroid masses.

At low masses, a wind can remove interstellar matter more effectively. It has been argued that in this regime mass loss controls the ratio of star to total mass. If there is a dark matter component, then the stripping of gas results in the fraction surviving in the form of stars being proportional to σ. Hence in the mass loss regime, $L/M \propto \sigma$.

5. Spheroid Formation and the Cosmic Infrared Background

While luminous starbursts are rare in the nearby universe, I have made the plausible conjecture that spheroid formation may be identified as a similar phenomenon. Spheroids are believed to form at high redshift according both to population synthesis models and the HST observations of ellipticals at redshift of order unity described in section 2. Although no examples of forming ellipticals have yet been identified, the cosmic infrared background acts as an ideal reservoir for detecting the cumulative diffuse background light generated by forming spheroids. One can most simply estimate the cosmic infrared background from forming galaxies by normalizing to the iron abundance.

Galaxy clusters are dominated by early–type galaxies. The intracluster medium, some 10-20 percent of the cluster mass, contributes more by about a factor of 2 to the total cluster mass then do the stars. In term of mass-to-light ratios, M/L for the Coma cluster is $\sim 300h$, whereas M/L for an elliptical galaxy to within its half-light radius is $\sim 10h$. The intracluster gas is enriched to about one-half the solar value. A convenient measure of the intracluster gas iron abundance is the iron mass-to-visible light ratio, measured to be ~ 4 times higher in clusters than for the Milky Way.

Current supernova rates do not account for the observed iron abundance of the intracluster gas. One infers the occurrence of early winds driven when the supernova rate was substantially higher than the present day value. This most plausibly occurred during the spheroid formation stage, when there was a substantial number of SN II. This leads to the prediction of SN II yields in the intracluster gas as recently seen in ASCA observations (8). It has been argued that the IMF must have been weighted towards more massive stars in spheroid formation in order to synthesize enough iron. What is certainly true is that the fundamental plane trend $M/L \propto M^{1/6}$, if due to compact stellar remnants, necessarily implies sufficient injection of heavy elements from luminous galaxies to account for the observed ICM abundance. Whether this is due to a top-heavy IMF or to star formation efficiency varying with galaxy mass is unclear.

The star formation efficiency interpretation implies that spheroids outside clusters drive winds of comparable strength to cluster galaxies, in order to account for the observed correlation between metallicity and velocity dispersion. A related clue is that nuclei of ellipticals generally show an enhanced abundance difference for α nuclei relative to iron (9), as expected for a SN II yield. This could again be attributed to a systematically top-heavy IMF or to systematically more efficient trapping of SN II ejecta during the starburst phase.

The contribution of forming galaxies to the diffuse background light may be estimated as follows. Denote the redshift of formation by z_f. The bolometric contribution to the diffuse background can be written as

$$\nu i_\nu = (4 \times 10^{-3} \mathrm{erg\, g^{-1}})\, c^2\, Z\, \rho_o\, \frac{c}{4\pi}\, (1 + z_f)^{-1}.$$

Scaling to disks gives

$$\nu i_\nu = 10 h_{h_2}^3 \left(\frac{\Omega_{disk}}{0.002}\right) \left(\frac{3}{1+z_f}\right) \left(\frac{M_{Fe}/M_*}{0.018}\right) \left(\frac{M_*/L_v}{3}\right) \mathrm{nwatts\, m^{-2} sr^{-1}}.$$

One would expect to observe the contribution from forming disks to the cosmic background at near infrared wavelengths.

The expected contribution from spheroids is about twice as large: rescaling, I obtain

$$\nu i_\nu = 20\, h_{1/2}^3 \left(\frac{\Omega_{spher}}{0.002}\right) \left(\frac{6}{1+z_f}\right) \left(\frac{M_{Fe}/M_*}{0.018}\right) \left(\frac{M_*/L_v}{12}\right) \mathrm{nwatts\, m^{-2} sr^{-1}}.$$

A major uncertainty is the predicted wavelength range where one might expect to observe the spheroid contribution. The starburst analogy suggests that spheroid formation is shrouded in dust, and the radiation is emitted in the far infrared spectral region.

In summary, elliptical galaxy formation most likely occurs early in the universe, and the luminous cores of these galaxies are likely to be shrouded in dust. Study of the diffuse far infrared background radiation offers a potential window for exploring the cumulative emission from these systems. Detection of individual starbursts at high redshift may be feasible at millimeter wavelengths. The predicted number of sources should exceed 10^3 per square degree, given that ultradeep galaxy counts are finding several hundred thousand galaxies per square degree and that the starburst duration at, say $z \sim 3$, must be a few percent of a Hubble time. Expected submillimeter luminosities are, as argued above, comparable to those of the ultraluminous IRAS galaxies at 100μ. A coordinated strategy of searching in deep fields with such instruments as the bolometer array at IRAM and SCUBA should be capable of setting useful constraints on this hypothesis.

References

1. Glazebrook, K., Ellis, R. S., Santiago, B. and Griffiths, R. 1995, MNRAS, 275, L119
2. Driver, S. P., Windhorst, R. A., and Griffiths, R. E. 1995, ApJ, in press.

3. Lilly, S. J., Tresse, L., Hammer, F., Crampton, D. and Le Fevre, O. 1995, ApJ, in press.
4. Glazebrook, K., Ellis, R. S., Colless, M., Broadhurst, T. J., Allington-Smith, J. and Tanvir, N. 1995, MNRAS, 273, 157
5. Griffiths, R. *et al.*1995, ApJ, 435, L19.
6. Dressler, A. and Gunn, J. E. 1992, ApJS, 78,1.
7. Bower, R. G., Lucey, J. R. and Ellis, R. S. 1995, MNRAS, 254, 601.
8. Mushotzky, R. 1994, in *Clusters of Galaxies*, ed. F. Durret *et al.*(Editions Frontieres: Gif-sur-Yvette), p. 167.
9. Worthey, G., Faber, S. M. and Gonzalez, J. J. 1992, ApJ, 398, 69.

MODELING THE COSMIC INFRARED BACKGROUND

Carol Lonsdale
IPAC 100-22
California Institute of Technology
and Jet Propulsion Laboratory
Pasadena, CA 91025

ABSTRACT

The contribution to the cosmic far-infrared and submillimeter background (CIBR) from extragalactic sources, including normal and active galaxies and protogalaxies, is not well constrained by direct observational number counts of galaxies, which are currently limited to the relatively local universe at these wavelengths. Therefore our estimates of their contribution to the background are based on model extrapolations. I review the principal modeling techniques currently in use, including Backward Evolution in time of a locally-observed luminosity function, and Forward Evolution in time of a high redshift galaxy population. The most modestly and passively evolving model extrapolations lie below the current FIRAS and DIRBE limits on the CIBR; however some of the more strongly evolving models are now in conflict with the best COBE limits, even though the DIRBE limits still notably contain some foreground zodiacal and galactic emission. The most important unknown parameters which govern the model predicted far-infrared and submillimeter emission of the integrated light of galaxies are the galaxy formation redshift; the dust temperature at high redshifts; the star formation history; and the variation of the dust optical depth with time, $\tau(t)$, which governs the fraction of the galaxy light that is reprocessed into the far-infrared. The next generation of spacecraft (ISO, WIRE, SIRTF), and of instruments on the new 10m class telescopes, will go a long way towards answering these questions, providing much insight into key aspects of galaxy formation and evolution.

1. INTRODUCTION

Why should we expect galaxies to produce enough infrared emission at early epochs that their integrated emission in this wavelength range will be both detectable and informative? The nucleosynthesis argument (*eg.* Peebles, these proceedings) indicates that we expect to find an integrated background light due to the formation of the metals we see in the local universe of about $10^{-8} W m^{-2} sr^{-1}$, of order a decade wide, somewhere in the electromagnetic spectrum. Such a background has yet to be detected in the UV-optical-NIR range. There are several lines of evidence which point to the conclusion that there was significant early metal and dust enrichment, at least in proto-elliptical galaxies, and that much of the early radiative output of galaxies emerged longward of 1μm and was subsequently reradiated to longer wavelengths. One piece of direct observational evidence for early dust and metal enrichment comes from the detection of both dust and molecular CO emission in IRAS F10214+4724 (Rowan-Robinson *et al.* 1991, Brown and Vanden Bout 1992, Solomon, Downes and Radford 1992) and the Cloverleaf quasar (Barvainis, Antonucci and Coleman 1992, Barvainis *et al.* 1994), at redshifts of z=2.36 and 2.56, respectively.

© 1996 American Institute of Physics

Although both of these objects are likely gravitationally lensed, the implied dust and CO masses are still large, indicating that in these objects at least, a very significant metal and dust enrichment has occurred. Not only do these objects illustrate enrichment, they also demonstrate high dust optical depth to optical/UV radiation since the far-infrared luminosities are high ($\gtrsim 10^{13} L_{\odot}$, after correction for lensing). Elbaz *at al.* (1992) and Mazzei and De Zotti (1994) have modeled IRAS F10214+4724 successfully as a very massive proto-elliptical galaxy of age about 1 Gyr; taking into account the probable lensing, their models will scale down to a match a more typical giant elliptical galaxy. (Alternatively, the very high infrared luminosity of these objects could arise in a dust torus around a luminous AGN (for which there is strong evidence in both cases), in which case their relevance to galaxy evolution in general is less obvious.)

We can also find some quite general evidence for a high far-infrared integrated background light from conditions in the local universe, where about 1/3 of the volume emissivity of all galaxies emerges in the far-infrared; this emission is dominated by ongoing massive star formation in disk galaxies. If the star formation rates in disk galaxies have undergone a relatively slow rate of change with time, as expected for isolated dissipatively-evolving systems and supported by direct measurements of current and recent star formation tracers in some nearby disks, it follows that their integrated far-infrared emission over the age of the universe is likely to be at least several percent of the Cosmic Microwave Background (Partridge and Peebles 1967). As emphasized by Mazzei and de Zotti (1994) and Franceschini *et al.* (1994), the integrated far-infrared emission of elliptical galaxies could be even higher than that of disks if ellipticals formed rapidly with a large initial star formation rate and a high optical depth, as in the models Elbaz *et al.* and Mazzei and de Zotti (1994) for F10214+4724. Support for the possibility of a high dust optical depth in protogalaxies comes from the fact that despite years of painstaking effort, protogalaxies have remained elusive in the optical and UV.

Another line of evidence argues that the integrated background in the far-infrared and submillimeter could be even higher than predicted by a model of slow dissipative evolution of isolated disk galaxies; the possibility that merging and starbursting may have been an important feature of galaxy evolution at intermediate times. A significant fraction of the local luminous galaxy energy density is generated in the far-infrared by starburst and/or AGN episodes associated with interacting and merging galaxies. The steeply rising IRAS 60μm number counts may imply an increase of this activity backwards in time, a scenario consistent with results deduced from submJy radio source counts and the optical and near-infrared color distributions of faint galaxy samples. Thus we might expect a very significant integrated light at far-infrared wavelengths from this phenomenon, in excess of that predicted under the assumption that disk galaxies have evolved quite passively as isolated systems.

2. MODELING THE INTEGRATED EMISSION OF GALAXIES

There are two basic types of model in use for understanding the evolution of galaxies with time and predicting the integrated backgrounds in the far-infrared (FIR) and submillimeter range. Both have their advantages and disadvantages but taken together they are strongly complementary.

The first modeling method I call Backward Evolution (BE) of the Luminosity Function (LF). In this class of model, a locally-derived luminosity function of galaxies is extrapolated backwards in time according to a simply parameterized evolutionary function, to match the observed number-magnitude, number-redshift, color-magnitude, *etc.*, distributions, as available. Since the IRAS-selected observational distributions are generally restricted to redshifts lower than about 0.3, large extrapolations of the evolutionary function backwards in time are required to make predictions about the integrated light of galaxies.

This class of model was orginally developed for application to radio source counts, and has also been extensively used in the interpretation of the evolution of QSOs. The method has the advantage of being empirical with a minimum of input and assumed parameters: the local luminosity function, the form of the evolution law, the spectral energy distribution(s) of galaxies (required in order to perform k-corrections as a function of z for the model galaxy population(s)) and a maximum redshift, z_{max}, at which the model integration is terminated. Such models provide an excellent context for a broad phenomenological overview of far-infrared galaxy evolution, but they can give limited physical insight into detailed galaxy evolution processes.

The second class of model, Forward Evolution, works the opposite way in time: a theoretical galaxy formation paradigm and evolutionary framework are first adopted, and then a population of galaxies is evolved forwards in time from the formation epoch to match the local $N(S)$, $N(z)$, and color distributions, *etc.* This class of model is more physical than backward LF evolution, since it can incorporate the detailed modeling of numerous physical parameters and processes. It also has the advantage that it can be highly multi-wavelength, drawing on data at many wavelengths to constrain the input parameters. Its advantages, of course, also highlight its disadvantages: the large number of assumptions and free parameters typically involved mean the models are often severely under-constrained.

2.1 Backward LF Evolution Models

In the simplest Backward LF Evolution models, the entire luminosity function is fixed in shape with time, and it translates *en masse* in luminosity and/or density at the rate specified by the adopted evolutionary law. The evolutionary law usually is either a single power law or exponential in z. Such models are called translational models, and they were originaly pioneered for the evolution of radio sources and quasars. The physical interpretation of such evolution involved an increasing luminosity of AGNs with lookback time for luminosity evolution, or an increasing fraction of galaxies possessing an AGN at earlier times for density evolution. In the context of far-infrared bright galaxies, power law translational evolution would be ideal for describing an increasing luminosity of starbursts and/or AGN with lookback time, or an increasing fraction of galaxies undergoing starburst episodes and/or AGN events with lookback time, as long as the rate of mergers is low. However, this kind of LF-invariant evolution is not well suited for describing true merging scenarios, in which smaller galaxies merge to form larger ones – such scenarios implicitly involve evolution of the shape of the luminosity function. It is also questionable for describing continuous passive evolution of stellar populations, which also involves an implicit evolution of the shape of the LF at some level (because galaxies of different shape and mass can be expected to brighten and fade with time at different rates), regardless of the merging process.

A sample of the parametric models of the far-infrared background that have been published based on translational evolution are shown in Figures 1 and 2. In these figures, α denotes pure power luminosity evolution, given by $L(t) = L(t_0)\,(1+z)^{\alpha}$; β denotes pure power law density evolution described as $\rho_{co}(t) = \rho_{co}(t_0)\,(1+z)^{\beta}$, where ρ_{co} is the co-moving density; and Q denotes the exponential luminosity evolution relation used by Oliver *et al.* (1992): $L(t) = L(t_0)\,\exp\{2/3Q[1-(1+z)^{-3/2}]\}$ (appropriate for Ω=1). Other published models include the power law models of Hacking and Soifer (1991), Weedman (1990), Franceschini *et al.* (1991) and Treyer and Silk (1993), which give similar results to those illustrated in Figures 1 and 2.

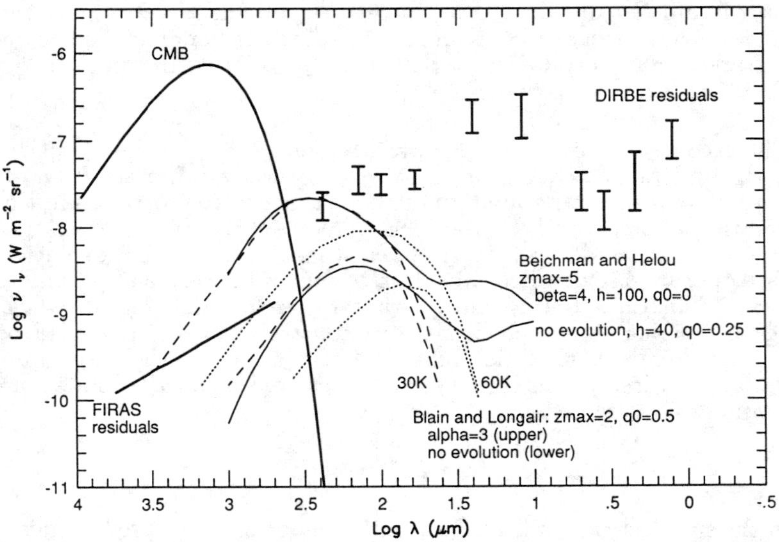

Figure 1: Comparison of the integrated CIBR predicted by representative Backward Evolution models of Beichman and Helou (1991) and Blain and Longair (1993) to the most recent FIRAS and DIRBE residual measurements, as summarized by Hauser (these proceedings).

For most of the BE models all galaxies comprising the LF are allowed evolve at the same rate. In the model of Franceschini *et al.* (1991), the three galaxies types evolve at different rates. The model of Treyer and Silk is different in nature from the others in that they do not allow the luminosity function of "normal" galaxies to evolve with time, but add a new population of dwarf galaxies whose characteristic space density, ϕ^* (Mpc^{-3}), is the parameter that evolves: $\phi^*_{dw} = 6.0\times10^{-2}((0.7/z)+1)^{-1}h^3$, where h=H$_0$ in units of 100 km/s/Mpc. This model is designed to explain the steepness of the observed blue number counts of galaxies with a population of dwarf galaxies which is present at $z = 0.7$ but has faded or merged to invisibility by the current epoch.

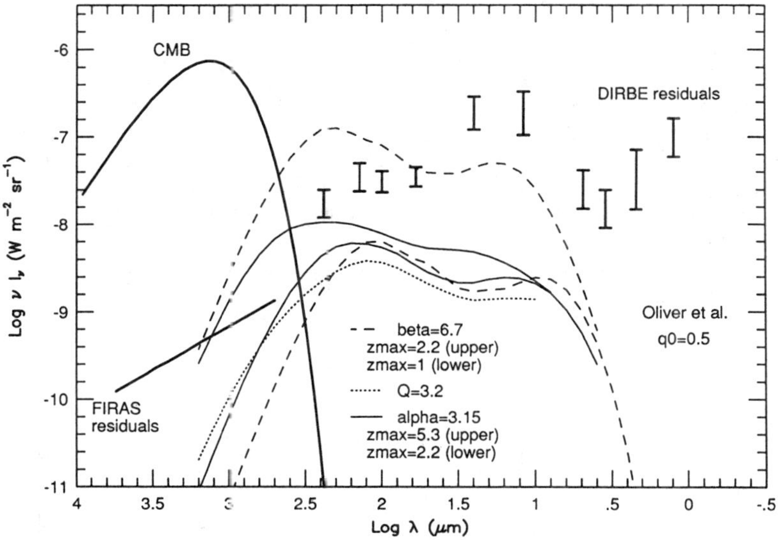

Figure 2: Comparison of the integrated CIBR predicted by the models of Oliver *et al.* (1992) to the most recent FIRAS and DIRBE residual measurements.

For all the models shown in Figures 1 and 2, except those labeled "no evolution", the evolutionary parameters (α, β or Q) have been selected by their authors to fit the slope of the locally-observed IRAS 60μm source counts. This is a very important aspect of the interpretation of these models which I will return to in Section 3.1.

2.2 Evolutionary Synthesis Models

The model of Franceschini *et al.* (1994) is based on the closed box chemical evolutionary population synthesis models of Mazzei, Xu and De Zotti (1992) and Mazzei, de Zotti and Xu (1994). Starting at a selected formation epoch, a population of model galaxies is given a range of masses and initial star formation rates and allowed to evolve forwards in time according to the adopted prescriptions for chemical and stellar evolutionary theory and a Salpeter IMF. The star formation rate, ψ, of each galaxy evolves with time as a power, n, of the fractional gas mass f_g, $\psi(t) = \psi_0 f_g^n M_\odot yr^{-1}$ where $f_g = M_{gas}/M_{gal}$ evolves as the changing stellar populations use up interstellar material and re-process some of it back for use by future generations. The far-infrared and submillimeter emission is modeled as the sum of a warm component heated by the massive stars and a cooler component heated by the older stars, depending on the evolving dust-to-gas ratio and dust optical depth. The galaxies with the highest initial star formation rates are elliptical galaxies, which use up most of their ISM very rapidly with an early, very far-infrared-luminous phase (a "protogalaxy" phase), and then settle down to a long passively evolving history of low far-infrared emissivity. Disk galaxies evolve more slowly, reaching an epoch of peak far-infrared emissivity at much later times. The latest galaxy types are still increasing in optical depth after a Hubble time.

As noted above, a young elliptical galaxy model can match the luminous z=2.3 IRAS galaxy F10214+4724 quite well, while the same model at 15 Gyr matches the UV-far-infrared energy distribution of local ellipticals well; see Mazzei and de Zotti (1994). Two integrated light models of Franceschini *et al.* (1994), differing in the value adopted for n, are shown in Figure 3; if their picture of galaxy evolution is correct then the integrated background light expected from galaxies in the wavelength range is clearly within close reach of being detected by COBE.

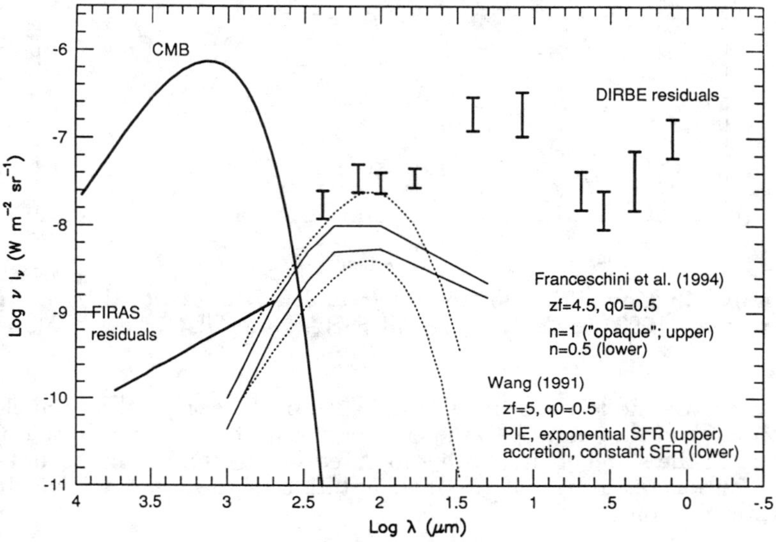

Figure 3: Comparison of the integrated CIBR predicted by the models of Franceschini *et al.* (1994) and Wang (1991) to the most recent FIRAS and DIRBE residual measurements.

Wang (1991a,b) takes a simpler approach to the evolving stellar energy output of his model galaxy population than Franceschini *et al.* (1994), and focuses on the chemical enrichment and evolution of the dust in molecular clouds in more detail. He investigates two enrichment/star formation rate scenarios: prompt initial enrichment with either a constant or exponentially declining SFR, and a model allowing continued accretion of material from intergalactic space coupled with constant star formation. Wang fixes the dust temperature at 30K, based on observations of local disks, rather than attempting a derivation of expected dust temperatures, and his galaxies are formed at a range of redshifts given by a Gaussian distribution with variable z_f and σ_z.

Wang finds that the dust content of young disk galaxies at early times can be up to 4 times larger than today, and the far-infrared luminosity can be two orders of magnitude greater. The PIE model predicts much stronger backgrounds than the accretion model because it shows strong evolution of the dust mass.

Wang's results are compared to those of Franceschini *et al.* (1993) in Figure 3, where it may be seen that for similar cosmology and z_f the "opaque" (n=1) model of Franceschini *et al.* (1994) is very similar to the PIE model of Wang with exponentially declining star formation, at least at $\lambda > 200\ \mu$m. At shorter wavelengths the models are expected to diverge because Wang, focusing only on the far-infrared emission, considers only a single temperature dust component.

Both the Franceschini *et al.* (1994) and the Wang models assume conservation of galaxy number density. Blain and Longair (1993) derive the integrated background for a hierarchical clustering model with merging based on the Press-Schecter formalism for the growth of structure. In their model the far-infrared and submillimeter emissivity of galaxies is proportional to the SFR, which varies as a gaussian with time during each galaxy merger event. They ran models adopting two different assumed dust temperatures, 30 and 60K, and constrained the models to generate the correct heavy element abundance and to fit local 60μm source counts. Their models are shown in Figure 4.

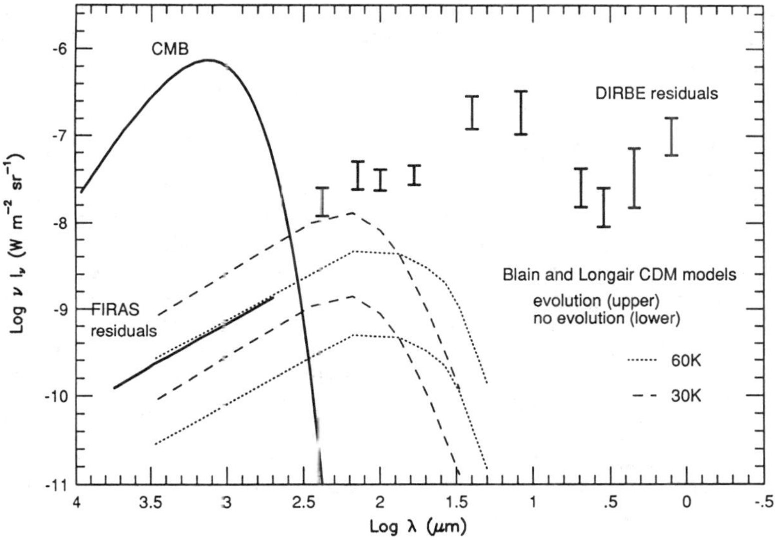

Figure 4: Comparison of the integrated CIBR predicted by the CDM models of Blain and Longair (1993) to the most recent FIRAS and DIRBE residual measurements

3. MODEL RESULTS AND CONCLUSIONS

Several of the models shown in Figures 1-4 are in conflict with the FIRAS residuals, and even with some of the current DIRBE limits, in spite of the fact that the DIRBE limits clearly still contain significant levels of foreground zodiacal and galactic emission. This confluence between theory and data means that we can begin to learn something about galaxy formation and evolution from the details of the modeling processes. In particular, it is very exciting to note that the COBE limits are getting down to the level predicted by the general nucleosynthesis argument of $\sim 10^{-8} Wm^{-2} sr^{-1}$, the level we expect if a large percentage of the total stellar energy density of the universe emerged in the FIR, therefore a detection at close to these levels may show that it actually did do so. The Desert *et al.* tentative detection (these proceedings) is therefore tantalizing.

3.1 Backward Evolution Models: Redshift Cutoff or Lower Evolutionary Rates?

The first conclusion that can be made from a consideration of the various models of Figs 1-4 is that for similar assumptions (cosmology, the galaxy formation epoch and the spectral energy distribution of the model galaxies), the strongly evolving power law Backward Evolution models tend to predict higher integrated backgrounds than the Forward Evolution models. Compare, for example, the α=3.15, z_f=5.3, Ω=1 BE luminosity evolution model of Oliver *et al.* with the z_f=5, Ω=1 models of Wang, which assume similar galaxy spectral energy distributions.

In general, all of power law the translational BE models that can fit the slope of the IRAS local 60μm number counts are in serious conflict with the COBE limits, unless a low redshift cutoff is imposed on them (the more complex model of Treyer and Silk is an exception and is discussed further below). The translational density evolution models, in particular, require an especially low redshift cutoff because higher evolutionary rates are nominally required to fit the local 60μm counts using density evolution rather than luminosity evolution (Hacking, Condon and Houck 1987; Saunders *et al.* 1990; Lonsdale *et al.* 1990). For luminosity evolution (and Ω=1) z_{max} is about $2-3$, and for density evolution it is about $1-2$, depending on the details of the adopted spectral energy distributions (see Rowan-Robinson, these proceedings, an update to the models of Oliver *et al.*). Of course, a traditional translational power law evolution model will always require a redshift cutoff at some point because such a model otherwise diverges at high redshift. Power law density evolution models in particular become meaningless at high redshifts, because at some point the mass of the total number of galaxies involved in the backward extrapolated series of mergers will exceed the total galactic mass of the universe. On the other hand, the exponential luminosity evolution model of Oliver *et al.* does not suffer from this problem because it converges at high redshift; indeed this model is not in conflict with the COBE limits.

How physically plausible is a redshift cutoff at $z \sim 2$ for the strongly evolving BE translational evolution models? Recall that this class of translational evolution model was developed as an aid to understanding the evolution with redshift of radio source counts and QSOs. As such they have been quite successful, and have a reasonably simple and physically plausible interpretation in terms of an increase with lookback time of the luminosity and/or frequency and/or duration of AGN events in galaxies. The character of the AGN events themselves are not expected to change much with lookback time, thus a redshift-invariant LF is not unreasonable. A turnover in redshift is also seen for QSOs, and is interpreted in terms of the formation epoch of QSOs. For the FIR number counts and integrated background an analogous scenario could be considered, since it is very possible that a significant fraction of the far-infrared luminosity of galaxies ultimately derives from dust-enshrouded AGN, especially in the most luminous galaxies (*e.g.*. Lonsdale, Smith and Lonsdale 1995), and from starbursts caused by interaction and merger events, and is therefore reasonably modeled to first order by translational LF evolution. But there remains a large fraction of the far-infrared emissivity of galaxies which derives from passive stellar evolutionary processes, in galaxy disks, which *are* expected to change continuously with cosmological epoch, and which *are not* well modeled by translational evolution of a fixed LF.

An alternative solution to the strongly evolving models with a redshift cut-off is to question whether the rate of evolution assumed in these models is simply too high. In fact there is good reason to question this, and the argument is very similar to that which has arisen around the "faint blue galaxy question". The focus of research in optical and near-infrared observational cosmology has recently revolved around the fact that redshift distributions of faint field galaxy samples routinely fail to find enough galaxies at high redshifts to agree with evolving models for the number-magnitude relations (see *e.g.*, Koo and Kron 1992). Several explanations for this failure have been put forward; the most popular invoke some sort of population of low luminosity galaxies that were present at moderate redshifts, and therefore contributed to steepening the number-magnitude relation, which have either faded to invisibility or merged with other galaxies by the present time (*eg.* Broadhurst, Ellis and Shanks 1998). A simpler explanation is that we have simply got either the shape of the local luminosity function (in particular, the slope of the faint end) or the normalization of the local number-magnitude relations wrong (*eg.* Lonsdale and Chokshi 1993; Gardner, these proceedings).

It now seems that there is a similar conflict between the steepness of the number-magnitude relation and redshift distributions at 60μm; Ashby, Houck and Hacking (1992) fail to find the high redshift galaxies expected for a translational evolution model that can fit the count slope. Therefore we should likewise question both the adopted shape of the 60μm LF and the normalization of the models to the number-magnitude relation, and we should consider more complex evolutionary forms. The far-infrared integrated background model of Treyer and Silk (1993) is based on a fading/merging dwarf scenario, such as those put forward to explain the faint blue galaxy problem, and indeed it has no conflict with the CIBR limits (it predicts a CIBR similar in intensity and distribution to the lower of the two models of Beichman and Helou illustrated in Figure 1). Alternatively, if we suspend the requirement that the BE models in general must match the 60μm $N(S)$ slope, then the more slowly evolving models of for example, Beichman and Helou, come into consideration.

3.2 The FE Models

To gain more physical insight into galaxy evolution than possible from the BE models, we can turn to the more detailed FE modeling approaches. The "opaque" (n=1) stellar population synthesis model of Franceschini *et al.* (1994) can be considered an illustration of a classical non-CDM galaxy formation model in which a large fraction of all the light of the forming elliptical protogalaxies emerges in the FIR. Indeed, a large fraction of all galaxy light emerges in the FIR in this model, leaving little of the approximately $10^{-8} W m^{-2} sr^{-1}$ expected on general grounds from the nucleosynthesis prediction to emerge at other wavelengths. The PIE model with exponential star formation of Wang is likewise a good illustration of a non-CDM universe in which most of the energy density of galaxies emerges in the FIR, as it also accounts for essentially all of the nucleosynthesis energy. The Blain and Longair evolving models illustrate an alternative CDM merging scenario, some of them also accounting for much of all nucleosynthesis energy. The fact that these models are so closely constrained by the new CIBR data means that they are clearly worthy of more detailed refinement, in anticipation of further observational progress in both the CIBR limits and in number count, redshift and color distributions at intermediate redshifts. It is worth emphasising, however, that there is room for uncertainty in those models which have been designed to fit the steep 60μm source counts. This is because, following the arguments in the previous section, such models may give a biased picture of far-infrared evolution rates in the local universe. Therefore further modeling should take care to fit the available FIR redshift distributions and take full account of the current uncertainties in the shape of the local LF and the absolute normalization of the N(S) relation.

The long wavelength shape of the integrated model spectra is the most critical aspect for comparison to the FIRAS limits (and the tentative FIRAS detection of Desert *et al.*). It's shape is principally governed by the convolution of the dust temperatures and the redshifts of the dominant contributing galaxies to the integrated background. At one extreme, a dominant contribution from very cool dust in very high redshift galaxies will cause the integrated background to greatly exceed the FIRAS limits, as in the 30K evolving CDM model of Blain and Longair, while at the other extreme a dominant contribution from relatively low redshift and/or relatively warm galaxies, such as the accretion model with low z_f of Wang, will predict an integrated emissivity at FIRAS wavelengths that is much lower.

4. FUTURE DIRECTIONS

Where do we go from here? The prospects for progress in understanding galaxy evolution via modeling of the Cosmic Infrared Background are obviously very exciting at present, since several different modeling approaches result in CIBR predictions that lie within reach of detection, and some plausible scenarios are already ruled out. Therefore more refined modeling approaches would clearly be very timely and important.

By itself, a detection of the CIBR will not distinguish between the variety of galaxy formation and evolution paradigms represented by the models discussed in this paper, but will tell us only in general terms what fraction of the total energy generated over the history of the universe was re-radiated into the FIR by dust; *ie.* it will give us an integral of the stellar energy and dust optical depth over time. The spectral shape of the CIBR will give us information about the integral of the temperature of that dust, and the redshift range at which it was emitted, over the history of the universe.

Therefore, of equal important to the refinement of the various models to agree with the CIBR measurements will be model predictions and detailed observations of galaxy populations at a variety of wavelengths and a range of redshifts which can be used to critically test these different model approaches. Redshift surveys of large samples of the faintest IRAS galaxies will help to solve the dichotomy between the steep slope of the 60μm number counts and the shallow redshift distribution (*eg.* Smith *et al.* 1996), while deep ISO and WIRE (Wide Field Infrared Explorer) surveys and their subsequent follow-up with redshift determinations will push these studies to redshifts of unity or above. High resolution HST or Keck imaging of the intermediate and high redshift ISO galaxies will shed much light on the overall role of mergers and starbursts in galaxy evolution. The new submillimeter SCUBA instrument promises to be extremely valuable because submillimeter number counts at intermediate redshifts are expected to be especially sensitive diagnostics of galaxy evolution models as the k-correction redshifts the strong FIR peak into the submillimeter band (see Blain and Longair 1993; also Guiderdoni, these proceedings).

ACKNOWLEDGEMENTS

This paper was supported by the Jet Propulsion Laboratory, California Institute of Technology, under a contract with the National Aeronautics and Space Administration.

REFERENCES

Ashby, M., Houck, J. R., & Hacking, P.B, AJ, 104, 980
Barvainis, R., Antonucci, R. & Coleman, P. 1992, ApJL, 399, L19
Barvainis, R., Tacconi, L., Antonucci, R., Alloin, D., & Coleman, P. 1994, Nat, 371, 586
Beichman, C. A., & Helou, G. 1991, ApJL, 370, L1
Blain, A. W., & Longair, M. S. 1993, MNRAS, 264, 509
Broadhurst, T. J., Ellis, R. S., & Shanks, T. 1988, MNRAS, 235, 827
Brown, R. L., & Vanden Bout, P. 1992, ApJ, 397, L19
Elbaz, D., Arnaud, M., Casse, M., Mirabel, I. F., Prantzos, N., & Vangioni-Flam, E. 1992, A&A, 265, L29
Franceschini, A., Toffolatti, L., Mazzei, P., Danese, L., & De Zotti, G. 1991, A&A, 89, 285
Franceschini, A., Mazzei, P., De Zotti, G., & Danese, L., 1993, ApJ, 427, 140
Hacking, P. B., & Soifer, B. T. 1991, ApJL, 367, L53
Hacking, P. B., Condon, J. J., & Houck, J. 1987, ApJ, 316, L15
Lonsdale, C. J., & Chokshi, A. 1993, AJ, 105, 1333
Koo, D. C., & Kron, R. G. 1992, ARA&A, 30, 613
Lonsdale, C. J., Smith, H. E., & Lonsdale, C. J., 1995, ApJ, 438, 632)
Lonsdale, C. J., Hacking, P. B., Conrow, T. P., & Rowan-Robinson, M. 1990, ApJ, 358, 60
Mazzei, P., & De Zotti 1994, ApJ, 426, 97
Mazzei, P., De Zotti, G., & Xu, C. 1994, ApJ, 422, 81
Mazzei, P., Xu, C., & De Zotti, G. 1992, A&A, 256, 45
Oliver, S. J., Rowan-Robinson, M., & Saunders, W. 1992, MN, 256, 15P
Partridge, B., & Peebles, P. J. E. 1967, ApJ, 148, 713
Rowan-Robinson, M., et al. 1991, Nat, 351, 719
Saunders, W., et al. 1990, MN, 242, 318
Smith, H. E., Lonsdale, C. J., Landy, B., Beichman, C. A., & Condon, J. J., 1995, in preparation)
Solomon, P. R., Downes, D., & Radford, S. J. E. 1992, ApJ, 398, L29
Treyer, M., & Silk, J. 1993, ApJL, 408, L1

Wang, B. 1991a, ApJ, 374, 456
Wang, B. 1991b, ApJ, 374, 465
Weedman, D. W. 1990, in Massive Stars in Starbursts, ed. C. Leitherer, N. R. Walborn, T. M. Hackman, & C. A. Norman, (Cambridge University Press: Cambridge), 317

IR GALAXY EVOLUTION AND CONTRIBUTIONS TO THE BACKGROUND FLUX

A. Franceschini, G. Granato, P. Mazzei, L. Danese, G. De Zotti
Osservatorio Astronomico di Padova
Vicolo Osservatorio 5
I-35122 Padova, Italy

ABSTRACT

We discuss plausible contributions from evolving galaxies to the cosmic infrared background (CIRB). These estimates are based on two alternative evolutionary paradigms, roughly consistent with currently available data. One scheme is a combination of luminosity and density evolution, and assumes that galaxy evolution from optical through radio wavelengths is driven by substantial *merging* events, increasing the average galactic mass with cosmic time and triggering star formation in the gas-rich systems.

The other extreme pattern assumes a constant mass function up to the formation epoch and pure luminosity evolution. In such a case, early-formed dust is assumed to hide from optical searches the expected bright star-formation phase of spheroidal galaxies. The corresponding dust re-radiation in the far-IR may have been, during these early phases, orders of magnitude larger than today, as possibly indicated by long-wavelength observations of hyperluminous IRAS galaxies, high redshift radio galaxies and quasars.

We show that the rates of luminosity/density evolution, which determine the amount of energy radiated by galaxies in the IR, are very roughly constrained by the IRAS galaxy counts (also complemented by faint radio and K-band counts, and, hopefully, by the forthcoming deep ISO surveys). On the other hand, the precise shape of the CIRB spectrum depends on evolution effects in the average dust temperature, which is likely to increase during the past active star-forming phases as a consequence of the increased radiation field.

Galaxies are the likely dominant sources of the IR background. A further significant contribution is expected in the mid- and far-IR from dust-enshrouded AGNs, which in the framework of unified models are expected to be numerous and strongly evolving.

Altogether, most of the observed background level, after subtraction of the obvious Galactic and Zodiacal foregrounds, appears to be due to the integrated contribution of known source populations. In the near-IR cosmological window in particular, any residual signals – e.g. due to radiation processes occurring at very high redshifts – are bound not to exceed relatively low intensity levels.

1. INTRODUCTION

Most part of the redshift interval after the recombination ($5 < z < 1500$, the so-called *dark* ages) is just matter of speculation, as no direct astrophysical information has been obtained from current surveys in any wavebands. It would be of key interest, of course, to at least partially fill in the observational gap between the almost undifferentiated primeval plasma mapped in the microwaves and the well structured universe we see at low redshifts in the optical. Also for $1 < z < 5$, which corresponds to a time interval at least half of the Hubble

© 1995 American Institute of Physics

time and during which most of the cosmic structures are believed to form, our information is very scanty, confined as it is to a restricted number of peculiar active sources (radiogalaxies and quasars).

Observations in the wide spectral domain between 1 μm and 1000 μm are ideal to investigate high-redshift phenomena and to explore the *dark ages*, for obvious reasons related to the effect of redshift, bringing optical-UV photons to longer wavelengths, and of dust, degrading the emitted light to even lower energy photons. Indeed, the presence of dust is almost ubiquitously found in currently known high-z galaxies.

Unfortunately, IR observations are severely hampered by atmospheric absorption and emissivity, and by the limited sensitivity of space telescopes, quickly running into problems of confusion because of the small collector sizes. In this situation, IR background observations, at least in the near-IR and sub-mm cosmological windows where the foregrounds are at the minimum, provide an excellent tool to probe the past history of the universe.

This paper is devoted to a brief discussion of the contribution to the CIRB by *known* classes of sources, *i.e.* galaxies and active nuclei. We also emphasize how CIRB observations may be informative about their past activity.

2. THE LUMINOSITY DENSITY OF LOCAL GALAXIES FROM 1 TO 1000 μm AND THE MINIMUM EXPECTED IR BACKGROUND

A precise knowledge of the local IR luminosity functions of galaxies is essential for several reasons. Basically, it allows a fully model-independent way to estimate a minimal contribution of known sources to the extragalactic IR background. This is a sort of *known* baseline for the CIRB, any (positive) evolution effects adding some amount of flux to such a baseline. On quite general grounds, we may write the contribution of a galaxy population to the background intensity at a given IR wavelength as

$$I_{IR} = \frac{1}{4\pi}\frac{c}{H_0}\int_0^{z_{max}} dz(1+z)^{-5}(1+\Omega_0 z)^{-1/2}\ j(z), \qquad (1)$$

where $\Omega_0 = 2q_0$ and where the redshift-dependent volume emissivity is given by $j_{IR}(z) = \int_{L_{min}}^{L_{max}} d\log L_{IR}\ L_{IR}\ \rho(L_{IR})\ E(L_{IR},z)\ K(L_{IR},z)$, an expression which convolves the local luminosity function (LLF) $\rho(L_{IR})$ (in units of $Mpc^{-3}/\Delta \log L$), the evolutionary correction $E(L,z)$ and the K-correction $K(L,z)$. The latter simplifies to $(1+z)^{-\alpha}$ for power-law spectra $L_\nu \propto \nu^{-\alpha}$. Assuming no evolution for the population objects ($E[L,z]=1$), one can see that, unless the K-corr is positive and very high, the CIRB flux is mostly contributed by local ($z<1$) objects. Then the volume emissivity simplifies to the local value $j_{IR}(z) \rightarrow j_{IR}(0) = \int d\log L\ L\ \rho_0(L)$.

Since it is unlikely that any populations could have faded by z of several tenths (corresponding to several Gyr back in time), and since the population emissivity is expected to be larger in the past because of the exhanced star-formation activity, then a lower limit to the contribution of the population to the CIRB is found by inserting $j_{IR}(0)$ into eq.(1). Note that this limit only depends on a reliable definition of the LLF.

We collect in Figure 1 some relevant information on the galaxy's LLFs in various optical, IR and mm bands, where existing surveys allow direct estimates.

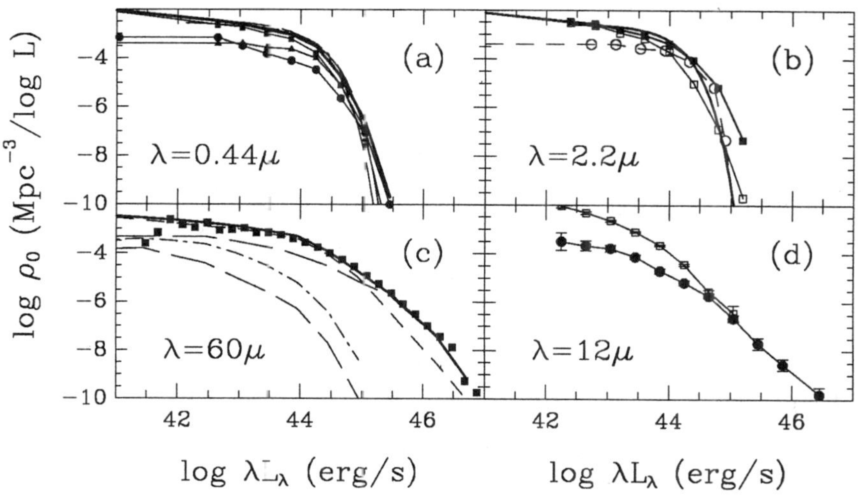

Figure 1. Multi-wavelength local luminosity functions of galaxies differentiated by galaxy types for $H_0 = 50\ Km/s/Mpc$ (see text for more details).

Since, as discussed in Sects. 3 and 4 below, the evolutionary histories of galaxies depend very significantly on the morphological class (being more extreme for spheroid- than for disk- dominated objects), we thought essential to keep the LLFs differentiated by morphological type.

Our adopted LLFs in Figure 1 are largely based on a complete sample of 1671 galaxies selected from the UGC by Franceschini *et al.* (1988a), whose depth is optimized with respect to the conflicting criteria of having the maximum sampled volume and the best morphological information. Fig.1a compares the B-band LLFs derived from this sample with published results by Efstathiou *et al.* (1988), De Lapparent *et al.* (1989) and Loveday *et al.* (1992) (note a revision of the latter by Ellis 1995). Values of the LLFs in the original paper by Franceschini *et al.* have been scaled here so as to make the corresponding global function (sum of the contributions from the various morphological components) to fit into the Efstathiou *et al.* and De Lapparent *et al.* best estimates, as well as to be consistent with LLFs based on K-band and IRAS selected samples, as explained below. The result of this optimization process shows good consistency with the other optical estimates for $M_B < -17$.

The K-band LLF (Fig.1b) is obtained from the optical one by applying average $B - K$ colour corrections. It compares well with a recent estimate by Glazebrook *et al.* (1995a) based on a K-selected sample of 93 galaxies with $z < 0.4$ (with just a marginal inconsistency for the E/S0 bin at $M_K = -27$). Note that this LLF is markedly different from, and updates, a previous estimate by Mobasher *et al.* (1993) based on an optically selected sample.

The 60 μm LLFs of Spiral and Starburst galaxies (Fig.1c) are scaled from the optical by means of bivariate luminosity distributions and have been adapted from Franceschini *et al.* (1988a) to take into account a more recent compilation by Saunders *et al.* (1990) based on IRAS selected samples. In particular, the contribution of ultra-luminous IRAS galaxies ($L_{IR} > 10^{12}\ L_\odot$) is fitted extrapolating the Starburst galaxy distribution. For the E and S0 galaxies, so elusive at

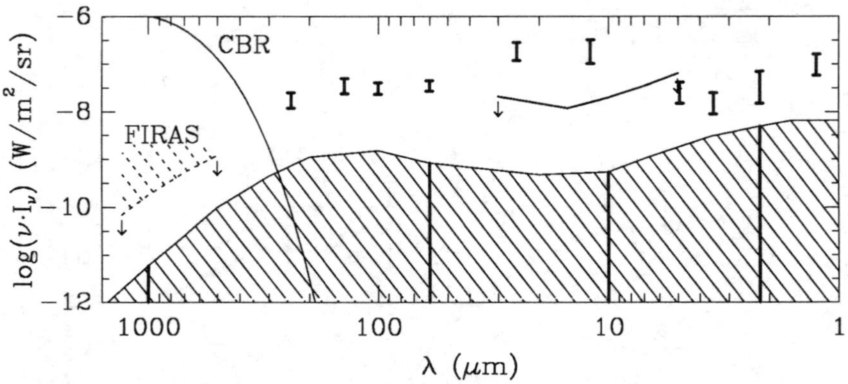

Figure 2. Minimum expected IR background from known galaxies, compared with the residual brightnesses estimated by local foreground removal from DIRBE all-sky maps (Hauser, 1994). Four wavebands where the galaxy emissivity has been directly estimated are shown.

far-IR wavelengths, we have transformed the B-band LLFs to 60 μm by adopting average flux ratios (see Franceschini *et al.*, 1994, for more details).

The 12 μm function in Fig.1d is from Rush *et al.* (1993). It is roughly consistent with the 60 μm one and with a bivariate L_{12}/L_{60} distribution. We compare here the LLF of normal galaxies to that of AGNs (Seyfert 1 + 2 galaxies) to emphasize that the latter contribute a much larger fraction ($\sim$ 12%) of the total volume emissivity in the band than found at the other wavelengths. Consequences of this will be discussed in Sect. 4 below.

The 1.3 mm LLF (not reported in Fig.1) is an extrapolation of the 60 μm function through a bivariate IR/mm luminosity distribution based on a recent millimetric survey by Franceschini and Andreani (1995) of a complete sample of 28 IRAS-selected galaxies. Admittedly, this function may lack contributions of galaxies dominated by very cold dust emission and under-represented in the IRAS samples.

It is now straightforward to integrate the volume emissivities in the five wavebands through eq.(1). This is done by adopting an upper limit to the redshift integral of $z_{max} = 1$. Interpolations between the 5 points are made using average spectra of local galaxies. The minimal expected CIRB thus found is compared in Figure 2 with the residual sky brightnesses in the ten DIRBE bands estimated by Hauser (1994) after removal of the local Zodiacal and Galactic foregrounds. This expected background is a conservative limit for various reasons. One is that IRAS 60 μm counts significantly require evolution. The other is related to the effect of large-scale clustering in the estimate of the LLFs: an underdensity by a factor two in the local universe has been invoked to interpret the very steep optical counts of galaxies at $B < 18$ (*e.g.* Ellis, 1995).

We see from Fig. 2 that there is room for a substantial cosmological evolution, except in the near-IR, where the contribution of local non-evolving galaxies is already within a factor 2 of the estimated total CIRB.

3. GALAXY EVOLUTION IN THE OPTICAL-IR-RADIO

Having defined the contribution to the CIRB of local galaxies, let us now consider possible effects of distant evolving objects.

3.1 GENERAL OBSERVATIONAL TRENDS

Optical searches have constituted for long time the primary tool to investigate the distant universe. A remarkable excess over a smooth extrapolation from brighter magnitudes was observed in the B-band galaxy counts. Counts, colors and spectroscopic surveys indicated a substantial increase with look-back time of the surface density of blue galaxies (Tyson, 1988). The observed redshift distributions down to $B \simeq 24$ (Colless *et al.* 1993; Glazebrook *et al.* 1995c), however, show that the source excess is due to blue objects at medium-redshifts ($0.3 < z < 1$), which cannot be accounted for by pure luminosity evolution (PLE) models.

Crucial constraints, concerning in particular the effects of dust in high-z galaxies (see Franceschini *et al.* 1994), come from observations at infrared wavelengths. Near-IR observations, in particular, have benefited by the impressively fast advances in infrared array technology, now able to compete with the optical ones in terms of limiting achievable sensitivity (Cowie *et al.* 1994; Gardner *et al.* 1993). Unlike optical counts, which are highly responsive to young stellar populations, hence prone to very uncertain K-corrections and biased against early-type galaxies, near-infrared counts are much less sensitive to the very recent star-formation history. K-band observations are setting tight constraints on galaxy evolution, also because the K-band light is usually a very good tracer of the baryonic content of galaxies. An extremely interesting result was that counts in the K band show a remarkable flattening fainter than $K \simeq 20$, which is at variance with the steep counts observed in the optical.

The first indication for an evolving IRAS galaxy population came from faint 60 μm counts (Hacking and Houck 1987; Franceschini *et al.* 1988b; Saunders *et al.* 1990). More recent redshift surveys (Oliver *et al.* 1995) suggest a strongly evolving luminosity function (with an evolution rate comparable to those of quasars, $L(z) \simeq L_0[1+z]^{3.3}$). Although the limited depth of the IRAS surveys implies a substantial uncertainty affecting this estimate, evolution of far-IR galaxies is also inferred from evolutionary studies of sub-mJy radio sources (Benn *et al.* 1993) and from the universal radio-to-far/IR luminosity correlation.

Two general evolutionary schemes proposed to interpret this phenomenology are briefly outlined in the next two Sects.

3.2 EVOLUTION THROUGH MERGING-INTERACTIONS

The basic question posed by observations of faint galaxies is why we do not see them at $z > 1$ in optical searches, in spite of their extreme sensitivity. One often discussed possibility is that at such redshifts luminous galaxies simply did not exist, because they formed only more recently (at $z < 1$) through accretion of smaller subunits, as envisaged in the *merging* picture of galaxy formation (Carlberg 1992; Broadhurst *et al.* 1992). The overall effect is a combination of luminosity evolution, due to galaxy-galaxy interactions triggering star-formation in the gas-rich systems, and of a systematic drift in the galaxy mass function increasing the average mass at the expense of the spatial number density.

This interpretation is attractive because it allows to simultaneously explain various observational evidences. The excess number of faint blue objects at medium redshifts observed in the optical, and the evolution of far-IR and radio selected galaxies are explained as the effect of interaction-induced star-formation. These numerous star-forming blue galaxies are not sampled in K, whose quickly converging counts are rather dominated by typically red L^* Sp and E/S0 galaxies. The absence of $z > 1$ galaxies is explained with the progressive merging making the mass function to change quite significantly already by $z \simeq 1$.

This model naturally fits into the general scheme of structure formation through gravitational instability proceeding from the smaller to the largest scales. In addition, galaxy interactions *are seen* in the local universe, and certainly bound to play a bigger role in the past. However, this *merging-interactions* scenario is not free of some serious problems. The basic one is that L^* galaxies *are seen* well above redshift one in various ways. In particular, absorption-line statistics in the high-z QSO's line-of-sight show that "normal field galaxies exhibit strikingly little evolution in their space density, luminosity, and optical/IR colors to redshifts as high as $z \sim 1.5$" (Steidel and Dickinson, 1994). But several additional problems have been mentioned: the uncomfortably high merging rate implied for local galaxies, the existence of thin stellar disks in spirals, the evidence that early-type galaxies are essentially coeval and formed well above $z = 2$, the existence of high-z radio galaxies and quasar-host galaxies.

3.3 LUMINOSITY EVOLUTION IN DUSTY ENVIRONMENTS

An alternative possibility is that the bulk of galaxies were indeed already formed by $z \simeq 1$, but cannot be detected because the large amount of optical-UV photons produced by newly born massive stars are suppressed by dust present in the diffuse medium from which stars themselves are formed. This is physically motivated by observations of star-forming regions in nearby galaxies, which are usually buried in heavily obscured molecular clouds. Even a small fraction of early produced metals, condensing into dust grains, may yield appreciable extinction and dust re-radiation, during a phase in which much of the baryonic content was still in a gaseous form, such a phase also coinciding with that of major energy production by stellar nucleosynthesis.

This second interpretation is maybe less elegant than the previously mentioned, as it requires a somewhat *ad-hoc* population of faint blue galaxies turning on at $z \leq 1$ and dominating the optical counts at the faint magnitudes. It makes, however, some testable predictions, in particular that the signs of the formation and early evolution of galaxies should be observable at long (IR and radio) wavelengths: not only dust extinction becomes negligible longward of few microns, but dust itself is expected to become a luminous source of IR photons.

Subsequent Sections will be devoted to test these evolutionary scenarios on existing data.

3.4 BASIC RECIPES FOR WIDE-BAND POPULATION SYNTHESIS

Near- and far-IR selections obviously sample quite different physical processes in galaxies: photospheric emission from red giant stars in the near-IR, and dust re-radiation from young stellar populations at far-IR wavelengths. To best exploit the information content of the two channels, we need a tool to relate properties of the ISM (gas fraction, metal abundance and dust content) with those of the stellar populations.

We make use of evolutionary population synthesis models by Mazzei et al. (1992, 1994) accounting for dust effects. Given the large spectral coverage, from the UV to 1 mm, these models allows a synoptic view of galaxy properties allover their emission wavebands. They provide a self-consistent description of chemical and photometric evolution incorporating stellar emission, internal extinction and dust re-radiation. Dust includes a cold component, heated by the general radiation field, a warm one associated with HII regions, and emission from PAHs and from circumstellar dust shells.

The star formation rate (SFR) and the initial mass function (IMF) are the basic input ingredients, ruling the evolution of metallicity, hence the far–IR output. The most critical parameter is the initial value of the SFR, SFR_0. The local spectral energy distributions (SEDs) and chemistry of late-type, disk-dominated, systems are reproduced by low values of SFR_0, followed by a SFR slowly decreasing with time. The evolution of the bolometric output and of the broad-band spectral shape is correspondingly very modest.

High values of SFR_0 provide a drastically different behaviour, due to the fast depletion of the gaseous content: the standard scenario for early-type, bulge-dominated, galaxies. The local SEDs are also explained, but, obviously, this is not enough to univocally constrain the evolutionary histories. If the early metal enrichment and dust contamination of the ISM were quick enough, as they are expected to be, then the fractional output $\frac{L_{IR}}{L_O+L_{IR}}$ in the IR (10-1000 μm) with respect to the bolometric $L_O + L_{IR}$ (0.1-1000 μm) may be a strong function of cosmic time, *i.e.* close to 100% at early epochs and almost zero locally. A possible example of this story may be IRAS F10214 (see below).

3.5 OBSERVATIONAL TESTS OF THE EVOLUTION

We show in this Section that current data do not warrant yet a definite choice between the above two extreme evolutionary patterns, both of them allowing (at least marginally) acceptable solutions, that are discussed in the following. Promising (and perhaps decisive to solve this ambiguity) observational improvements are soon expected from the forthcoming ISO mission (in particular to test the dusty spheroid idea of Sect. 3.3), by HST imaging of very faint galaxies, spectroscopic surveys with 10-m class telescopes, and identifications of QSO line-of-sight absorbers.

3.5.1 Optical- and near-IR-selected galaxy samples

Faint galaxy observations in the optical provide a primary motivation for the *merging* hypothesis. The large number of medium-luminosity objects of small angular dimensions (0.2″ – 0.4″) found at $z \sim 0.5$ dominating the faint B and, to a less extent, the I counts, and the morphological information from high-resolution imaging (Glazebrook *et al.* 1995b), all seem to imply the optical activity to be triggered in low-mass systems by interactions. The most natural way to make them disappear by z=0 is by merging into larger galactic aggregates. The alternative is to assume that they fade away by a quick consumption of the thermonuclear fuel and wind production (Babul and Rees 1992).

Indications much less in favour of a *merging*-dominated evolution are coming instead from near-IR sensitive surveys. Impressive results on counts and (B,I,K) colour distributions down to K=21.9 have been produced by the IfA team at Mauna Kea (*e.g.* Cowie *et al.* 1994). Spectroscopic observations nearly complete to $K \leq 18$ and $\sim$ 70% complete at K=19-20 have also been carried out (Songaila *et al.* 1994). These spectroscopically confirmed galaxies are distributed in redshift up to $z \approx 1$, with one object at $z = 2.35$. Glazebrook *et al.* (1995a) analyse a complete spectroscopic sample K=17.3. The results of these surveys may be summarized as follows. Any luminosity evolution in K is modest, *i.e.* <1 mag at $z < 1$, and also the space density evolution is moderate, but it implies a *decrease* of the source number with redshift. Thus the over-

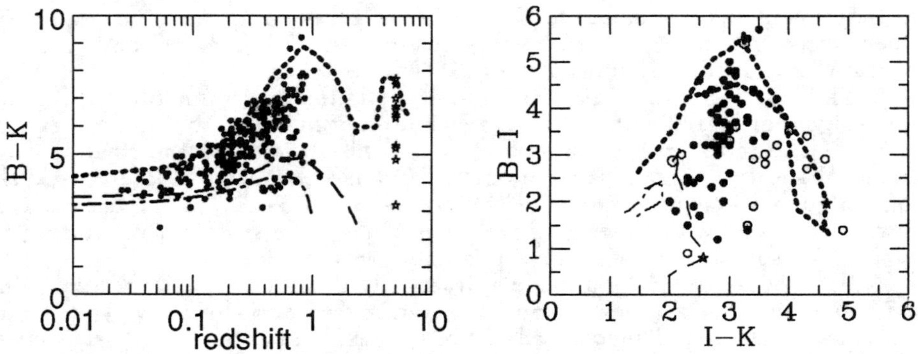

Figure 3. *(a)* Colour-redshift relationships predicted by the PLE model (Sects. 3.3 to 3.5), compared to data by Songaila *et al.* (1994). *(b)* Colour-colour plots for intermediate ($0.5 < z < 1$) galaxies (filled circles) and of candidate high-z galaxies (open circles), compared to model predictions.

all evolution seems opposite to that expected for a merging-dominated model (Glazebrook *et al.*). Finally, spectral features show an enhanced star-formation activity at z=1 compared to the present time.

The two leading groups in the field seem to disagree in the interpretation of these data. Indeed, Songaila *et al.* and Cowie *et al.* suggest that the merger model is favoured by their analysis of the dependence of the median redshift on K magnitude (showing a turnover at $K > 18$) and by the scarcity of very red objects which might lie at $z > 1$. Their conclusions, however, are based on estimates of photometric redshifts for objects at $K > 18$ and on colour analyses of very faint galaxies, both performed assuming template SEDs from local galaxies.

We point out, instead, that evolutionary effects in the spectra of early–type systems, in particular, cannot be neglected at $z \gtrsim 1$. The use of template SEDs from local objects may lead to wrong interpretations. A younger stellar population implies bluer colours at $z > 1$ in comparison with simply reshifted local SEDs.

Indeed, we find that by taking into account the combined spectral effects of the evolving stars and of the ISM's dust extinction (see Sect. 3.4), all basic statistical properties of K-selected galaxies can be explained by pure luminosity evolution of the observed LLFs (see also Gronwall and Koo 1995; Gardner 1994; Elston 1994). Hence we explain, in particular, the blueing trend observed at $K > 19$ and $z \gtrsim 1$ as an evolutionary effect in the stellar populations, rather than a change in the galaxy's population mixture.

Figure 3 compares the colour and redshift data by Songaila et al. (1994) with predictions based on a PLE model. We have also analysed the colours of the 16 unidentified galaxies in the K selected sample by Songaila *et al.* and found that most of them may well correspond to $z > 1$ objects. We then suggest that several high-z ellipticals during an early phase of evolution could be found among these K-selected objects.

A world model with $q_0 = 0.15$ and $z_{form} = 7$ for early–type and $z \simeq 2$ for late-type galaxies is indicated by a match of the observed counts, LLFs, redshift and colour distributions.

On the other hand, this is by no means a unique solution, a merging-interaction evolution certainly being also consistent. We have made a comparative analysis of optical and near-IR survey data through a simplified approach in which the luminosity functions of various morphological classes evolve following the change with time of the average galaxy mass, and consequently of the number density, and following a luminosity evolution due to the enhanced star-formation triggered by interactions. We find that acceptable solutions in which the merging rate is such that the average galactic mass is reduced to 30% of the local value by $z = 1$ and the luminosity per unit baryonic mass is $\simeq 3$ times higher at that time in the B band ($q_0 = 0.15$) for spiral and starburst galaxies. For early-type galaxies we assume canonical evolution with moderate dust extinction (see Sect. 3.4) and an evolution of the mass function similar to that of late-type galaxies.

This model is also consistent with the above-mentioned K band data, in the limited statistics of the available samples, and within the large uncertainties about the fraction of high-z objects in complete K samples.

3.5.2 Far-IR and radio-selected samples

The available information on distant galaxies is orders of magnitude less detailed here than in the optical-K band. As for the far-IR, and at least until the advent of ISO, we just have a few areas covered by IRAS to moderate sensitivities and with limited spectroscopic information. Similar problems affect the faint radio source surveys. Again a crucial question here concerns the high-z tail of the redshift distribution: these surveys usually suffer by at least 30% incompleteness in the redshift information (the farther the selection wavelength from the optical, the worse this incompleteness is).

In such a situation, one can take the two extreme attitudes: *a)* to assume that the missing redshifts have basically the same distribution of the measured ones, and this would bring to support a merging evolution; or *b)* to postulate that a fraction of the missing objects would be at $z > 1$, something in favour of PLE evolution. By definition, measuring redshifts for dusty high redshift galaxies will be a difficult, maybe impossible, task, so that alternative criteria based on multi-wavelength optical-IR-mm-radio broad-band observations are probably to be defined to solve the ambiguity. This will need sensitive observations from space (ISO + future missions as ISAS, SIRTF, FIRST, EDISON, etc.) and from ground.

At the moment, we have just verified that the same two models fitting the optical-K band data also reproduce survey data at longer (IR and radio) wavelengths.

As a second step, one may consider some high-z galaxies, selected in particular at long wavelengths, and try to understand the basic physics ruling their e.m. output. From the limited information on broad-band spectra of a few of them (cfr. Figure 4) we see that a dusty starburst may contribute most of their luminosity. Note that current evidences for a lensing effect in IRAS $F10214+4724$ could just scale down its luminosity and baryonic mass by roughly a factor ten.

Several other distant IR- and radio-selected galaxies and optical quasars have been found to contain large amounts of dust (Isaak *et al.* 1994; Carilli *et al.* 1995), hence a very rich ISM. But again, it is just a matter of attitude, at the moment, to consider these objects as peculiar and extremely rare monsters in the far tail of the galaxy phenomenology, or, instead, truly representative examples of the evolutionary behaviour of high redshift spheroids.

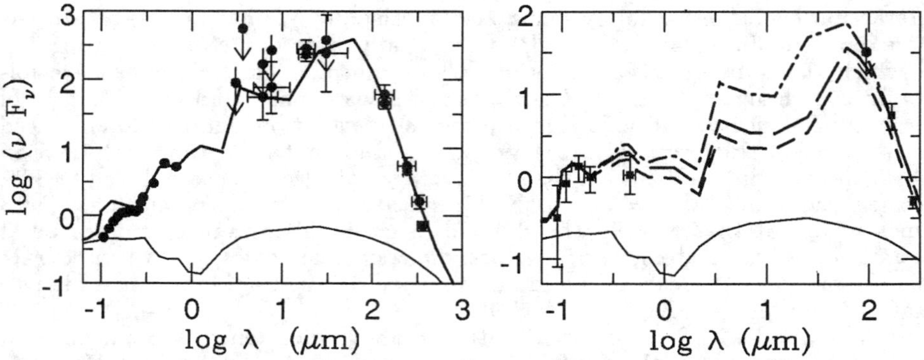

Figure 4. Optical through *mm* SEDs of two high-z primeval galaxy candidates. *(a)*: IRAS F10214 (Rowan-Robinson et al. 1991); *(b)*: 4C41.17 (Chambers *et al.* 1990). Predicted SEDs are based on a model of primeval dusty spheroids (see Sects. 3.3 and 3.4 and Mazzei and De Zotti 1994).

To summarize, we argue that, on the basis of currently available data, both the merging-dominated model and the luminosity evolution model in dusty environments are viable interpretations. Both of them will be considered in the final discussion of Sect. 5.

4. IR EMISSION FROM ACTIVE GALACTIC NUCLEI

The infrared luminosity is often a significant portion of the bolometric emission of AGNs. In the case of radio quiet AGN the origin of the IR emission is still debated, although evidence has recently accumulated in favour of dust reprocessing of the primary optical–UV light. Arguments in favour of dust emission are: *(i)* the observed SEDs showing a steep rise of the submillimetric continuum between 1000 and 100 μm (*e.g.* Chini, Kreysa and Biermann 1989), far too steep for any non–thermal component, and a local minimum in νF_ν at $\lambda \simeq 1\mu$m, which is naturally explained by dust radiating at a limiting sublimation temperature $T_s \sim 1500$ K; *(ii)* the IR variability and its relationship with those in the UV and optical with time delays which are easily interpreted with dust models; and *(iii)* the observed low level of polarization.

The presence of optically thick dust with azimuthal symmetry has been also advocated to explain the observed difference between broad– and narrow–line objects in the framework of unified AGN theories. In addition to the broad lines observed in polarized optical light in narrow-line nuclei, broad components of hydrogen IR lines have more recently been seen in direct near-IR light, suggestive of the possibility of a direct view of the broad–line regions in the IR. Further indications for dust around AGN has been found through studies of ionization cones in several Seyfert 2 galaxies, which demonstrate that the gas in the hosting galaxies sees anisotropic ionizing sources. Also X–ray observations revealed very high absorbing columns in many Seyfert 2 galaxies, where Seyfert 1 X-ray spectra exhibit moderate to low or even no absorption.

The observed nuclear SEDs of Seyfert 1 are now rather well defined and show a large bump peaking at $\lambda \sim 10$ to 30 μm. The optical components exhibit relatively moderate amounts of extinction in the optical–UV ($E[B-V] \lesssim 0.3$). By contrast the available information on the SEDs of narrow line AGN is rather scanty. Only for NGC 1068 a reasonably good nuclear IR spectrum exists: it exhibits a maximum at $\lambda \sim 30$ μm and falls off at shorter wavelengths much more rapidly than in the case of broad–line objects, as expected in the unified scheme.

Various kinds of dust configurations have been fitted to the available data, with various equatorial depths, covering factors, shapes, radial and axial dust density distributions. Standard galactic composition, size distribution and sublimation temperatures are assumed for dust grains. These models can be broadly separeted in two classes. A first one predicts extremely high optical depths ($A_V \gtrsim 500$) and very compact structures for tori (Pier and Krolik 1993), but then additional dust components from the galaxy are needed to fit the observed spectra and surface brightness.

The second family of models require less extreme optical depth ($A_V \sim 100$) and more extended tori (Granato & Danese 1994), in which case the torus reproduces the entire IR spectrum and extends in the mid-IR over several tens of parsecs, in agreement with the observations. Also the predicted anisotropy of the 10μm emission is within the limits required by the comparison of the Seyfert 1 and 2 LLFs.

Statistics on the number of optically hidden, dust-extinguished, type-2 AGNs have been tried through various arguments. One in particular compares the 12 μm luminosity functions of Seyfert 1 and 2 with the emissivity properties in hard X–rays (which are a good tracer of the nuclear non-thermal activity and minimally affected by gas/dust on the line-of-sight). These analyses provide however discrepant results. It then seems worthwhile to keep the ratio of type-2 to type-1 objects running from unity to roughly 5.

5. CONTRIBUTIONS TO THE IR BACKGROUND FLUX

The past integrated activity of galaxies, AGNs and other hypothetical primeval objects has been actively searched in the IR through background measurements. Certainly, a phase that even very deep searches have failed to reveal is that corresponding to the early metal enrichment of gas from which the oldest stars were formed. The corresponding bolometric background intensity is

$$I_{bol} = 1.44\ 10^{-8} \left(\frac{\Omega_b}{0.01}\right) \left(\frac{f_m}{0.1}\right) \left(\frac{5}{1+z_F}\right) (W/m^2/sr), \tag{2}$$

where Ω_b is the baryon density parameter, z_F is the galaxy formation redshift, and f_m is the fraction of processed baryons. It is interesting to note that the presence of dust (either primeval, or along the line-of-sight) would produce a precise anti-correlation signal between CIRB intensities in near- and far-IR wavebands, the former directly contributed by the early stellar generations, the latter by dust reprocessing.

Three main observational channels have been followed to test the CIRB, in particular in the near-IR and sub-mm cosmological windows where the foregrounds are at the minimum: rocket flights of dedicated instrumentation, all-sky surveys by space platforms (IRAS and COBE), and indirect estimates based on spectral observations of extragalactic γ-ray sources (Stecker and de Jager 1993). To avoid overcrowding, we report in Figure 5 (as in Fig. 2) only the most recent results from the dedicated DIRBE experiment on COBE (Hauser 1994), and an indirect estimate based on γ-ray observations of AGN in the mid-IR (Dwek and

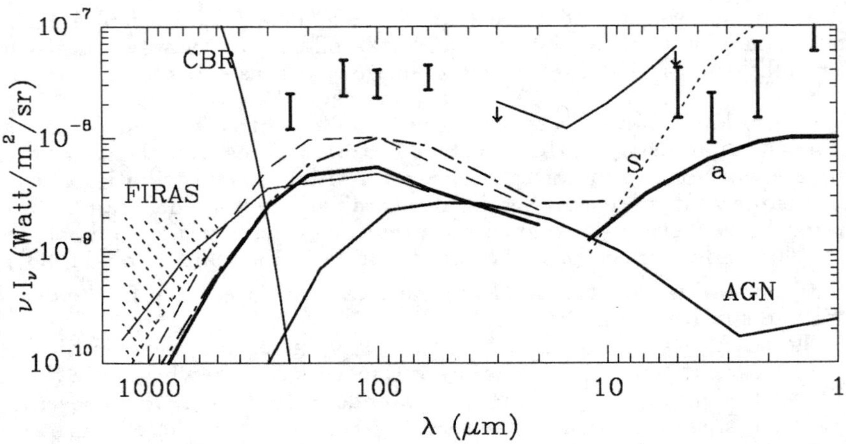

Figure 5. Extragalactic background light at IR to sub-mm wavelengths. The residual brightnesses estimated by Hauser (1994) are compared with predictions discussed in Sect. 5.

Slavin 1994), where the other methods are weak because of the overwhelming brightness of the IPD.

5.1 The near-IR background

For $\lambda < 10\ \mu m$ we report in Figure 5 as a thick line the contribution of galaxies estimated from the PLE evolution model mentioned in Sect. 3.5. Note that this estimate is quite a robust one, constrained as it is by existing deep counts of galaxies in K band. The predicted spectral shape is also constrained by photometric evolution models. We estimate that uncertainties related to this prediction amount to few tens of percent in this energy range.

Two observational limits (the Hauser's estimate in Fig. 5 and one by Stecker and De Jager (1993) are already quite close to such a predicted background. Altogether, the current information (from both direct observational limits, indirect estimates and theoretical predictions) is consistent with the near-IR background mostly originating from ordinary stellar processes in galaxies (Stecker and De Jager, 1995).

5.2 The far-IR/sub-mm background

In the long-wavelength part of Fig. 5 we compare predictions of various evolutionary models discussed in Sect. 3. We see an overall factor three uncertainty in such predictions. The most variant model corresponds to the thin line and is based on a simple kinematical prescription fitting the 60 μm statistics (Franceschini *et al.* 1991). This implies an higher flux level at the long wavelengths (possibly already inconsistent with FIRAS estimates), because of the dust temperature distribution assumed constant with time. The lower mm background predicted by all other more physical models is due to an increase of the dust temperature in the past, following the increased SFR and average radiation field in galaxies.

Within the physical models, the spread in the predicted levels are partly due to various degrees of evolution allowed by the IRAS survey data (the thick continuous and the dashed lines corresponding to a moderate and to a strong

PLE evolution), in part to the average redshift at which the CIRB is produced. PLE models predict a fairly large redshift for the main phases of galaxy activity ($z \gtrsim 3$), whereas typical *merging* schemes assume most of it confined to $z \lesssim 1$ (the dot-dashed line, see details on the model in Sect. 3.5). How this is reflected in the CIRB spectrum is clearly shown in the Figure.

5.3 A mid-IR background from hot dust in AGNs?

At the minimum between the predicted ***starlight*** background in the near-IR and the ***dust re-radiation*** feature in the far-IR, there is a spectral region ($10 < \lambda < 20\mu m$) in which properties of even local galaxies are very poorly known (essentially because of the poor sensitivity of IRAS at these wavelengths). Existing data suggest, in any case, that the integrated emission of normal galaxies should keep below $3\ 10^{-9}\ W/m^2/sr$, unless dust with implausibly high temperature would characterize the early evolution phases.

On the other hand, preliminary estimates based on $\gamma-$ray observations of MKN 421 indicate an CIRB level at 10 μm larger than that (Stecker and De Jager, 1995; see also Dwek and Slavin, 1994). Although not conclusive, these results warrant further investigation of alternative radiation processes with respect to normal stellar processes in galaxies. As previously discussed, dust hot enough to emit at these wavelengths is likely to be present close to the nuclear environments in AGNs. The corresponding contribution to the mid-IR background is set by the convolution of the local IR volume emissivity of these AGNs and their evolution rate. We report in Figure 5 (line marked AGN) a crude estimate based on the 12 μm luminosity function of type-1 and -2 Seyferts (Fig.1d) and an evolution rate equal to that of optical QSOs. The adopted IR spectra for the two AGN types are average SEDs of Seyfert's 1 and that SED of NGC 1068 in Granato and Danese (1994). The IR LF of Seyferts 2 has been renormalized upwards by a factor 4 to account for the fraction of optically undetected, dust extinguished AGNs predicted by the unified AGN model. With these somewhat extreme ingredients, the contribution of dusty AGNs in the range 10 to 50 μm may exceed those of other galaxy populations (see Granato *et al.*, these Proceedings, for a more detailed analysis).

6. PERSPECTIVES

Current observational limits on the CIRB in the two most important cosmological windows (the near-IR and the sub-millimetric one) have already come so close to the level expected for the integrated contribution of extragalactic sources to allow very significant achievements for observational cosmology.

When confirmed, the available limits in the near-IR will severely constrain any radiation processes occurring at high redshifts – *e.g.* from pre-galactic stellar generations. In the sub-mm, there are good chances soon to probe – after a delicate removal of the Galactic foreground – the integrated emission from dust during early evolution phases in galaxies, then allowing a fundamental test of galaxy formation and evolution. The measured shape of the background here will provide an integrated information on the formation epoch and on the average intensity of the radiation field in the sources.

CIRB measurements in the mid-IR would be extremely relevant to test the integrated emission from dust-enshrouded Active Galactic Nuclei and Quasars, but they will possibly have to wait for missions flying the outer Solar System

REFERENCES

Babul, A., & Rees, M. 1992, MNRAS 255, 346
Broadhurst, T.J., Ellis, R., and Shanks, T. 1992, Nature, 355, 55
Carilli, C..L., Owen, F.N., & Harris, D.E. 1994, AJ 107, 480
Carlberg, R.G. 1992, ApJ , 399, L31
Chambers, K., Miley, G., & Van Breugel, W. 1990, ApJ 363, 21
Chini, R., Kreysa, E., & Biermann, P.L. 1989, A&A 219, 87
Cowie, L.L., Gardner, J.P., Songaila, A., Hodapp, K.W., Wainscoat, R.J. 1994, ApJ 434, 114
De Lapparent, V., Geller, M., and Huchra, J. 1989, ApJ , 343, 1
Dwek, E., & Slavin, J. 1994, ApJ 436, 696
Efstathiou, G., Kaiser, N., Ellis, R.S., & Frenk, C.S., 1990, MNRAS 242, 318
Ellis, R. 1995, Preprint
Elston, R. 1994, in "IR astronomy with Arrays", Kluwer, 33
Franceschini, A., Danese, L., De Zotti, G., & Toffolatti, L. 1988a, MNRAS , 233, 157
Franceschini, A., Danese, L., De Zotti, G., & Xu, C. 1988b, MNRAS , 233, 175
Franceschini, A., Toffolatti, L., Mazzei, P., Danese, L., & De Zotti, G. 1991, ApJS 89, 285
Franceschini, A., Mazzei, P., De Zotti, G., & Danese, L., 1994, ApJ 427, 140
Franceschini, A., & Andreani, P. 1995, ApJ 440, L5
Gardner, J.P., Cowie, L.L., & Wainscoat, 1993, ApJ 415, L9
Gardner, J.P. 1994, in "IR astronomy with Arrays", Kluwer, 29
Glazebrook, K., Peacock, J., Miller, L., & Collins, C. 1995a, MNRAS , in press
Glazebrook, K., Ellis, R., Santiago, B., Griffiths, R. 1995b, MNRAS , in press
Glazebrook, K., *et al.* 1995, MNRAS 273, 157
Granato, L., Danese, L., 1994, MNRAS 268, 235
Gronwall, C., & Koo, D.C. 1995, ApJ 440, L1
Hacking, P., Houck, J.R., 1987, ApJS 63, 311
Hauser, M. 1994, in "Examining the Big Bang and Diffuse Background Radiations, IAU Symp. 168, The Hague
Isaak, K.G., McMahon, R.G., Hills, R.E. & Withington, S.: 1994, MNRAS 269, L28
Mazzei, P., Xu, C., De Zotti, G., 1992, A&A 256, 45
Mazzei, P., De Zotti, G., Xu, C., 1994, ApJ 422, 81
Mazzei, P., & De Zotti, G. 1994, MNRAS 266, L5
Mobasher, B., Ellis, R., & Sharples 1993, MNRAS 263, 560
Oliver, S.J., *et al.* 1995, in "Wide-Field Spectroscopy and the Distant Universe", eds. S.J. Maddox and A. Aragon-Salamanca, World Scientific
Pier, E.A., & Krolik, J.H. 1993, ApJ 401, 99
Rowan-Robinson, M., et al. 1991, Nature, 351, 719
Rowan-Robinson, M., Benn, C., Lawrence, A., McMahon, R., & Broadhurst, T. 1993, MNRAS 263, 123
Rush, B., Malkan, M.A., Spinoglio, L., 1993 ApJS 98, 1
Saunders, W., *et al.* 1990, MNRAS 242, 318
Songaila, A., Cowie, L.L., Hu, E.M., Gardner, J.P., 1994, ApJS 94, 461
Stecker, F., & de Jager, O. 1993, ApJ 415, L71
Stecker, F., & de Jager, O. 1995, Space Science Review, in press
Steidel, C., & Dickinson, M. 1994, in "Wide Field Spectroscopy and the Distant Universe", eds. S.J. Maddox and A. Aragon-Salamanca, World Scientific
Tyson, A. 1988, AJ , 96, 1

FIRST STARS IN THE UNIVERSE

L.M. Ozernoy[1,2] and V.F. Mukhanov[3†]

[1] Institute for Computational Sciences and Informatics & Physics and Astronomy Department, George Mason University, Fairfax, VA 22030-4444

[2] Laboratory for Astronomy and Solar Physics, NASA/GSFC, Greenbelt, MD 20771

[3] Meudon Observatory, France

ABSTRACT

Soon after decoupling (redshift $z \approx 1000$), the opacity of contracting primordial clouds is determined mainly by Rayleigh scattering off neutral hydrogen. This process becomes essential for a contracting primeval cloud after its density exceeds $n \approx 10^8$ cm^{-3} when the cooling due to *f-b* (free-bound) emission dominates over Compton cooling off microwave background radiation. Lyman alpha photons released in the course of *f-b* emission are absorbed again within the cloud until they eventually split up, due to two-photon emission, into lower energy quanta. As the density of the cloud increases, the opacity due to Rayleigh scattering increases until the diffusion time of photons becomes larger than the free-fall time. At this moment, the Jeans mass becomes $M_J = (\pi^4/3)(kT/m_{\rm H})^{5/2}(\sigma_R/cG^2 m_{\rm H})$, where $\sigma_R = \sigma_T[\nu_{\rm eff}^2/(\nu_{\rm eff}^2 - \nu_0^2)]^2 \approx 4 \cdot 10^{-3}\sigma_T$; σ_R and σ_T are the Rayleigh and Thomson scattering cross sections, correspondingly; $\nu_{\rm eff}$ is an effective frequency of *f-b* emission (close to the Balmer continuum edge); and ν_0 is the resonant frequency corresponding to the Lyman continuum edge.

Evaluation of M_J requires calculation of thermal balance in the cloud. Energy gains include shock heating due to (i) collisions of fragments in the turbulent collapsing cloud and (ii) collisions of the clouds themselves. However, since the Compton cooling implies a conservative constraint $T \leq 10^4$ K, this immediately results in a stringent upper limit to $M_J \leq 0.1$ M$_\odot$. In fact, M_J might even be as small as $\sim$ a few 10^{-3} M$_\odot$.

Although M_J does not depend explicitly on n, "locking" of photons due to an inequality $t_{\rm diff} \geq t_{\rm free-fall}$ sets in at densities much exceeding those at which the contracting atomic gas is converted to molecular form so that a proper account of Rayleigh scattering should be done numerically. However, this cannot alter the conclusion that the first stars in the Universe might form before the production of metals and be of very low mass, such as brown dwarfs and "jupiters". Dark galactic halos can contain a significant amount of these low mass baryonic objects formed in protogalactic gas before formation of galaxies.

---

† Permanent address: *Institute for Nuclear Research, Moscow*

© 1996 American Institute of Physics

GENERAL PROPERTIES OF GALAXY FORMATION FROM TRIAXIAL PROTOGALAXIES WITH MASSIVE STAR FORMATION.

Peter Berczik
Main Astronomical Observatory, Ukrainian National Academy of Sciences
Golosiiv, Kiev-022, 252650, Ukraine

Email ID
berczik@mao.gluk.apc.org

ABSTRACT

This is the second article in the series of papers about the 3D models of triaxial galaxies formation carried out by Smoothed Particle Hydrodynamics (SPH) method (Berczik & Kolesnik (1993), Berczik & Kolesnik (1994), Berczik & Kolesnik (1995)). We use the most common triaxial model for initial conditions of protogalaxies. The body of protogalaxies we model by triaxial ellipsoids with semiaxies A, B and C. Initial velocity fields is defined by initial angular velocity $\mathbf{\Omega}_0$. The total angular momentum is not coincide with semiaxies. Therefore the initial velocity field can be describe with formule: $\mathbf{V}_0(x, y, z) = [\mathbf{\Omega}_0 \times \mathbf{r}]$. In this paper we discuss the series of galaxy formation models with initial massive Star Formation (SF) and try to find its parameters.

1. Introduction

We not discuss here in detail the process leading to the formation of triaxial protogalaxy. This is the independent cosmological problem that needs the separate investigation. But, as we know, the protogalactic perturbation in early universe is triaxial (Berdeen, Bond, Kaiser & Szalay (1986), White & Silk (1979)). Therefore, when the galaxy is formed through the evolution of triaxial protogalaxy fragment the resulting galaxy must be triaxial.

The problem of origin of the initial angular momentum of protogalaxy have many different solutions (Hiotelis & Voglis (1989), Chernin (1993)). But, in common cases, all models shows that there exists probability that the protogalaxy have the initial angular momentum vector which doesn't coincide with any of main axes of triaxial initial protogalactic figure. Then, in a result of dynamical collapse we can expect a triaxial galaxy having

© 1996 American Institute of Physics

complicated internal kinematics (Curir & Diaferio (1994)).

The number of data, that galaxies are triaxial in their majority, increases steadily. First of all, elliptical galaxies are triaxial (Illingworth (1977), Binney (1978), Binney (1985)). This conclusion, in many cases, follows from the analysis of shape and brightness distribution of images of elliptical galaxies. Still more convincing data on the triaxial shape of ellipticals give their kinematical properties. New investigations confirm this idea and now we may make the conclusion that all ellipticals are triaxial (Franx, Illingworth & Heckman (1989), Kormendy & Djorgovski (1989), de Zeeuw & Franx (1991), Franx, Illingworth & de Zeeuw (1991)).

The warping processes in the disk system also possible to examine as evidence of general triaxiality of this system (Binney (1992)). Asymmetries in the distribution of brightness and neutral hydrogen also often observed in disk galaxies (Richter & Sancisi (1994)). On the base of data about the bar - galaxies we conclude that in a considerable part such systems have essentially triaxial gravitational potential (Pfenniger (1984)). Now the number of data about triaxiality of bulges in spiral galaxies steadily increases (Bertola, Vietori & Zeilinger (1991), Weinberg (1992)). The Milky Way galaxy and M31 also have extended triaxial bulges (Blitz & Spergel (1991a), Blitz & Spergel (1991b)).

Great number of triaxial shaping among galaxies generates the idea that triaxial configuration initially exists on a protogalactic stage of evolution. We expect that this triaxial form already exists on the early state of evolution of protogalaxies, when the star formation still not start and protogalaxies have only pure gas components. From our point of view, the triaxiality of protogalaxies is the main generating factor of present triaxial shape of galaxies.

In this paper we consider a case, when in initial state a protogalaxy is a homogeneous triaxial ellipsoide having an angular momentum directed arbitrary in space. Simulations were carried out by our, own modificated, SPH method (Berczik & Kolesnik (1993), Berczik & Kolesnik (1995)). In this method we include to standard SPH algorithm the process of SF and SN explosions. The properties of early galaxies with initial massive SF, that can be formed in a result of collapse of triaxial protogalaxies, are considered. In the result of model we try to analyze the SF parameters in forming galaxies.

2. Basic parameters of model

We suppose the common initial conditions for protogalaxy. The initial body of protogalaxy we model by homogenious triaxial HI ellipsoids with semiaxis A, B and C. In the model we cary out the calculation with $N = 4169$ gas particles. The initial center of particles we take in space with equal step for three coordinat axis. Therefore the initial configuration of particles form the system of homogenious cubic cells. In this paper we present the model when the mass of HI gas protogalaxy, is $M_{HI} = 5 \cdot 10^{11}\ M_{\odot}$. The initial temperature of gas $T_0 = 10^4\ K$. The initial orientation of ellipsoid semiaxis $A = 100\ kpc$, $B = 75\ kpc$ and $C = 50\ kpc$ is coincide with the axis of coordinate system (x, y, z). The initial velocity fields are defined through the angular velocity $\mathbf{\Omega}_0(x, y, z)$.

$$\mathbf{V}_0(x, y, z) = [\mathbf{\Omega}_0(x, y, z) \times \mathbf{r}]$$

The initial angular velocity we take constant in space. The orientation of initial angular velocity we defined by its components $\Omega_0 = (0.75,\ 0,\ 0.75) \cdot \Omega_{cir}$. Here the $\Omega_{cir} = V_{cir}/A$ and $V_{cir} = \sqrt{G \cdot M_{HI}/A}$.

The standard spin parameter is defined as (Hiotelis, Voglis & Contopoulos (1991), Hiotelis & Voglis (1991)):

$$Spin = \frac{|\ \mathbf{L}_0\ | \cdot \sqrt{|\ E_0^{gr}\ |}}{G \cdot M_{HI}^{5/2}}.$$

Where the $\mathbf{L}_0$ is the total initial angular momentum and the E_0^{gr} is the total initial gravitational energy of protogalaxy. The value of spin parameter in our model is $Spin \approx 0.3$. This value is typical for spiral galaxies.

In the calculation of our model we introduce the simple cooling function $\Lambda(T)$ in the form presented in works by Rees & Ostriker (1977), Navarro & Benz (1991), Dalgarno & McCray (1972). We use the standard cooling function for an optically thin plasma with a primordial composition in ionization equilibrium. This includes thermal bremsstrahlung on hydrogen and helium (H^+ and He^{++}), radiative recombination, dielectronic recombination and excitation of discrete levels.

The calculation were carried out by modificated standard SPH method (Berczik & Kolesnik (1993), Berczik & Kolesnik (1995)). To this SPH algorithm we added the process of SF and SN explosions. The SF injects a lot of mass and thermal energy in to the galactic system through SN explosions and stellar winds (see e.g. Leitherer, Robert & Drissen (1992)). Provided that we are only concerned with first order effects, we use the most resonable simplified way of treating SF with SPH (see also Katz (1992)). We assume what each particle consists of the gas and star component, e.g. $m_i \equiv m_i^{gas} + m_i^*$. At the begining for all particles we assume what $m_i^* \equiv 0$.

The first step in our SF algorithm is the criteria to find the gas particle in which the stars are formed. Some possible SF criteria in numerical simulation have been examined by Friedli & Benz (1993). We use the idea that the SF is start if the absolute value of gravitational energy of the particle greater then its thermal energy e.g. $| E_i^{gr} | > E_i^{th}$.

For gravitational and thermal energy of the gas particle i, we have the expressions:

$$E_i^{gr} = -\frac{3}{5} \cdot G \cdot \frac{(m_i^{gas})^2}{h_i}, \qquad E_i^{th} = \frac{3}{2} \cdot m_i^{gas} \cdot c_i^2.$$

Where the $c_i^2 = \frac{\Re \cdot T_i}{\mu}$ is the isothermal sound speed in second power and h_i is the smoothed radius of particle i. Therefore, the critical temperature of SF for particle i we define as:

$$T_i^{crit} = \frac{8\pi\mu}{15\Re} \cdot G \cdot \rho_i \cdot h_i^2.$$

If temperature of particle T_i less then the critical temperature T_i^{crit} the gas particle i will be subjected to SF.

When the collapsing particles i have been defined we update the m_i^{gas} and m_i^*. The first step is the SF process:

$$(m_i^*)^I := \eta \cdot m_i^{gas} + m_i^*, \qquad (m_i^{gas})^I := (1-\eta) \cdot m_i^{gas}.$$

The second step is the SN explosions:

$$(m_i^*)^{II} := (1-R) \cdot \eta \cdot m_i^{gas} + m_i^*, \qquad (m_i^{gas})^{II} := (1-(1-R) \cdot \eta) \cdot m_i^{gas}.$$

Where the η is the fraction of gas converted into stars in accordance with the initial mass function (IMF). The parameter R is the fraction of mass going back to interstellar medium (ISM) by SN and winds. We define $\epsilon = (1-R) \cdot \eta$ as the global efficiency of star formation. Typical values are

$R \approx 0.45$, $\eta \approx 0.02 \div 0.5$ which are poorly known quantities and consequently $\epsilon \approx 0.01 \div 0.25$. In presented model we use the value for $R = 0.45$, and for $\eta = 0.02$.

For injection the energy by SN explosions we use the model in which the explosions energy converted totaly to thermal energy of gas particle. The total energy produced by SN for particle i is:

$$\Delta E_i^{SN} = R^{SN} \cdot \eta \cdot m_i^{gas} \cdot \Delta E_{SN}^{mean}.$$

Where the R^{SN} is the number of SN created per unit mass of stars formed. The ΔE_{SN}^{mean} is the mean energy produced by one SNII. In our calculations we use the value $\Delta E_{SN}^{mean} = 10^{44}\ J$. If the IMF is of the type $\psi(m) = A \cdot m^{-\gamma}$ then

$$R^{SN} \equiv \frac{\int_{m_{SN}}^{m_u} \psi(m) \cdot dm}{\int_{m_l}^{m_u} m \cdot \psi(m) \cdot dm} = \frac{(2-\gamma) \cdot (m_{SN}^{1-\gamma} - m_u^{1-\gamma})}{(1-\gamma) \cdot (m_l^{2-\gamma} - m_u^{2-\gamma})},$$

where m_l is the lowest stellar mass, m_u the upper stellar mass, and m_{SN} is the mass threshold to form SN. Many (often controversial) papers are devoted to the IMF problem (see for instance reviews by Scalo (1986), Scalo (1989), Maeder (1993)). Standard values in the solar neighbourhood are for the IMF slope $\gamma \approx 2.7$ (Salpeter's value is 2.35), $m_l \approx 0.1\ M_\odot$, $m_u \approx 100\ M_\odot$, and $m_{SN} \approx 8\ M_\odot$, i.e. $R^{SN} \approx 0.002\ M_\odot^{-1}$. There are evidences in favour of top - heavy IMF in starbursts (Scalo (1989), Silk (1993)) that is to say IMF with smaller $\gamma < 1.5$ and higher $m_l > 2\ M_\odot$, i.e. $R^{SN} \approx 0.03\ M_\odot^{-1}$. A decrease of $m_u < 30\ M_\odot$ has also been suggested (Joseph (1991)). In presented calculation we use the standard value for $R^{SN} = 0.002\ M_\odot^{-1}$.

Therefore for specific internal energy of i - th particle after SN is:

$$(u_i)^{II} := u_i + \frac{\Delta E_i^{SN}}{(m_i^{gas})^{II}} = u_i + \frac{R^{SN} \cdot \eta \cdot m_i^{gas} \cdot \Delta E_{SN}^{mean}}{(1-(1-R) \cdot \eta) \cdot m_i^{gas}}.$$

Finally we have:

$$(u_i)^{II} := u_i + \frac{R^{SN} \cdot \eta \cdot \Delta E_{SN}^{mean}}{(1-(1-R) \cdot \eta)}.$$

In our calculations we use this formula for updating the specific internal energy after process of SF and SN explosions for gas particles.

3. Conclusions

Using our calculation we may make several general conclusions about the properties of forming galaxies and the properties of early SF:

- In the result of collapse of our triaxial model, the complex triaxial star - dominated galaxy with extended inclined, gas - dominated halo structure is formed.
- The resulting stellar galaxy have the initial intense global spiral structure and typical long - living triaxial central bar.
- Even very moderate SF rate (with parameter $\eta = 0.02$) produce the stellar component of galaxy, which have ≈ 80 % of mass of the total system. In the central 20 kpc the mass of stars is $\approx 2 \cdot 10^{11}$ $M_{\odot}$.
- The SN explosions produce in gas component of galaxy with very inhomogenuous, fractal like, distribution of temperature from 10^4 K to 10^6 K.

References

Berczik P. & Kolesnik I.G. 1993, Kinem. i Fiz. Nebes. Tel, **9**, No 2, 3

Berczik P. & Kolesnik I.G. 1994, in THE FORMATION OF THE MILKY WAY, ed. Alfaro E.J., Cambridge Univ. Press.

Berczik P. & Kolesnik I.G. 1995, A&A, in press

Berdeen J.M., Bond J.R., Kaiser N. & Szalay A.S. 1986, ApJ, **304**, 15

Bertola F., Vietori M. & Zeilinger W.W. 1991, ApJ **374**, L13

Binney J. 1978, MNRAS, **183**, 501

Binney J. 1985, MNRAS **212**, 767

Binney J. 1992, ARA&A **30**, 51

Blitz L. & Spergel D.N. 1991a, ApJ **370**, 205

Blitz L. & Spergel D.N. 1991b, ApJ **379**, 631

Chernin A.D. 1993, A&A, **267**, 315

Curir A. & Diaferio A. 1994, A&A, **285**, 389

de Zeeuw T. & Franx M. 1991, ARA&A **29**, 239

Dalgarno A. & McCray R.A. 1972, ARA&A **10**, 375

Franx M., Illingworth G. & Heckman T. 1989, ApJ **344**, 613

Franx M., Illingworth G. & de Zeeuw T. 1991, ApJ, **383**, 112
Friedli D. & Benz W. 1993, A&A **268**, 65
Hiotelis N. & Voglis N. 1989, A&A, **218**, 1
Hiotelis N., Voglis N. & Contopoulos G. 1991, A&A, **242**, 69
Hiotelis N. & Voglis N. 1991, A&A, **243**, 333
Illingworth G. 1977, ApJ, **218**, L43
Joseph R. 1991, in MASSIVE STARS IN STARBURSTS, eds. Leitherer C., Walborn N.R., Heckman T.M., Norman C.A., Cambridge Univ. Press., 259
Katz N. 1992, ApJ **391**, 502
Kormendy J. & Djorgovski S. 1989, ARA&A **27**, 235
Leitherer C., Robert C. & Drissen L. 1992, ApJ **401**, 596
Maeder A. 1993, in FIRST LIGHT IN THE UNIVERSE, eds. Rocca - Volmerange B., Dennefeld M., Guideroni B., Tran Thanh Van J., Editions Frontières, Gif sur Yvette
Navarro J.F. & Benz W. 1991, ApJ **380**, 320
Navarro J.F. & Benz W. 1991, ApJ **380**, 320
Pfenniger D. 1984, A&A, **134**, 373
Rees M. & Ostriker J.P. 1977, MNRAS **179**, 541
Richter O.G. & Sancisi R. 1994, A&A, **290**, 9
Scalo J.M. 1986, Fund. Cosmic Physics **11**, 1
Scalo J.M. 1989, in WINDOWS ON GALAXIES, ASSL no. 160, eds. Fabbiano G., Gallagher J.S., Renzini A., Kluwer, Dordrecht, 125
Silk J.I. 1993, in THE FEEDBACK OF CHEMICAL EVOLUTION ON THE STELLAR CONTENT OF GALAXIES, eds. Alloin D., Stasinska G., Observatorie de Paris, Meudon, 299
Weinberg M.D. 1992, ApJ **384**, 81
White S.D.M. & Silk J. 1979, ApJ, **231**, 1

ON THE DETERMINATION OF THE IR BACKGROUND FROM EXTRAGALACTIC TeV GAMMA-RAY ABSORPTION

F.W. Stecker
Laboratory for High Energy Astrophysics
NASA Goddard Space Flight Center
Greenbelt, MD 20771

ABSTRACT

Following our previously proposed technique, we have used the recent γ-ray observations of Mkn 421 to place theoretically significant constraints on the magnitude of the intergalactic infrared radiation field (IIRF). Our 2σ upper limit of ~ 10 nw m^{-2} sr^{-1} in the 2 to 10 μm range is consistent with normal IR production by stellar processes in galaxies. They rule out exotic mechanisms proposed to produce a larger IIRF. The data on the spectrum of Mkn 421 also hint at a possible absorption cutoff which could be produced by an IIRF of the magnitude expected from stellar emission and reprocessing in galaxies. The background flux in the 10 to 40 μm range is both estimated and theoretically expected to be in the range between about 2 and 20 nw m^{-2} sr^{-1}. Such a flux may be eventually detectable with the DIRBE experiment on COBE, provided that the foreground radiation from zodiacal light can be modeled well enough to be subtracted out. Using models for the low energy intergalactic photon spectrum from microwave to ultraviolet energies, we have also calculated the opacity of intergalactic space to γ-rays as a function of energy and redshift.

1. INTRODUCTION

The potential importance of the photon-photon pair production process in high energy astrophysics was first pointed out by Nikishov (1961). Unfortunately, his early paper overestimated the energy density of the low energy IIRF by about three orders of magnitude. However, with the discovery of the cosmic microwave background radiation, Gould and Schreder (1966) and Jelley (1966) were quick to point out the opacity of the universe to photons of energy greater than 100 TeV. Stecker (1969) and Fazio and Stecker (1970) generalized these calculations to high redshifts, showing that photons originating at a redshift z will be absorbed above an energy of $\sim 100(1+z)^{-2}$ TeV. Following the detection of the blazar 3C279 at energies of 0.1 to 5 GeV by the EGRET team (Hartman, *et al.* 1992), we (Stecker, De Jager and Salamon 1992, Paper I) proposed that one look for pair-production absorption features in blazar spectra at TeV energies to determine the intensity of the IIRF.

As we discussed in Paper I, an EGRET– strength "grazar" (observed γ-ray blazar) with a hard spectrum extending to multi-TeV energies is potentially detectable with ground-based telescopes. Pair-production interactions with the cosmic microwave background radiation will not cut off the γ-ray spectrum of a nearby (low redshift) source below an energy of ~ 100 TeV (Fazio and Stecker 1970). The techniques involved in making ground-based observations of TeV

© 1996 American Institute of Physics

γ-ray sources have been reviewed by Weekes (1988).

Shortly after Paper I was published, the Whipple Observatory group reported the discovery of $\sim$ TeV γ-rays coming from the nearby blazar Mkn 421, a BL Lac object (Punch *et al.* 1992). This source has a hard, roughly E^{-2} γ-ray spectrum extending to the TeV energy range (Lin, *et al.* 1992; Mohanty, *et al.* 1993) where the pair production process involving infrared photons becomes relevant. Following this discovery, we made use of the TeV observations of Mkn 421 to carry out our suggestion of using such sources to probe the IIRF (Stecker and De Jager 1993; De Jager, Stecker and Salamon 1994). This has also been done by Dwek and Slavin (1994).

There are now over 50 grazars which have been detected by the EGRET team (Thompson, *et al.* 1995). These sources, optically violent variable quasars and BL Lac objects, have been detected with a redshift range extending beyond 2. Of these objects, only the low redshift BL Lac object, Mkn 421 was seen by the Whipple telescope. The fact that the Whipple team did not detect the much brighter EGRET source 3C279 at TeV energies (Vacanti, *et al.* 1990, Kerrick, *et al.* 1993) is consistent with our previous predictions (Paper I; Stecker 1993) because this source is at a much higher redshift (z=0.54). However, the interpretation is somewhat complicated by the significant time-variability of this source at GeV energies and the fact that simultaneous measurements were not made. The very recent observation by the Whipple group of another nearby BL Lac object, Mkn 501 (Quinn, *et al.* 1995), whose redshift of 0.034 is almost identical to that of Mkn 421, supports the idea that low redshift sources are more observable at high energies because there is less absorption by pair production interactions with the IIRF.

We have calculated the opacity of the universe as a function of redshift to γ-rays extending from multi-GeV energies to energies in the PeV region using semiempirical models of the low energy intergalactic radiation fields from the microwave range to the UV range (Stecker and De Jager 1995). We have shown that if the intrinsic spectrum of an extragalactic source extends beyond 1 TeV, then the IIRF should cut off the observed spectrum between $\sim$ 10 GeV and a few TeV, depending on the redshift of the source.

As we suggested in Paper I, our more recent results confirm the idea that using sub-TeV and multi-TeV observations of grazars, one can derive useful information about the IIRF, provided that the intrinsic source spectrum does not cut off in this energy range. This appears to be the case with Mkn 421. There is possible evidence for such a sharp drop in the spectrum of Mkn 421 above 3 TeV where intergalactic absorption is expected to occur. We have used this evidence to estimate the intensity of the IIRF.

We have also pointed out that by combining a measurement of the opacity of the universe to high energy γ-rays with a direct measurement of the IIRF, one can determine the value of the Hubble parameter H_0 at cosmologically significant distances, independent of the distance ladder (Salamon, Stecker and De Jager 1994). Such a determination of H_0 has become even more critical with the recent, relatively "local" measurement of H_0 using Cepheids in the Virgo cluster (Pierce, *et al.* 1994; Freedman, *et al.* 1994). These determinations using observations from the Hubble Space Telescope give values of $H_0 \sim 80$ to 90 km s^{-1} Mpc^{-1}, resulting in a *prima facie* conflict with age determinations of old stars in globular clusters. A determination of a smaller value for H_0 sampled at truly cosmological redshifts may resolve this conflict.

2. THE OPACITY OF THE UNIVERSE TO GAMMA-RAYS

Consider the interaction of a TeV γ-ray of energy $E(z) = (1+z)E$ with a soft photon of energy $\epsilon(z) = (1+z)\epsilon$ at a redshift z, where E and ϵ are the presently observed (z=0) photon energies. Pair production is expected above the threshold energy condition $E\epsilon(1+z)^2 x > 2(mc^2)^2$, where $x = (1-\cos\theta)$ and θ is the angle between the photon directions. The cross section (Heitler 1960) is maximized when the soft photon energy is in the infrared range:

$$\epsilon(E)(1+z) \simeq \frac{2(mc^2)^2}{E(1+z)} \simeq 0.5(\frac{1 \text{ TeV}}{E(1+z)}) \text{ eV} \tag{1}$$

For a ~ 1 TeV γ-ray interacting at a low redshift, this corresponds to a soft photon wavelength near the K-band ($\sim 2\mu$m). Thus, infrared photons with wavelengths around 2μm will contribute most to the absorption of TeV γ-rays. If the photons with energy ϵ have a number density $n(\epsilon)d\epsilon$ cm^{-3}, the optical depth for γ-rays will be

$$\tau(E) = \frac{c}{H_o} \int_0^z dz\ (1+z)(1+\Omega z)^{-1/2} \int_0^2 dx \frac{x}{2} \int_{\epsilon_1}^{\infty} d\epsilon\ n(\epsilon)\sigma\left[2x\epsilon E(1+z)^2\right] \tag{2}$$

where $\epsilon_1 = 2m^2c^4/Ex(1+z)^2$ and σ is the cross section for pair production as a function of the square of center-of-momentum energy (Stecker 1971).

The intergalactic background radiation from galactic starlight and dust reradiation has recently been modeled in some spectral detail (Franceschini, *et al.* 1994). Using their models for stellar and dust emission, together with the COBE results on the cosmic microwave-FIR background radiation and using the UV results of Martin and Bowyer (1989) extrapolated to the Lyman limit, we have modeled the entire range of the low energy photon background spectrum (Stecker and De Jager 1995). It should be kept in mind that this UV estimate is very uncertain and represents an upper limit on the UV flux. This flux may drop off significantly at energies below the Lyman limit.

By making use of this modeled intergalactic spectrum, one can calculate the opacity of the universe at various redshifts over a wide range of γ-ray energies. Fig. 1 shows the optical depth of the universe to γ-rays originating at redshifts corresponding to the grazars designated.

We have used the opacity curve shown in Fig. 1 to fit the Whipple data on Mkn 421 as shown in Fig. 2. The Whipple spectrum is well fitted by a simple E^{-2} source spectrum with absorption included ($\chi^2 = 7.7$ for $8-2=6$ degrees of freedom). Fig. 2 also shows the latest totaled EGRET spectra (Lin, *et al.* 1994). It can be seen that an extrapolation of our Mkn 421 fit on the TeV data will adequately agree with the EGRET results.

Again, using the opacity curves shown in Fig. 1, we have calculated predicted turnovers in other EGRET observed grazars at various redshifts, assuming power-law spectra with the indices fitted by the EGRET team (Fichtel, *et al.* 1994). Our results are shown in Fig. 3. It can be seen that we predict turnovers in such spectra which should be observable with sensitive enough multi-GeV and sub-TeV detectors and used in concert to test our model of intergalactic absorption.

3. ANALYSIS OF THE TeV SPECTRUM OF Mkn 421

Since Mkn 421 is presently the only blazar for which a TeV spectrum exists, it takes on special significance as our one data tool for probing the IIRF. For a low redshift source ($z \ll 1$) such as Mkn 421, the calculation of the optical depth, τ, simplifies considerably. Taking the soft photon spectrum as $n(\epsilon) = \kappa\epsilon^{-\alpha}$, for small redshifts , z, (with $\Omega = 1$)

$$\tau(E) = \frac{2^{\alpha}cz}{(\alpha+1)H_o}(\frac{2(mc^2)^2}{E})n[\frac{2(mc^2)^2}{E}]\int_1^{\infty} u^{-\alpha}\, du\ \sigma[\beta = \sqrt{1-1/u}] \tag{3}$$

Even though the EGRET and Whipple detections were not simultaneous, they can be cautiously combined. Although the intensity of Mkn 421 may occasionally vary by large amounts on a timescale of days (Kerrick, *et al.* 1995), it was shown by the Whipple group that averaging on a timescale of several weeks results in a 20% variation at most. Such a timescale is the one relevant to the above results.

Using the spectral data for Mkn 421, the Whipple group (Mohanty *et al.* 1993) were able to fit a single power law between their data set below 5.1 TeV and the EGRET data set at $\sim$ GeV energies (Lin, *et al.* 1992). The spectrum that they obtained was

$$dN/dE = (1.02 \pm 0.14) \times 10^{-11} E^{-2.06\pm0.04}\ \ \mathrm{cm^{-2}s^{-1}TeV^{-1}} \tag{4}$$

We (Stecker and De Jager 1993) used the Whipple data on the TeV energy spectrum from Mkr421 to put limits on τ, and therefore to put limits on the infrared radiation density. We generalized a power-law fit (*e.g.*, eq. (4)) to include absorption by assuming spectra of the form

$$dN/dE = KE^{-\Gamma}\exp\left[-\tau(E)\right] \tag{5}$$

with K, Γ and κ as free parameters. Then the best χ^2 fit is for

$$dN/dE = (1.11 \pm 0.21) \times 10^{-11} E_{\mathrm{TeV}}^{-2.04\pm0.04} \exp\left(-\tau(E)\right) \tag{6}$$

For normal galaxies (Stecker *et al.* 1977) with $n(\epsilon) = \kappa_1 \epsilon^{-2.55}$, in units of $\mathrm{eV}^{-1}\mathrm{cm}^{-3}$ with ϵ in eV, we find

$$\tau(E) = 143\kappa_1 E_{\mathrm{TeV}}^{1.55} \tag{7}$$

and, for ϵu_ϵ as constant (or $n(\epsilon) = \kappa_2 \epsilon^{-2}$), we find that

$$\tau(E) = 120\kappa_2 E_{\mathrm{TeV}}. \tag{8}$$

Expressions (7) and (8) for $\tau(E)$ should roughly hold up to $E \sim 10$ TeV if the simple power-law form for $n(\epsilon)$ taken by Stecker *et al.*(1977) holds.

Using $\tau(E)$ of the forms given by eqs.(7) and (8) and assuming $H_o = 75$ km/s/Mpc) $\tau(E) = 0.048 E_{\mathrm{TeV}}$ (for $\alpha = 2$) and $\tau(E) = 0.043 E_{\mathrm{TeV}}^{1.55}$ (for $\alpha = 2.55$) (again valid only up to $E \sim 10$ TeV) with a minimum $\chi_8^2 = 4.9$ for $11 - 3 = 8$ degrees of freedom for both models for $n(\epsilon)$. The upper limits on τ are obtained by allowing all free parameters to vary simultaneously, which allows us to construct confidence ellipsoids. The 1σ and 2σ upper limits on κ_1 in Eq. (6) (i.e. for $\alpha = 2.55$) are $\kappa_1 < 9 \times 10^{-4}$ and $\kappa_1 < 1.9 \times 10^{-3}$ respectively, whereas the corresponding numbers for κ_2 in Eq. (7) (i.e. for $\alpha = 2$) are $\kappa_2 < 2.1 \times 10^{-3}$ and $\kappa_2 < 4.0 \times 10^{-3}$ respectively. The 2σ or 95% upper limits on $\tau(E)$ are then $\tau(E) < 0.27 E_{\mathrm{TeV}}^{1.55}$, which implies $n(\epsilon) < 1.9 \times 10^{-3} \epsilon^{-2.55}$ $\mathrm{eV}^{-1}\mathrm{cm}^{-3}$ for $\alpha = 2.55$. For $\alpha = 2$ we find $\tau(E) < 0.48 E_{\mathrm{TeV}}$, which implies $n(\epsilon) < 4.0 \times 10^{-3} \epsilon^{-2}$ $\mathrm{eV}^{-1}\mathrm{cm}^{-3}$. This photon density corresponds to an upper limit on the flux of $\nu I_\nu < 10$ nw $\mathrm{m}^{-2}\mathrm{sr}^{-1}$.

In the above analysis, the spectral information *above* 3.4 TeV was neglected since the highest energy points were only upper limits. However, we note that using the data from Mohanty *et al.* (1993) and extrapolating the $E^{-2.06}$ differential spectrum to energies above 3 TeV, we expect ~ 58 ON-source events (source γ-rays and cosmic-ray background) between 3.4 TeV and 5.1 TeV, whereas only 26 *on*-source events were observed, which is consistent with the 27 *off*-source cosmic ray background events collected with similar exposure. Furthermore, above 5.1 TeV we expect ~ 32 *on*-source events, whereas only 10 were observed against the 12 background events. These data may be interpreted as indicating the presence of an absorption feature in the *integral* spectrum of Mkn 421 manifested at the highest energies. Such an interpretation allows us to obtain an estimate of the IIRF, rather than an upper limit (De Jager, Stecker and Salamon 1994). This empirical estimate is consistent with theoretical estimates of the IIRF (Stecker, *et al.* 1977).

There are three main caveats to consider with regard to the possible cutoff in the high energy end of the observed spectrum of Mkn 421 being considered as an absorption effect of the IIRF: (1) The source spectrum may have an *intrinsic emission* cutoff above ~ 3 TeV. However, one may then ask why the intrinsic spectrum would turn over at the energy naturally expected for an IIRF absorption feature. We also note that emission models involving Compton scattering of high energy electrons in the source would not produce such an apparently sharp cutoff (Ghisellini, Maraschi and Treves 1985; Dermer and Schlickheiser 1993; Bloom and Marscher 1993). (2) *Intrinsic absorption* in the source may produce a high–energy cutoff. However, we consider it highly improbable that the infrared characteristics of the source would conspire to exactly mimic that which would be produced by a reasonably estimated IIRF. Also, if the observed IR flux from Mkn 421 (Makino, *et al.* 1987) is from the galaxy, the implied optical depth would not be large enough to produce an observable absorption effect, while if it originates in the jet and is beamed, the intrinsic flux may be considerably reduced as well as the interaction rate. We note, however, that if the absorption *is* intrinsic, then our estimate becomes the best upper limit extant on the IIRF. (3) Perhaps, *most importantly*, there is a systematic uncertainty in the event rate derived by Mohanty, *et al.* (1993) in the energy range above 5 TeV, which is as yet poorly determined. A possible *oversubtraction* of assumed cosmic ray produced events, which can become more confused with γ-ray produced events at the higher energies, might have artificially supressed the highest energy event rate reported by Mohanty, *et al.* (1993) (R. Lamb, priv. comm.). We note, however, that the Crab nebula, a much stronger source, was observed by the Whipple group to have a power–law spectrum with *no* such turnover up to $\sim$ 10 TeV (Vacanti, *et al.* 1991).

We also note that intergalactic electron-photon cascading (Protheroe and Stanev 1993; Aharonian, Coppi and Volk 1993) is not expected to significantly influence the TeV spectrum of Mkn 421, since the source is observed within a point-spread function with HWHM of $\sim$ 0.2 degrees, and it can be shown that an intergalactic magnetic field of 10^{-20} G or greater would deflect the cascade TeV electrons away from the source direction, creating a γ-ray halo much larger than the point-spread function. Since the measured magnetic field strengths in superclusters are of the order of a μG (Kronberg 1994 and references therein), it is highly unlikely that the intergalactic field would be as weak as 10^{-20} G.

Thus, our interpretation of the Mkn 421 spectrum (De Jager, *et al.* 1994) may indeed be the first determination of the IIRF at a level which is expected from stellar processes in galaxies (Stecker, *et al.* 1977).

4. DETERMINATIONS OF THE INTERGALACTIC IR BACKGROUND

A theoretical upper limit for the infrared energy density from starlight in normal galaxies was estimated by Stecker, *et al.* (1977), assuming that all of the energy from nucleosynthesis in stars is emitted in the infrared. Estimates of the intergalactic and optical radiation densities were obtained by Yoshii and Takahara (1988), Takahara (1990), and Tyson (1990), based on galaxy counts and other modelling considerations. Cowie, *et al.* (1990) have estimated the IIRF using a K-band deep galaxy survey.

Direct measurements of the extragalactic infrared and optical radiation are plagued by the difficulty of separating the true extragalactic component from zodiacal light and galactic radiation. The observational upper limits obtained in the near infrared by Matsumoto, *et al.* (1988) and the optical (Dube, *et al.* 1979) and the residuals obtained from the ongoing analysis of the COBE-DIRBE data (Hauser 1995) all lie above the theoretical estimates discussed above.

The TeV data from Mkn 421 have provided both estimates and upper limits which are already of theoretical significance. Our conservative upper limit on the background in the 2 to 10 μm range of about 10 nw m^{-2}sr^{-1}, obtained from the Whipple data on Mkn 421 below 3 TeV (Stecker and De Jager 1993), rules out various exotic mechanisms for producing larger fluxes, such as some exploding star, decaying particle, massive object and black hole models (Carr 1988). They are consistent with the extragalactic near infrared background originating from ordinary stellar processes in galaxies (Stecker, *et al.* 1977; Franceschini *et al.* 1994). Our estimates (De Jager, Stecker and Salamon 1994) indicate that at least 30% of the energy emitted by stars and dust in galaxies ends up in the infrared. Fig. 4 shows our previously obtained upper limit (Stecker and De Jager 1993), indicated by the horizontal line labled SD93, and our IIRF estimate (De Jager, Stecker and Salamon 1994), indicated by the dashed-line parallelogram labled DSS94. It can be seen that our estimate from the Mkn 421 data lies somewhat above the model of Franceschini, *et al.* (1994). This can be from either of three causes: (1) uncertainty in the interpretation of the Mkn 421 event rate (R. Lamb, private communication), (2) some intrinsic absorption in Mkn 421 (see previous discussion), or an underestimate in the modeled IIRF owing to the omission of contributions from infrared bright extragalactic sources (De Zotti, *et al.* 1994). The upper limits given by the COBE-DIRBE residuals (Hauser 1994) are also shown in Fig. 4.

The background flux in the 10 to 40 μm range is both estimated here and theoretically expected to be in the range of 2 to 20 nw m^{-2} sr^{-1}. Such a flux may be eventually detectable with the DIRBE experiment on COBE, provided that the foreground radiation from zodiacal light can be modeled well enough to be subtracted out (Mather, 1982; Hauser 1990; Mather, *et al.* 1990). Indeed, the minimum mean residual sky brightnesses reported recently by the COBE-DIRBE team are in the range of $\sim$ 10 to 15 nw m^{-2} sr^{-1} in the 2 to 5 μm range and $\sim$ 10 to 30 nw m^{-2} sr^{-1} between 60 and 240 μm, with significant foreground emission still remaining (Hauser 1994). These residuals are comparable with our upper limits in the 2 to 10 μm range and not too far above the cosmic background estimates which we have obtained in the 10 to 40 μm range. Further analysis of the COBE-DIRBE data may thus result in a direct measurement of the IIRF. In principle, such a measurement may be used in combination with high energy γ-ray absorption measurements to determine the value of the Hubble parameter at cosmologically significant redshifts (Salamon, Stecker and De Jager 1994).

ACKNOWLEDGMENTS

As referenced in the text, this paper is based on work undertaken jointly with my colleagues O.C. De Jager and M.H. Salamon.

REFERENCES

Aharonian, F.A., Coppi, P. & Volk, H.J. 1993, Proc. 23rd Int. Cosmic Ray Conf. (Calgary: Univ. of Calgary Press), 1, 451

Bloom, S.D. & Marscher, A. 1993, in Compton Gamma Ray Observatory, ed.

M. Friedlander, N. Gehrels, and D.J. Macomb (New York: AIP), 578
Carr, B.J., 1988, in Comets to Cosmology, ed. A. Lawrence (Berlin: Springer-Verlag), 265
Cowie, L.L., Gardner, J.P., Lilly, S.J. and McLean, I. 1990, Ap. J. Letters 360, L1
De Jager, O.C., Stecker, F.W. & Salamon, M.H. 1994, Nature, 369, 294
Dermer, C.D. & Schlickheiser, R. 1993, Ap. J. 416, 458
De Zotti, G. *et al.* 1994, preprint.
Dube R.R., Wickes, W.C. & Wilkinson, D.T. 1979, Ap. J. 232, 333
Dwek, E. & Slavin, J. 1994, Ap. J. , 436, 696
Fazio, G.G. & Stecker, F.W. 1970, Nature 226, 135
Franceschini, *et al.* 1994, Ap. J. 427, 140
Freedman, W. *et al.* 1994, Nature 371, 757
Gould, R. J. & Schreder, G.P. 1966, Phys. Rev. Letters 16, 252
Ghisellini, G.& Maraschi, L. 1989, Ap. J. 340, 181
Hartman, R.C. *et al.* 1992, Ap. J. Letters 385, L1
Hauser, M.G. 1995 in Examining the Big Bang and Diffuse Background Radiations, Proc. I.A.U. Symp. No. 168, in press
Hauser, M.G. *et al.* 1990, in After the First Three Minutes, AIP, New York, 222, 161
Heitler, W. 1960, The Quantum Theory of Radiation (London: Oxford Press)
Hurley, K. *et al.* 1994, Nature 372, 652
Jelley, J.V. 1966, Phys. Rev. Letters 16, 479
Kerrick, A.D. *et al.* 1993, in Proc. 23rd Int. Cosmic Ray Conf. (Calgary: Univ. of Calgary Press), 1, 405
Kerrick, A.D. *et al.* 1995, Ap. J. Letters 438, L63
Kronberg, P.P. 1994, Rep. Prog. Phys. 57, 325
Lin, Y.C. *et al.* 1992, Ap. J. Letters 401, L61
Lin, Y.C. *et al.* 1994, in The 2nd Compton Symposium, ed. C.E. Fichtel, N. Gehrels, and J.P. Norris (New York: AIP), 582
Makino, F. *et al.* 1987, Ap. J. 313, 662
Martin, C, & Bowyer, S. 1989, Ap. J. 338, 677
Mather, J.C. 1982, Optical Engineering 21, 769
Mather, J.C. *et al.* 1990, Ap. J. Letters 354, L37
Mohanty, G. *et al.* 1993, Proc. 23rd Int. Cosmic Ray Conf. (Calgary: Univ. of Calgary Press), 1, 440
Matsumoto, T., Akiba, M. & Murakami, H. 1988, Ap. J. 332, 575
Mattox, J.R., *et al.* 1993, Ap. J. **410**, 609
Nikishov, A.I. 1962, Sov. Phys. JETP, 14, 393
Pierce, M.J. *et al.* 1994, Nature 371, 385
Protheroe, R.J. & Stanev, T. 1993, Mon. Not. R. Astron. Soc. 264, 191
Quinn, J. *et al.*, 1995, I.A.U. Circ. No. 6178
Salamon, M.H., Stecker, F.W. & De Jager, O.C. 1994, Ap. J. Letters 423, L1
Stecker, F.W. 1969, Ap. J. 157, 507
Stecker, F.W. 1971, Cosmic Gamma Rays (Baltimore: Mono Book Co)
Stecker, F.W. 1993, in Particle Astrophysics, Proc. Rencontres de Blois, ed. J. Tran Thanh Van (Gif-sur-Yvette: Editions Frontiers), 111
Stecker, F.W. & De Jager, O.C. 1993, Ap. J. Letters 415, L71
Stecker, F.W. & De. Jager, O.C. 1995, Space Sci. Rev., in press
Stecker, F.W., De Jager, O.C. & Salamon, M.H. 1992, Ap. J. Letters 390, L49 (Paper I)
Stecker, F.W., Puget, J.L. & Fazio, G.G. 1977, Ap. J. Letters 214, L51

Takahara, F. 1990, Prog. Theor. Phys. 83, 1071
Thompson, D.J. *et al.*1995, Ap. J. Suppl., in press
Tyson, J.A. 1990, in The Galactic and Extragalactic Background Radiation, ed. S. Bowyer and C. Leinert (Dordrecht: I.A.U.), 245
Vacanti, G. *et al.* 1990, in Proc. 21st Int. Cosmic Ray Conf. (Adelaide: Univ. of Adelaide), 2, 329
Vacanti, G. *et al.* 1991, Ap. J. 337, 467
Weekes, T.C. 1988, Phys. Rep. 160, 1
Yoshii, Y. & Takahara, F. 1988, Ap. J. 326, 1

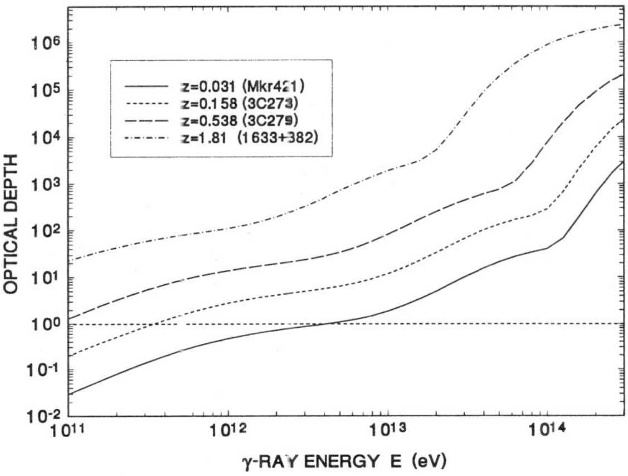

Figure 1: Optical depth versus energy for photons from four EGRET grazars obtained using the our modeled photon spectrum (see text) and taking $H_o = 50$ km s^{-1}Mpc^{-1} (Stecker and De Jager 1995).

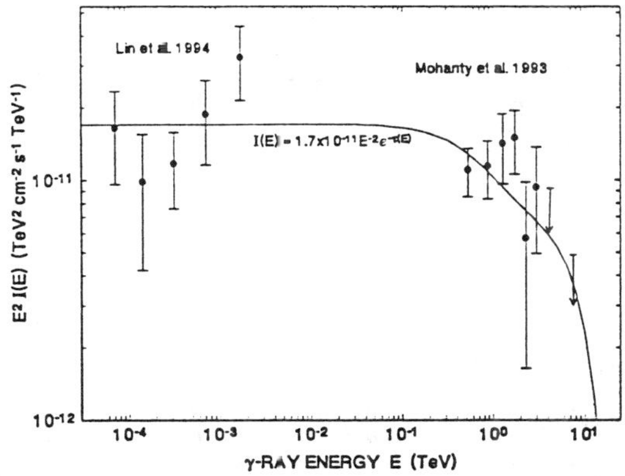

Figure 2: Mkn 421 spectrum fitted to the Whipple data (Mohanty, *et al.* 1993) with absorption calculated using the model spectrum and taking $H_o = 50$ km s^{-1}Mpc^{-1} (Stecker and De Jager 1995). The 2σ differential upper limits at 3 and 5 TeV derived by using the expected absorption spectrum. The EGRET data (Lin, *et al.* 1994) are also shown.

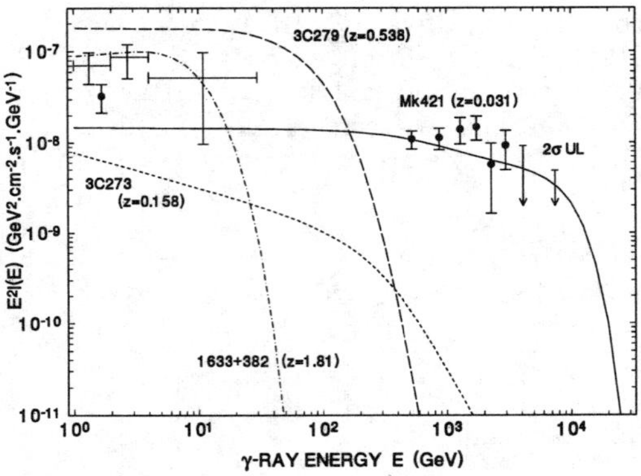

Figure 3: Power-law spectra extapolated from EGRET data for four observed grazars (Stecker and De Jager 1995). The absorption calculations assumed $H_o = 50$ km s^{-1}Mpc^{-1}. The solid circles are data points for Mkn 421. The plain crosses are data points for the high redshift grazar 1633+382 (Mattox, *et al.* 1993).

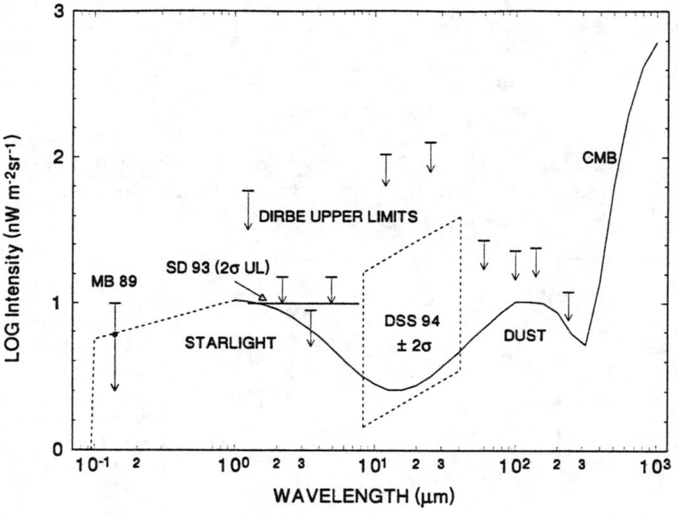

Figure 4: The intergalactic photon spectrum modeled as discussed in the text (Stecker and De Jager 1995). The results of Martin and Bowyer (1989) are indicated as 'MB 89'. We also show our previous upper limit (Stecker and De Jager 1993, 'SD 93') and our estimate of the IIRF (De Jager, Stecker and Salamon 1994, 'DSS 94'), obtained from the TeV data on Mkn 421. The DIRBE upper limits (Hauser 1994) are indicated by the arrows.

DUST IN DAMPED LYMAN-ALPHA SYSTEMS AND THE CIBR

S. M. Fall
Space Telescope Science Institute
3700 San Martin Drive, Baltimore, MD 21218

e-mail id
fall@stsci.edu

ABSTRACT

Dust has been detected in damped Lyα systems by two independent methods: the reddening of background quasars and the gas-phase depletion of Cr relative to Zn. Both methods give a typical dust-to-gas ratio at $z \approx 2.2$ roughly 10% of that in the present-day solar neighborhood. In this talk, I will review some of the consequences of these observations. First, dust causes obscuration and hence incompleteness in optically selected samples of quasars. The "missing" fraction at $z = 3$ is estimated to be 10-70%. Corrections for this effect help to reconcile the ionizing UV background inferred from the proximity effect at high redshifts with that expected from quasars alone. Second, since the samples of damped Lyα systems were obtained primarily from optically selected samples of quasars, they too must be incomplete at some level. This effect biases the statistics of the damped Lyα systems, such as the distribution of HI column densities $f(N, z)$, and especially the comoving density of gas $\Omega_g(z)$. The bias has important implications concerning the histories of star formation and metal production in the damped Lyα systems. In particular, it resolves an apparent paradox that has arisen in attempts to understand the chemical evolution of the damped Lyα systems (the so-called "G-dwarf problem"). Third, the dust is likely to absorb most of the Lyman-alpha photons produced in the damped Lyα systems, which may account for the null results of most searches for Lyα emission from them.

© 1996 American Institute of Physics

The Evolution of the Far Infrared Galaxy Population and the Predicted Far Infrared Background

Michael Rowan-Robinson and Chris Pearson

Astrophysics Group, Blackett Laboratory, Imperial College, Prince Consort Rd, London SW7 2BZ

Abstract. Direct estimates of the rate of evolution of IRAS galaxies from redshift surveys are reviewed. There is now good evidence for strong evolution of the starburst galaxy population and this is supported by the interpretation of the 60 μm source-counts. The sub-mJy radio source population at 1.4 GHz has also been found to be essentially identical to the IRAS starburst galaxy population and faint radio source-counts support luminosity evolution models.

A new model is presented for source-counts and integrated background radiation from radio to optical wavelengths. For normal spirals, only 30% of their energy, on average, is emitted in the far infrared, so they must be optically thin. For starburst galaxies we estimate that 95% of their optical and uv radiation is absorbed by dust and reemitted at far infrared wavelengths. Our predicted far infrared integrated background intensity lies well below the DIRBE upper limits, but interestingly close to the FIRAS limit at 500 μm. It also lies in the range estimated from absorption of Tev γ-rays in AGN.

1. DIRECT ESTIMATES OF THE RATE OF EVOLUTION OF IRAS GALAXIES

The evolution of the 60 μm luminosity function can be characterized by:

$$\eta(L,z) = \phi(z)\ \eta_0(L/L_*(z))$$

where, for density evolution we take

$$\varphi(z) = (1+z)^P\ ,\ z < z_f$$
$$= 0\ ,\ z > z_f\ ,$$

and for luminosity evolution we take

$$L_*(z) = (1+z)^Q\ ,\ z < z_*\ ,$$
$$= (1+z_*)^Q\ ,\ z_* < z < z_f\ ,$$
$$= 0\ ,\ z > z_f\ .$$

The most direct evidence for evolution comes from large redshift surveys. Saunders et al (1990) found that for the QDOT 1-in-6 sparsely sampled IRAS galaxy redshift survey to S(60) = 0.6 Jy:

© 1996 American Institute of Physics

$$Q = 3.1 \pm 1.0$$
$$\text{or} \quad P = 6.7 \pm 2.3.$$

After correction for redshift errors, for the effects of Malmquist bias, and for non-linearity in the IRAS PSC flux-scale, Oliver et al (1995) found that this should be revised to:

$$P = 4.2 \pm 2.3.$$

From the shallower but larger 1.2 Jy survey, Fisher et al (1992) found no clear evidence for evolution:

$$P = 2 \pm 3$$

but the uncertainty is such that the QDOT values are not inconsistent with this.

The much deeper IRAS FSS redshift survey of Oliver et al (1995), which consists of 1400 galaxies in 700 sq deg with S(60) ≥ 0.2 Jy, after correction for Malmquist bias, and with a K-correction based on a 2-component (starburst+cirrus) fit to the 100/60 μm colours, yields strong evidence for evolution:

	P	Q
V/V_{max} method	5.5 ± 1.8	3.7 ± 0.9
binned V/V_{max} method	4.9 ± 1.4	2.6 ± 0.8
max. likelihood	6.3 ± 1.5	3.5 ± 0.7
average	5.6 ± 1.6	3.3 ± 0.8

2. EVIDENCE FOR EVOLUTION FROM 60 μm COUNTS

Several groups have found that the 60 μm source-counts can only be understood if the IRAS galaxy population is subject to strong evolution (Hacking and Houck 1987, Hacking et al 1987, 1989, Danese et al 1987, Lonsdale and Hacking 1989, Lonsdale et al 1990, Hacking and Soifer 1991, Oliver et al 1992). In Fig 1 of Oliver et al (1992), for example, observations from the IRAS PSC, the IRAS FSS and the pointed observations of Hacking and Houck (1987), are compared with predictions for the case of no evolution, and for strong density and luminosity evolution. The no evolution case is clearly inconsistent with the observations, while either density or luminosity evolution give consistent results. Counts with ISO will allow the latter two models to be distinguished.

3. SUB-mJy RADIO-SOURCES AS STARBURST GALAXIES

The bright radio source-counts (S(1.4 GHz) >> 1 mJy) are due to radio-galaxies and quasars undergoing strong evolution, which is approximately of the form of luminosity evolution (eg Condon 1984, Dunlop and Peacock 1990), though some models involve a small amount of density evolution also. Below 1 mJy at 1.4 GHz the slope of the counts steepens again and there is evidence of a new population of blue radio-emitting galaxies (Mitchell and Condon 1985, Windhorst et al 1987, Thuan and Condon 1987, Franceschini et al 1988, Condon 1989, Lonsdale and Harmon 1991).

Benn et al (1993) have carried out spectroscopy of a sample of 112 identifications of sources with S(1.4 GHz) $\geq$ 0.1 mJy and shown that below 1 mJy most of the sources are starburst galaxies very similar to those seen by IRAS. The luminosity function for these galaxies agrees well with that at 60 μm, shifted by the radio-fir relation

$$S(60\ \mu m) \ = \ 90\ S(1.4\ GHz).$$

Fits to the sub-mJy source-counts then show that strong evolution is required in this population, with luminosity evolution strongly favoured over density evolution (Rowan-Robinson et al 1993a).

4. A NEW MODEL FOR COUNTS AND BACKGROUND RADIATION FROM RADIO TO X-RAYS

There have been a number of attempts to model the source-counts and background radiation at wide range of wavelengths (eg Franceschini et al 1991, Blain and Longair 1993). We describe here a new study by Pearson and Rowan-Robinson (1995) which gives predictions of source-counts at radio, submillimetre, far infrared, near infrared and optical wavelengths, and is being extended to X-rays. The galaxy populations included are: (a) normal spirals + cirrus, characterized by cool 100/60 μm coloours, (b) starburst and ultraluminous ir galaxies, characterized by warm 100/60 μm colours, (c) the new category of hyperluminous ir galaxies, with $L_{60} > 10^{13}$ L_o ($\Omega = 1$, $H_o = 50$), (d) Seyferts and quasars + dust tori, and (e) elliptical galaxies. Clearly the separation of poulations (a) and (b) is an oversimplification since there is some star formation proceeding in normal spirals.

For classes (b), (c), (d) we assume strong power-law luminosity evolution with $Q = 3.1$, $z* = 2$, $z_f = 5$, which gives a good fit to the optical data for QSOs (Boyle et al 1988), fits the bright radio source-counts for radio-galaxies and quasars (Dunlop and Peacock 1990), is consistent with the direct evidence for evolution in IRAS galaxies (section 1 above) and fits the the sub-mJy source-counts (Rowan-Robinson et al 1993a). The physical motivation for such a common evolution would be that interactions simultaneously drive star formation and feed gas into black holes in active galactic nuclei. Thus the evolution represents the evolution of the gas supply from interactions with time.

The spectral energy distributions we have assumed for the different populations are shown in Table 1:

TABLE1:

population	fir	opt/uv
(a) normal spiral	cirrus model (RR92)	Sbc (Coleman et al 1980, Yoshii et al 1988), 30% of light assumed absorbed by dust and reradiated in ir
(b) starburst	sb model (RRE93)	Sab, $\nu < \nu_B$ (Coleman et al 1980) HII, $\nu > \nu_B$ (Mk 36, Neugebauer et al 1976) 95% of light absorbed by dust and reradiated in ir
(c) hyperluminous	F10214+4724 continuum (RR et al 1993b)	
(d) Seyferts/quasars	torus model (RR95)	average QSO continuum (RR95)

With these assumed spectral energy distributions and rates of evolution, we then need to define the luminosity functions for each population at only one wavelength. For normal spirals and ellipticals we use the B-band Schecter functions of Efstathiou et al (1988). The assumption that 30% of the optical-uv light is absorbed by interstellar dust and reemitted in the far infrared then yields a good fit to the 60 μm luminosity function for galaxies with 'cool' far infrared colours (Saunders et al 1989). For the starburst and hyperluminous galaxies we use the 60 μm luminosity function for 'warm' galaxies (Saunders et al 1990). Finally for Seyferts and quasars we use the 12 μm luminosity function data of Rush et al (1993), refitted with a 2 power-law function (Lawrence et al 1986).

The resulting fits to the 1.4 GHz, 60 μm, K-band and B-band counts are extremely good (Figs 1a-d). The inclusion of a strongly evolving population of starburst galaxies accounts for the excess B-band counts down to B= 22 m., but not the fainter excess of blue galaxies. The increasing proportion of galaxies with emission lines in fainter optical galaxy redshift surveys finds a natural explanation in this picture. This analysis shows that the proportion of the optical-uv light from a typical starburst that is absorbed by dust and reemitted in the far ir can not be much less than 95%, otherwise the optical counts would be violated. On the other hand the figure can not be 100%, otherwise no emission lines or continuum colour changes would be seen in starburst galaxies. Inclusion of stellar evolution effects in the treatment of the B-band SED and evolution of normal galaxies will have only a small effect on these models.

We can now predict the far infrared background from these models: this is shown in Fig 2, with contribution from each population indicated. Fig 3 shows the predicted background compared with the upper limits derived from COBE. The predicted intensity of the background is well below the latest DIRBE limits (Hauser 1995), but is interestingly close to the FIRAS limit at 500 μm. It is also within the range calculated from absorption of Tev γ-rays (De Jager et al 1994).

5. DISCUSSION AND FUTURE WORK

The evidence that the starburst galaxy population is undergoing strong evolution seems very strong, both from IRAS galaxy redshift surveys and from the interpretation of the faint 1.4 GHz and 60 μm source-counts. The rate of evolution appears to be similar to that seen in optically selected QSOs and in radio-loud quasars and radio-galaxies. A possible mechanism to account for this similarity is that both processes are driven by galaxy interactions and mergers.

We have presented a model for source-counts and integrated background radiation from radio to optical wavelengths and have been able to link the far infrared luminosity functions and evolution to the behaviour of the corresponding populations at optical wavelengths. For normal spiral galaxies only about 30% of the total optical and uv light is absorbed and reemitted at far infrared wavelengths. These galaxies are therefore, on average, optically thin. This result has been clear since the work of Rowan-Robinson et al (1987) and Rowan-Robinson and Crawford (1989) (see also Rowan-Robinson 1992). It is only by ignoring the far infrared evidence that claims that normal spirals are optically thick (eg Disney et al 1989, Valentijn 1990, Burstein et al 1991) can be sustained. Obviously, though, some parts of normal spirals (eg giant molecular clouds, the nucleus) are highly optically thick.

We are also able for the first time to give a quantative estimate for the average proportion of optical-uv light in a starburst which is absorbed by dust and reemitted in the far infrared. We find a figure of 95%, which is consistemt with the very high 60 μm to Hα flux ratios found for starburst galaxies (Leech et al 1989). A figure much lower than this would lead to violation of the optical source-counts for all galaxies. At this figure, the increasing proportion of emission line galaxies seen in optical galaxy redshift surveys down to B = 23 m. finds a natural explanation.

Our treatment does not explicitly include star formation in elliptical galaxies, which is at a low rate at present epochs but will have been much more intense at earlier times. However the strong evolution of the starburst population back to z = 5 assumed in our evolution model must probably be interpreted as including the contribution of star formation in ellipticals.

Our predicted background intensities are similar to those obtained in earlier calculations (Beichman and Helou 1991, Franceschini et al 1991, Oliver et al 1992, Blain and Longair 1993). However direct estimates of the rate of evolution and improved models of the spectral energy distributions of the different populations should significantly reduce the uncertainties in our estimate.

Future work:
(1) We will shortly be able to test the evolution rate in the PSCz IRAS galaxy redshift survey, a total of 15000 galaxies with S(60 μm) ≥ 0.6 Jy (Saunders et al 1995a).
(2) We plan to greatly increase the sample size for our sub-mJy surveys using the Australia Telescope and the 2dF facility at the AAT (Mobasher, Cram and Rowan-Robinson 1995).

(3) We have been awarded 215 hours of ISO Open Time for deep surveys of about 20 sq deg of sky at 90 and 15 μm (Rowan-Robinson et al 1995a).
(4) We plan to carry out a survey at 850, 450 and 350 μm with the SCUBA instrument at the JCMT (Rowan-Robinson et al 1995b).

REFERENCES

Beichman C.A., and Helou G., 1991, ApJ 370, L1
Benn C. R., Rowan-Robinson M., McMahon R.G., Broadhurst T.J., Lawrence A., 1993, MNRAS 263, 98
Blain A.W., and Longair M.S., 1993, MNRAS 264, 509
Boyle B.J., Shanks T., and Peterson B.A., 1988, MNRAS 235, 935
Burstein D., Haynes M., Faber S.M., 1991, Nat 353, 515
Coleman G.D., Wu C., Weedman D.W., 1980, ApJS 43, 393
Condon J.J., 1984, ApJ 284, 44
Condon J.J., 1989, ApJ 338, 13
Danese L., deZotti G., Franceshini A., Toffolatti L., 1987, ApJ 318, L15
De Jager O.C., Stecker F.W., Salamon M.H., 1994, Nat 369, 294
Disney M., Davies J., Phillips S., 1989, MNRAS 239, 939
Dunlop J.S., and Peacock J.A., 1990, MNRAS 247, 19
Efstathiou G., EllisR.S., Peterson B.A., 1988, MNRAS 232, 431
Fisher K.B., Strauss M.A., Davis M., Yahil A., Huchra J.P., 1992, ApJ 389, 188
Franceschini A., Toffolatti L., Mazzei P., Danese L., de Zotti G., 1991, AAS 89, 295
Franceschini A., Danese L., de Zotti G., Xu C., 1988, MNRAS 233, 175
Hacking P.B., and Houck J.R., 1987, ApJS 63, 311
Hacking P.B., Condon J.J., Houck J.R., 1987, ApJ 316, L15
Hacking P.B., Condon J.J., Houck J.R., Beichman C.A., 1989, ApJ 339, 12
Hacking P.B. and Soifer B.T., 1991, ApJ 367, L49
Hauser M.G., 1995, in IAU Symposium 168 *Examining the Big Bang and Diffuse Background Radiations*
Lawrence A., Walker D., Rowan-Robinson M., Leech K.J., Penston M.V., 1986, MNRAS 219, 687
Lonsdale C.J. and Hacking P.B., 1989, ApJ 338, 712
Lonsdale C.J.,and Harmon R.T., 1991, Adv.Space Res. 11, 255
Lonsdale C.J., Hacking P.B., Conrow T.P., Rowan-Robinson M., 1990, ApJ 358, 60
Mitchell K.J., and Condon J.J., 1985, AJ 90, 1957
Neugebauer G., Becklin E.E., Oke J.B., Searle L., 1976, ApJ 205, 29
Oliver S.J., Rowan-Robinson M., Saunders W., 1992, MNRAS 256, 15p
Oliver S. et al 1995, in *Wide-Field Spectroscopy and the Distant Universe* eds S.J.Maddox, A.Aragon-Salamanca, World Scientific (Singapore)
Pearson C. and Rowan-Robinson M., 1995, MNRAS (in press)
Rowan-Robinson M., Helou G., Walker D., 1987, MNRAS 227, 589
Rowan-Robinson M. and Crawford J., 1989, MNRAS 238, 523
Rowan-Robinson M. 1992, MNRAS 258, 787
Rowan-Robinson M. Benn C R., Lawrence A., McMahon R.G., Broadhurst T.J., 1993a, MNRAS 263, 123
Rowan-Robinson M. et al 1993b, MNRAS 261, 513
Rowan-Robinson M., 1995, MNRAS 272, 737

Rowan-Robinson M., and Efstathiou A., 1993, MNRAS 263, 675
Rowan-Robinson M., Cesarsky C., Danese L., Franceshini L., Genzel R., Lawrence A., Lemke D., McMahon R., Miley G., Oliver S., Puget J.-L., Rocca-Volmerange B., 1995a
Rowan-Robinson M., Pearson C., Mobasher B., Griffin M., Gear W., Lawrence A., Dunlop J., Longair M.S., Blain A.W., 1995b
Rush B., Malkan M., Spinoglio L., 1993, ApJS 89, 1
Saunders W.S., Rowan-Robinson M., Lawrence A., Efstathiou G., Kaiser N., Ellis R.S., Frenk C.S., 1990, MNRAS 248, 318
Saunders et al 1995, in *Wide-Field Spectroscopy and the Distant Universe* eds S.J.Maddox, A.Aragon-Salamanca, World Scientific (Singapore)
Thuan T.X., and Condon J.J., 1987, ApJ 322, L9
Valentijn E., 1990, Nat 346, 153
Windhorst R.A., Dressler A., Koo D.C., 1987, in *Observational Cosmology*, eds A.Hewit, G.Burbidge, L.Z.Fang, Reidel (Dordrecht), p.573
Yoshii Y., Takahara F., 1988, ApJ 326, 1

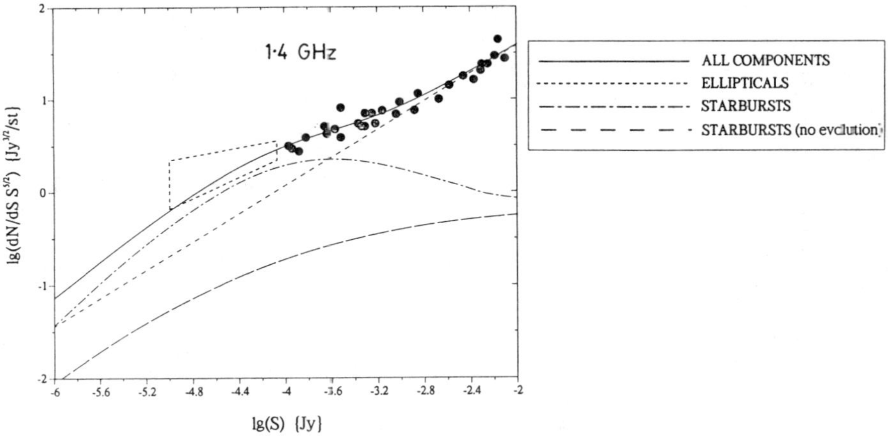

FIGURE 1a: Normalised differential source counts at 1.4 GHz, with predicted contributions of different populations.

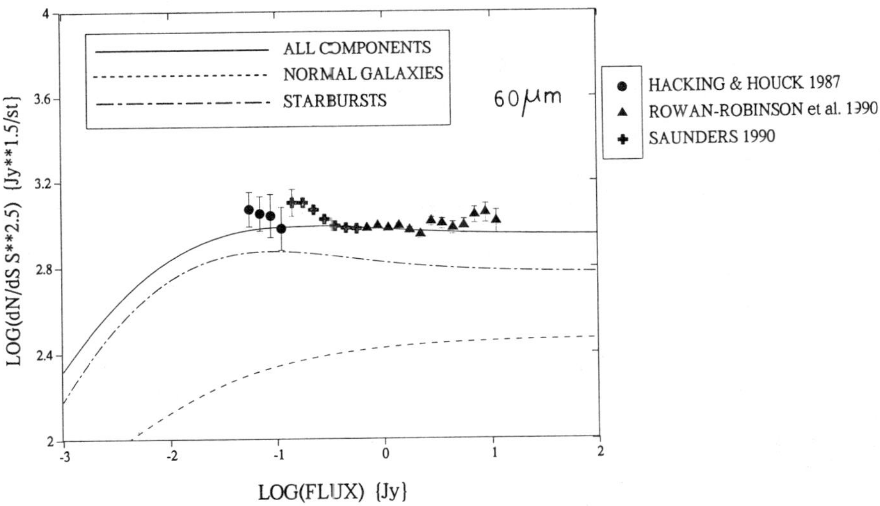

FIGURE 1b: Normalised differential source counts at 60 μm, with predicted contributions of different populations.

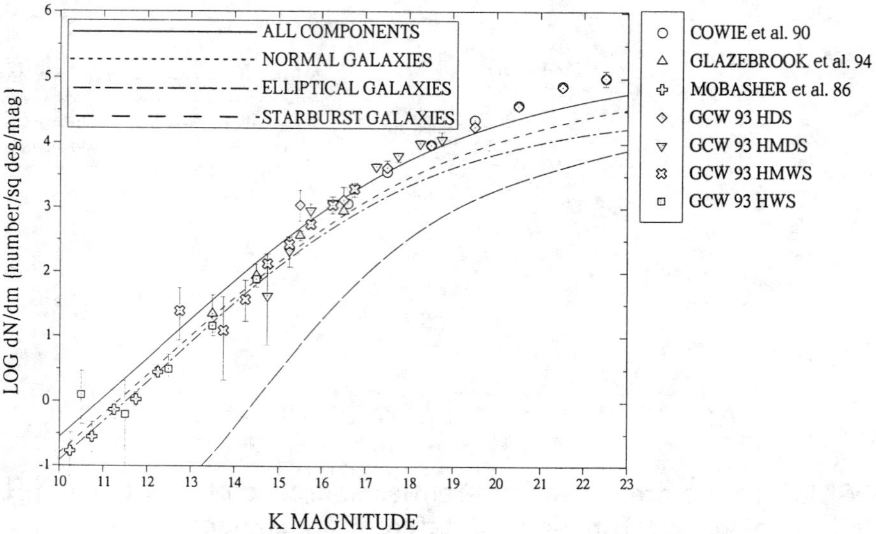

FIGURE 1c: Differential source counts in the K band, with predicted contributions of different populations

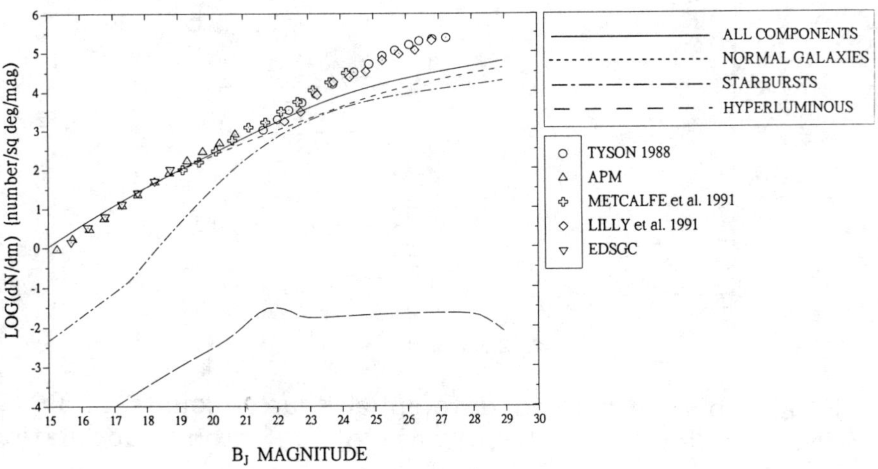

FIGURE 1d: Differential source counts in the B band, with predicted contributions of different populations.

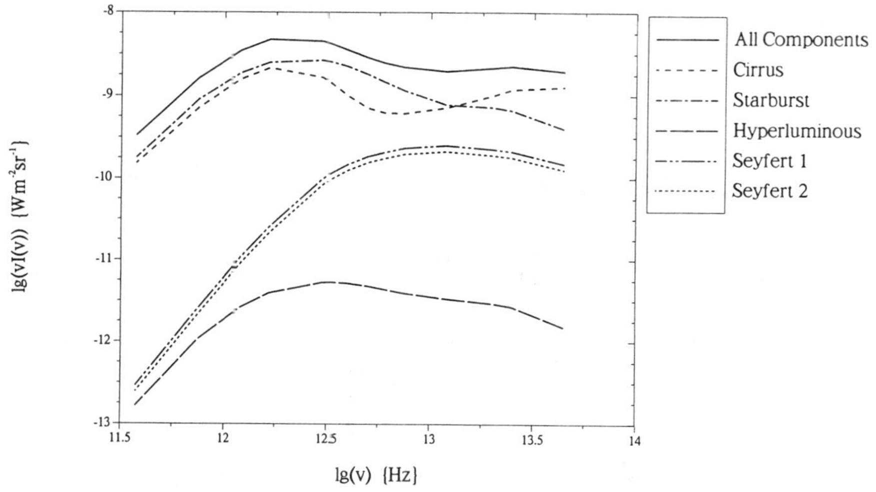

FIGURE 2: Predicted integrated background radiation intensity in the middle and far infrared, and submillimetre bands, showing contribution from different populations.

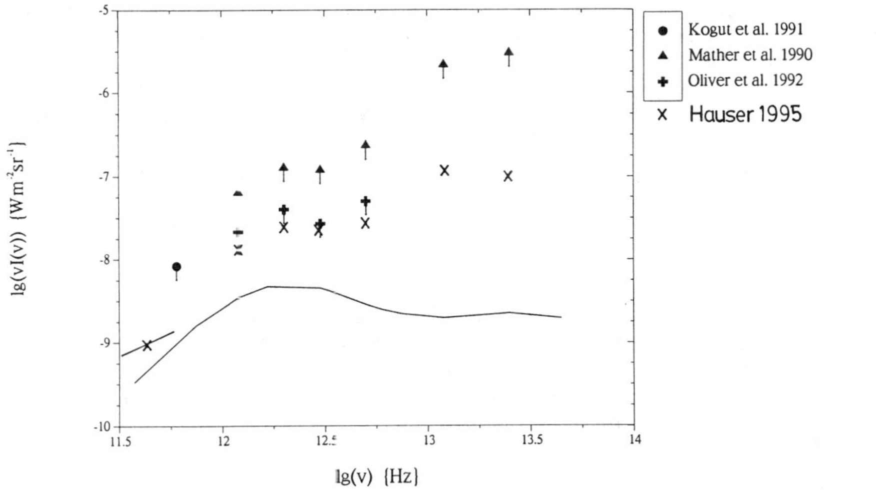

FIGURE 3: Predicted integrated background intensity compared with COBE upper limits.

FAINT GALAXY COUNTS AND DIFFUSE BACKGROUNDS

IN THE SUBMM/MM RANGE

Bruno Guiderdoni, Eric Hivon, François R. Bouchet
Institut d'Astrophysique de Paris, CNRS, 98bis Boulevard Arago F–75014 Paris

Bruno Maffei
Queen Mary and Westfield College, Mile End Road, London E1 4NS

& Richard Gispert
Institut d'Astrophysique Spatiale, Bât. 121, Univ. Paris XI, F–91405 Orsay Cedex

ABSTRACT

A second generation of satellites, like the COBRAS/SAMBA project studied at ESA, would observe the anisotropies of the CMBR on sub–degree scales, where the contamination of fluctuations due to various foregrounds and backgrounds is important. In particular, it is necessary to estimate the contribution of the diffuse background of unresolved galaxies. For that purpose, this paper proposes a new modelling of the evolution of the IR to mm properties of galaxies which includes the following series of physical processes: growth of linear perturbations (gravitationally dominated by dark matter), non–dissipative collapse and virialization of haloes, shock–heating and subsequent cooling and dissipative collapse of baryonic gas, formation of a rotationally–supported disk–like core, star formation and stellar evolution, stellar feedback, stellar emission, absorption of star light by dust and re–emission from the IR to the mm. The model reproduces the IRAS luminosity functions and deep galaxy counts, and gives predictions of the faint galaxy counts and diffuse background of unresolved galaxies in the IR and submm/mm ranges, which are compared with the present upper limits. Finally, we show exemples of sky simulations using this new model of galaxy evolution.

1. INTRODUCTION

After the great success of the COBE mission, new projects like the COBRAS/SAMBA satellite experiment aim at observing the anisotropies of the CMBR on scales below the degree (see the presentation by Bouchet, this Conference). On these scales, the separation of the various foregrounds and backgrounds which superimpose to the fluctuations of the CMBR is more difficult than for the 7° field of view of the COBE DMR experiment. In this perspective, the French RUMBA collaboration (standing for Research Upon Microwave Background Anisotropies) is developping simulations of the contributions of these various components to the fluctuations of the submm/mm sky in order to study the possible accuracy of their separation.

The diffuse background of unresolved galaxies is one of these components. This background has not been observed yet (but see Puget *et al.* 1995, and Désert *et al.*, this Conference). We should worry about the sensitivity of the numerous theoretical estimates proposed in the literature, to the unknown early

© 1996 American Institute of Physics

evolution of galaxies. This uncertainty has particularly strong consequences in the submm/mm range, as illustrated by Fig. 1 which shows observer–frame model spectra of a bright luminous IR galaxy at various redshifts. It turns out that there is a wavelength range, between $\sim 600\mu m$ and ~ 4 mm, where the distance effect is counterbalanced by the "negative k–correction" due to the huge rest–frame emission bump at $\sim 100\mu m$. In this range, the apparent flux of galaxies depends weakly on redshift. Moreover, this particular range brackets the maximum of emission of the CMBR. The modelling of the diffuse background of unresolved galaxies is consequently very sensitive to the early stages of galaxy evolution.

Two classes of models have been proposed to predict the amplitude of the background. In the first class, the submm/mm luminosity function undergoes luminosity and/or number evolutions which are simply parametrized as power laws of $(1 + z)$ (e.g. Beichman & Helou 1991; Hacking & Soifer 1991). In the second class, the photometric evolution is taken into account on more physical bases by computing the evolution of stellar populations, the chemical evolution of the gas, the absorption of the UV and visible star light by dust, and the re–emission of the energy in the IR and submm/mm range (Franceschini *et al.* 1991, 1994). Nevertheless, these models assume that all galaxies form at the same redshift z_{for} and that there is no number evolution.

As a matter of fact, the paradigm of the hierarchical growth of structures implies that there is no clear–cut redshift z_{for} since galaxy formation is a continuous process. In order to implement a model of photometric and chemical evolution of galaxies into a more realistic scenario of dynamical formation, we have used ideas which have been elaborated by various authors during the last ten years (White & Rees 1978; Schaeffer & Silk 1985; Evrard 1989; Cole 1991; Blanchard *et al.* 1992; White & Frenk 1991; Lacey & Silk 1991; Lacey *et al.* 1993, Kauffmann *et al.* 1993, 1994; Cole *et al.* 1994; Heyl *et al.* 1992). Indeed, the photometric evolution of stellar populations is the crucial link to relate the predictions of the theory to observations. It turns out that, in spite of differences in the details of the models, these approaches share the same spirit and lead to conclusions in the UV, visible and (stellar) near–IR which are remarkably similar. None of these models has been applied, up to now, to the prediction of the properties of galaxies in the far–IR and submm/mm ranges.

2. A SCHEMATIC VIEW OF GALAXY FORMATION

For that purpose, we have to identify the main physical processes which connect the mass of the initial perturbation to the IR and submm/mm radiation emitted by the end–product galaxy. If we assume that the universe is dominated by non–baryonic dark matter, the formation and evolution of a galaxy in its dark matter halo can be briefly sketched as follows: the initial perturbation, which is gravitationally dominated by non–baryonic dark matter, grows and collapses. During the (non-dissipative) collapse, the halo virializes, through the formation of a mean potential well seen by all particles, which consequently share the same velocity distribution. The shock–heated baryonic component subsequently cools and collapses at the center of the potential well. Since the baryons initially share the small rotation velocity of the halo created by tidal interactions with other haloes, and because of angular momentum conservation, their collapse stops when they reach rotational equilibrium. Stars begin to form in this disk-like baryonic core and evolve through the main stages of stellar evolution. The luminosity and spectrum of the stellar population strongly change during this evolution. At the end of their lifetime, stars explode and eject gas, heavy elements and energy back into the interstellar medium. This energy feedback is likely to alter the star formation efficiency. Part of the UV and visible light is absorbed by dust and the energy is re–emitted in the IR and submm/mm ranges. Finally, we have to conceive these evolutionary processes in the framework of the continuous, hierarchical growth of structures, in which small haloes merge to form bigger haloes. Moreover, galaxy merging in these merged haloes triggers bursts of star formation.

We need simplifying assumptions to build semi–analytic models including these evolutionary processes. In particular, the initial, linear perturbation is assumed to be spherical and homogeneous. This is the so–called "top–hat" model, which has the friendly property of being entirely defined by two parameters, for instance the size R and the density contrast δ_0 which are the linearly–extrapolated values at $z = 0$, or, equivalently, by its mass M and collapse redshift z_{coll}. If $\Omega_0 = 1$:

$$M = \frac{4}{3}\pi R^3 \rho_0 \qquad ; \qquad (1 + z_{coll}) = \frac{\delta_0}{\delta_c} \tag{1}$$

The current density is ρ_0 and $\delta_c = 1.68$ (for $\Omega_0 = 1$). The properties of the virialized halo are entirely computable as functions of M and z_{coll}. Its mass profile is a singular, isothermal sphere truncated at "virial radius" r_V. We have:

$$\rho_H(r) = \frac{V_c^2}{4\pi r^2} \tag{2}$$

$$r_V = \frac{3R}{10\delta_0} = 1.7\ 10^{-5}(\frac{M}{M_\odot})^{1/3}(1 + z_{coll})^{-1} h_{100}^{-2/3} \tag{3}$$

given in Mpc. The "circular velocity" (in km s^{-1}) is defined as:

$$V_c \equiv (\frac{GM}{r_V})^{1/2} = 1.6\ 10^{-2}(\frac{M}{M_\odot})^{1/3}(1 + z_{coll})^{1/2} h_{100}^{1/3} \tag{4}$$

Finally, the virial temperature of the baryonic gas is:

$$T_V \equiv \frac{1}{2}\frac{\mu m_p}{k} V_c^2 = 35.9(\frac{V_c}{kms^{-1}})^2\ K \tag{5}$$

with a mean molecular weight $\mu = 0.59$ corresponding to ionized gas with respective H and He mass fractions $X = 0.75$ and $Y = 0.25$. m_p is the mass of the proton.

3. FROM MASS TO LIGHT

Fig. 2 shows four panels which describe the main steps involved in relating the power spectrum of linear fluctuations $P(k)$ to the luminosity function $\phi(L)$ (hereafter LF). Panel *a* gives the variance $\sigma_0(M)$ of the density contrast $\delta\rho/\rho$ (linearly extrapolated to $z = 0$) for perturbations with mass scale M and size

R. The variance is computed from the power spectrum by smoothing with a top–hat window.

$$\sigma_0(M)^2 = \frac{1}{8\pi^3}\int P(k)|W_k(kR)|^2 4\pi k^2 dk \tag{6}$$

where $W_k(y)$ is the Fourier transform of the top–hat filtering function:

$$W(x) = \frac{1}{\frac{4}{3}\pi R^3} \qquad |x| < R \tag{7}$$

and 0 otherwise. The choice of $P(k)$ depends on the cosmological parameters, and $P(k)$ is normalized such as $\sigma_0(M) = b^{-1}$ for a smoothing radius $R = 8h_{100}^{-1}$ Mpc. The value of the bias parameter b is 2.5 for the classical CDM (which does not accomodate the large scale observations). In Panel b, the number density of collapsed haloes is easily computed from the Press–Schechter prescription (Press & Schechter 1974; hereafter PS).

$$n(M,z)dM = \sqrt{\frac{2}{\pi}}\frac{\rho_0}{M^2}\frac{\delta_c(1+z)}{\sigma_0(M)}\exp -\frac{\delta_c^2(1+z)^2}{2\sigma_0(M)^2}\left|\frac{d\ln\sigma_0(M)}{d\ln M}\right|dM \tag{8}$$

Efstathiou & Rees (1988) first noticed that this prescription gives a remarkably good fit of the number distribution of haloes in N–body simulations. Recent works (Bond *et al.* 1991; Bower 1991) settled this formalism on somewhat firmer theoretical grounds. If the power spectrum can be approximated by a power law $P(k) \propto k^n$, the low–mass end is $n(M)dM \propto M^{(n-9)/6}dM$. The CDM spectral index varies between -3 and 1, yielding a slope $\gamma = -2$ to -1.67, close to -2 on galaxy scales. We assume that new galaxies form at each generation of halo formation and that galaxies are not destroyed when the number density of haloes with low masses begins to decrease at low redshift. Our assumption is very similar to the one of Haehnelt & Rees (1993) who modelled QSO formation in haloes, and should be well fit to the description of "active" objects sensitive to the clustering process.

The baryonic gas cools down in the potential well of the dark matter halo, by a process identical to cooling flows observed at the center of rich clusters. The cooling time at halo radius r is:

$$t_{cool}(r) = \frac{3}{2}\frac{n_{tot}(r)kT_V}{n_e^2(r)\Lambda(T_V)} = \frac{3\pi G\mu_e^2 m_p^2}{\Omega_b}\frac{r^2}{\Lambda(T_V)} \tag{9}$$

with $\mu_e = 1.14$ for ionized primordial gas. Ω_b is the baryonic density. The equation $t_{cool}(r_{cool}) = t(z)$ defines a cooling radius r_{cool} as a function of redshift z. At this redshift, only gas inside r_{cool} (or r_V if $r_{cool} > r_V$) cools and is available for star formation. This cooling criterion introduces a high–mass cut-off in the mass distribution of cold baryonic cores. Haloes with masses larger than $\sim 10^{13}M_\odot$ (the exact value depends on Ω_b) are unable to produce cool gas and harbour a group or a cluster. Galaxies form in haloes with masses lower than this limit (Panel c). On the low–mass end, the cooling is so efficient that all the gas can cool, leading to a slope of the mass distribution of baryonic cores

$M_b\ \beta \simeq \gamma$. This is the so-called "overcooling" problem (Cole 1991; Blanchard *et al.* 1992).

When gas cools, the final radius is related to the initial radius of the gas by conservation of angular momentum: $r_D = \lambda \min(r_V, r_{cool})$ (Fall & Rees 1985), with the dimensionless spin parameter $\lambda \equiv J|E|^{1/2}G^{-1}M^{-5/2} \simeq 0.05 \pm 0.03$ (Barnes & Efstathiou 1987; Efstathiou *et al.* 1988; Zurek *et al.* 1988). A dynamical time scale in the disk-like core is $t_{dyn} \equiv r_D/V_c$. This gas settles in rotational equilibrium. Then stars form according to a star formation rate history $\tau_*(t)$. Locally, the star formation rate depends on numerous physical parameters. Nevertheless, many phenomenological studies seem to show that, on galaxy scales, it varies as a power of the total gas content (the so-called Schmidt law), with index 1–2 (see e.g. Kennicutt 1989). Here we assume that the star formation time scale t_* is proportional to the dynamical time scale of the disk t_{dyn} and we introduce a first fudge factor f_1.

$$\tau_*(t) = f_1 \frac{M_{gas}(t)}{t_{dyn}} \tag{10}$$

The explosion of massive stars can expel gas from the galaxies and stop star formation, leading to a strong increase of the mass/luminosity ratios in small objects. Observationally, HI holes and X-ray superbubbles are good evidence that such galactic winds are present in galaxies. By equating the gas binding energy to the thermal energy ejected by supernovae:

$$\frac{1}{2} M_{gas}(t) (\frac{V_{esc}}{V_c})^2 V_c^2 = f_2 \int_0^{t_W} \tau_*(t') \eta_{SN} E_{SN} dt' \tag{11}$$

where η_{SN} is the number of SNe per unit mass of stars, depending on the IMF, and $E_{SN} \sim 10^{51}$ erg is the output mechanical energy of a SN. Since much of this energy is subsequently radiated away, we inserted a second fudge factor $0 < f_2 \leq 1$. The mass fraction of stars forming before the triggering of the galactic wind at time t_W is $F_* = (1 + \beta(V_c))^{-1}$ with $\beta(V_c) \equiv (V_{hot}/V_c)^2$ and $V_{hot} \equiv 220 f_2^{1/2}$ km s^{-1} for the standard IMF used in Sec. 4.

4. SPECTRAL EVOLUTION FROM THE UV TO THE SUBMM/MM

A model of spectrophotometric evolution is used to compute the mass-to-luminosity ratios and the UV to NIR spectra of the stellar populations at each age. The stars are placed on the Zero-Age Main Sequence of the HR diagram according to the Initial Mass Function. The models use compilations of stellar evolutionary tracks taking into account the various stages of stellar evolution in order to compute at each time step the distribution of the stellar populations in the HR diagram. This distribution is combined with a library of stellar spectra. The apparent magnitudes and colours at redshift z are derived from the synthetic spectra of the stellar populations after coupling with the cosmological model and convolving with the response curves of filters. The model which is used here is described in Guiderdoni & Rocca-Volmerange (1987, 1988), and Rocca-Volmerange & Guiderdoni (1988), with upgraded stellar tracks from Schaller *et al.* (1992). We keep a standard IMF with slope $x = 1.7$ for stars more massive than $2M_\odot$ (observational value from Scalo 1986), $x = 1.35$ between 1 and $2M_\odot$, and $x = 0.25$ below.

These mass–to–luminosity ratios inserted in the baryonic core mass function of Panel *c* lead to the LF in Panel *d.* It turns out that, in spite of the variety of star formation rate histories and the strong variation of mass/luminosity ratios in time scales of a few Gyr, the *shape* of the LF is surprisingly similar to that of the baryonic mass function. The slope of the low–luminosity end is $\alpha \simeq \beta \simeq -1.6$ to -1.8, while the observed slope of the nearby field LF derived from wide–field surveys is $\alpha \simeq -1$ (see e.g. Loveday *et al.* 1992). Although this steep LF seems in disagreement with the nearby surveys, it is necessary to reproduce the deep redshift surveys (down to $B_J < 24$) and the faint galaxy counts (down to $B_J < 28$). Such a prediction is a robust result of this class of models (see Kauffmann *et al.* 1994; Cole *et al.* 1994, and e.g. the review by Guiderdoni 1995). Subtle selection effects due to surface brighness could explain the discrepancy between the nearby LF and the high–z one (Mc Gaugh 1994).

Part of this energy is absorbed by dust associated to the gaseous component and re–emitted in the IR, submm and mm ranges. As in Guiderdoni & Rocca–Volmerange (1987) and Franceschini *et al.* (1991, 1994), we assume a simple geometric distribution of the stars and gas in the disks. If $\tau_\lambda(t)$ is the optical thickness of the disk at time t, the extinction correction (averaged over inclination angle i) is:

$$A_\lambda(t) = -2.5 \log < \frac{1-\exp(-\tau_\lambda(t)/\cos i)}{\tau_\lambda(t)/\cos i} >_i \tag{12}$$

providing that the stars and gas have the same height scale. The optical thickness is given by:

$$\tau_\lambda(t) = 3.25(\frac{A_\lambda}{A_V})_{Z_\odot}(\frac{Z_g(t)}{Z_\odot})^s g(t) \tag{13}$$

where $g(t) \equiv M_{gas}(t)/M_b$. The extinction curve depends on the metallicity $Z_g(t)$ of the gas according to power–law interpolations based on the Solar Neighbourhood and the Magellanic Clouds, with $s = 1.35$ for $\lambda < 2000$ Å and $s = 1.6$ for $\lambda > 2000$ Å. The extinction curve for solar metallicity is taken from Mathis *et al.* (1983). The metallicity of the gas is estimated from the Instant Recycling Approximation $Z_g(t) = -y_Z \ln g(t)$ with a yield y_Z depending on the choice of the IMF. The optical thickness is maximal for $g(t) = e^{-s} \simeq 0.20$ with $s = 1.6$.

The total star light in the UV and optical is absorbed by dust and re–emitted at longer wavelengths with a "bolometric" IR luminosity L_{IR}. The emission spectra of galaxies are computed as a sum of various components, according to the method developped by Maffei (1994). The observational correlations of the IRAS flux ratios 12μm/60μm, 25μm/60μm and 100μm/60μm with L_{IR} (Soifer & Neugebauer 1991) are reproduced by adding template emission spectra (taken from the model by Désert *et al.* 1990) for Polycyclic Aromatic Hydrocarbons, very small grains and big grains to the near–IR stellar emission. The synchrotron emission is proportional to L_{IR}. The resulting spectra are computed from a few μm to several mm and evolve with L_{IR} such as more luminous galaxies emit preferentially at shorter wavelengths. It is possible to reproduce the spectrophotometry of various individual galaxies by this method (Maffei 1994). In spite of its shortcomings, this approach takes into account the observed spectral evolution in the IR and submm/mm ranges, while previous works (e.g. Franceschini *et al.* 1991, 1994) only use a single template spectrum in the whole spectral range.

5. THE FIR LUMINOSITY FUNCTION

The predictions of the model depend on (1) the cosmological parameters: the Hubble constant H_0, the density parameter Ω_0 (and the cosmological constant Λ), the baryonic fraction Ω_b, the choice of the power spectrum of linear fluctuations and its normalization b; (2) the free parameters f_1 and f_2 (or V_{hot}) associated with star formation and stellar feedback. Hereafter, we shall consider $H_0 = 50$ km s^{-1} Mpc^{-1}, $\Omega_0 = 1$ and $\Lambda = 0$, and focus on the "classical" Cold Dark Matter model with $b = 2.5$. The parameter f_1 is fixed by the necessity to reproduce the observed star formation time scales in disks (Kennicutt *et al.* 1994) and its order of magnitude is $f_1 \sim 100 - 400$. We take $V_{hot} = 150$ km s^{-1}. The baryonic fraction, which is used as a last parameter to fit the total luminosity in the LF, is strongly constrained by the primordial nucleosynthesis.

Fig. 3 shows the formation rates of galaxies versus redshift in haloes with various masses. Bigger haloes form later on an average, leading to a continuous process of galaxy formation. Fig. 4 gives predictions for the $z = 0$ luminosity functions at wavelengths 25 μm, 60 μm, and 100 μm, compared with the observational determinations drawn from the IRAS Bright Galaxy Sample (Soifer & Neugebauer 1991). Panel *a* shows the predicted LF for normal star formation in disks (involving 90 % of the available cold gas) while panel *b* gives the LF for an initial burst of star formation (involving 10 % of the cold gas) in an optically–thick region where all the energy emitted in the UV and visible is re–emitted in the IR and submm/mm. Panel *c* shows the result when the two modes of star formation are taken into account. The three LF are accomodated simultaneously with $\Omega_b = 0.05$. This model also yields predictions in the optical. As anticipated in Sec. 4, the predicted B–band LF is steeper than the nearby LF (e.g. Loveday *et al.* 1992). We did not attempt to model the surface–brightness effects in the optical. On the contrary, these selection effects do not seem to be important in the IR since the data do show the steep slope predicted by this class of models.

6. FAINT GALAXY COUNTS AND DIFFUSE BACKGROUND

The model proposed in the previous section is "minimal", meaning that most of the star formation occurs in a quiet mode, in disks, while only the brightest galaxies of the IRAS Bright Galaxy Sample are seen during a burst episode. We hereafter assume that this "regular" behaviour is similar at high z. Fig. 5 and 6 respectively give the galaxy counts at 60 μm based on the evolving LF, with the observational values drawn from the IRAS Faint Galaxy Sample, and the predicted counts for various wavelengths between 20 μm and 2 cm. Clearly there are three regimes: at short wavelengths, the effect of curvature and the positive "k–correction" produce the bend of the faint counts; in the submm/mm range, the negative "k–correction" produces a strong bump in the faint counts, which reflects the passage of the 100 μm emission bump into the observing bands; finally, at larger wavelengths, the luminosity is so small that all galaxies at these flux levels are nearby, in the euclidean zone, with $N(> S) \propto S^{-3/2}$.

These faint counts can be summed up and yield predictions for the amplitude of the diffuse background, as displayed in Fig. 7, with other models and the current upper limits given by FIRAS and DIRBE. It turns out that, because of the non–conservation of the number density of galaxies with time (due to the continuous process of galaxy formation), the background is smaller than the predictions without number evolution (up to $z_{for} = 5$), but with luminosity evolution (Franceschini *et al.* 1991, 1994), and even than the predictions without evolution, obtained by integrating the IRAS LF up to $z_{for} = 3$ to 9. Moreover, in spite of the redshift range contributing to the background, and of the spectral evolution in the dust emission, the shape of the background spectrum is rather similar to that of the cold dust.

The isotropic component extracted by Puget *et al.* (1995) from a re–analysis of the FIRAS data is also plotted in Fig. 7. The "minimal" model introduced previously is unable to accomodate this component, because the redshifts of galaxy formation in an $\Omega_0 = 1$ CDM universe are too low to shift the 100 μm emission bump in the mm range. In the models by Franceschni *et al.* (1991, 1994), all galaxies form at $z_{for} = 4.5 - 5$, in contrast with our predictions for a CDM model. We tested another model in which all star formation at $z > 1$ occurs in optically–thick bursts, and the proportion of mass involved in bursts increases as $0.05(1+z)^{4.5}$ for $z < 1$. This power law comes from the increase of the galaxy pair rate observed by Zepf & Koo (1989). This "strong burst" model gives a fair fit of the IRAS LF and leads to a stronger diffuse background in the far–IR, but is not sufficient in the mm range. The "strong burst" model with a bias parameter $b = 1$ and a more extreme model in which all star formation occurs in bursts (whatever the redshift) lead to similar failures. Only a model with $\Omega_b = 0.2$, which violates the constraints on nucleosynthesis and strongly overestimates the IRAS LF, is able to reproduce the isotropic component.

As a matter of fact, an increase of the IR luminosity (due, for instance, to an IMF with only massive stars) leads to spectra peaking at shorter wavelengths, if the correlation of the spectrum with L_{IR} still holds at high z, and has little influence on the mm range. We are left with *ad hoc* explanations (high baryonic fraction, or non–standard dust properties at high z), or with the evidence that galaxies form at higher z than predicted by the $\Omega_0 = 1$ CDM universe.

7. SIMULATIONS OF THE SKY OF UNRESOLVED GALAXIES

Thess models have been used to create maps of the faint galaxy contribution to the CMBR anisotropies as the COBRAS/SAMBA experiment would observe them. For each redshift z, the number of galaxies located in the cone of observation is given by the model number density. Their positions are picked at random or extracted from an external catalog, and their spectral luminosities are Monte–Carlo realizations of the predicted luminosity function. We then compute the flux received from each of these galaxies on 500×500 pixels with an invidual size of 1.5×1.5 arcmin, for a set of frequencies lying between 15 and 15000 GHz.

Spatial correlation between sources would have several effects on observed map. Of course, it would have a signature on correlation between pixels, but it would also promote the contribution of faintest galaxies to flux anisotropies (Barcons 1992). To implement source clustering as it occurs at cluster scales, we performed an N-body simulation for an initial power spectrum $P(k) \propto k^{-2}$ with about 17 million particles (256^3) using a PM code (Bouchet *et al.* 1985; Moutarde *et al.* 1991) with 256^3 cells. The initial conditions were imprinted using Zeldovich approximation on a "glass"-like particle distribution (White, private communication). The comoving box length was fixed to be $280h_{100}^{-1}$ Mpc, and today ($z = 0$) is identified to the expansion factor when the variance on a sphere of radius $R = 8h_{100}^{-1}$ Mpc is equal to 1. At each time step of the simulation, one records positions of particles located inside the cone of observation with redshift z. This catalog yields the positions of galaxies implied in simulations of anisotropies.

The maps obtained with these simulations can be degraded both in spatial and spectral resolution to meet COBRAS/SAMBA specifications. The beam angular size lies between 4.5 and 30 arcmin, and the relative bandwidth is about 0.2 (see the presentation by Bouchet, this Conference). Instrumental noise can be added before being compared to similar maps for other contributions (CMBR, dust, free–free emission).

Fig. 8 shows two maps of CMBR fluctuations due to faint galaxies for $0 \leq z \leq 5$, at 299 GHz, with clustering (left panel) and without clustering (right panel). Each map is 2.5 deg. large and the spatial resolution is 1.5 arcmin. Two isocontours of brightness are indicated : $\sigma/2$ and σ, where σ is the standard deviation measured on each map. In the clustering case, the variance is larger than for a random distribution of sources, because of some bright pixels which contain either close sources or a superposition of remote sources. These rare events are selected on the left–hand map as compared to the right–hand one.

Fig. 9 shows the variance of CMBR, dust and faint galaxies as a function of frequency for a spatial resolution of 1.5 arcmin. For the $\Omega_0 = 1$ CDM models, the fluctuations of the diffuse background of galaxies at 1.5 arcmin scales are larger than those of the faintest Galactic dust only at $\lambda < 150\mu$m.

8. CONCLUSIONS

The separation of the sub–degree anisotropies due to the CMBR and the various foregrounds/backgrounds, which would be observed by the second generation of submm/mm satellites like the COBRAS/SAMBA project, can be prepared by an improved modelling and realistic sky simulations. In this paper, we proposed preliminary results on the diffuse background of unresolved galaxies. We introduced a new model of galaxy evolution which takes into account the main physical processes from the collapse of the density fluctuations to the absorption of UV and visible star light by dust and the re–emission in the IR and submm/mm ranges. The predictions of this new model are consistent with the dynamical process of continuous galaxy formation predicted by the hierarchical growth of structures in an $\Omega_0 = 1$ CDM universe, and represent a significant progress with respect to previous phenomenological models. The predicted number density evolution and the relatively low redshifts of galaxy formation have strong consequences on the predictions of the diffuse background. Galaxies with normal dust properties in an $\Omega_0 = 1$ CDM universe seem unable to emit significantly in the mm range and cannot explain the isotropic component which has been extracted from the FIRAS data by Puget *at al.* (1995).

The influence of the star formation rate history, the geometry of the dust and star distribution, and the cosmological parameters on the submm/mm properties of galaxies can be studied by this new approach. Finally, the amplitude of the mean value and fluctuations of the diffuse background of unresolved galaxies can be estimated. The information on the various backgrounds will help to optimize the design of the instrument, the choice of the observing channels and strategy, and the algorithm of component separation.

ACKNOWLEDGEMENTS

We are pleased to thank the organizers of the COBE Workshop on "Unveiling the Cosmic Infrared background" for the stimulating and fruitful discussions. Many thanks to Jean–Loup Puget and François–Xavier Désert for their comments and suggestions.

REFERENCES

Barcons, X. 1992. ApJ, **396**, 460
Barnes, J., & Efstathiou, G. 1987. ApJ, **319**, 575
Beichman, C.A., & Helou, G. 1991. ApJL, **370**, L1
Blanchard, A., Valls–Gabaud, D., & Mamon, G. 1992. A&A, **264**, 365
Bond, J.R., Cole, S., Efstathiou, G., & Kaiser, N. 1991. ApJ, **379**, 440
Bouchet, F. R., Adam, J.-C., & Pellat, R. 1985. A&A, **144**, 413
Bower, R.G. 1991. MNRAS, **248**, 332
Cole, S. 1991. ApJ, **367**, 45
Cole, S., Aragón–Salamanca, A., Frenk, C.S., Navarro, J.F., & Zepf, S.E. 1994. MNRAS, **271**, 781
Désert, F.X., Boulanger, F., & Puget, J.L. 1990. AA, **237**, 215
Efstatiou, G., & Rees, M.J. 1988. MNRAS, **230**, 5P
Efstathiou, G., Frenk, C.S., White, S.D.M., & Davis, M. 1988. MNRAS, **235**, 715
Evrard, A.E. 1989. ApJ, **341**, 26
Fall, S.M., & Rees, M.J. 1985. ApJ, **298**, 18
Franceschini, A., Toffolatti, L., Mazzei, P., Danese, L., & De Zotti, G. 1991. ApJSS, **89**, 285
Franceschini, A., Mazzei, P., De Zotti, G., & Danese, L. 1991. ApJ, **427**, 140
Guiderdoni, B., & Rocca–Volmerange, B. 1987. A&A, **186**, 1
Guiderdoni, B., & Rocca–Volmerange, B. 1988. A&ASS, **74**, 185
Guiderdoni, B., & Rocca–Volmerange, B. 1990. A&A, **227**, 362
Guiderdoni, B. 1995, in *Wide–Field Spectroscopy and the Distant Universe*, S. Maddox *et al.* (eds), World Scientific
Hacking, P.B., & Soifer, B.T. 1991. ApJL, **367**, L49
Haehnelt, M.G., & Rees, M.J. 1993. MNRAS, **263**, 168
Hauser, M.G. 1995, in *Proceedings of the IAU Symp. n° 168, Examining the Big Bang and Diffuse Background Radiation*, The Hague, August 1994
Heyl, J.S., Cole, S., Frenk, C.S., & Navarro, J.F. 1995. MNRAS, **274**, 755
Kauffmann, G.A.M., & White, S.D.M. 1993. MNRAS, **261**, 921
Kauffmann, G.A.M., White, S.D.M., & Guiderdoni, B. 1993. MNRAS, **264**, 201
Kauffmann, G.A.M., Guiderdoni, B., & White, S.D.M. 1994. MNRAS, **267**, 981
Kennicutt, R.C. 1989. ApJ, **344**, 685
Kennicutt, R.C., Tamblyn, P., & Congdon, C.W. 1994. ApJ, **435**, 22
Lacey, C., & Silk, J. 1991. ApJ, **381**, 14
Lacey, C., Guiderdoni, B., Rocca–Volmerange, B., & Silk, J. 1993. ApJ, **402**, 15
Lonsdale, C.J., Hacking, P.B., Conrow, T.P., & Rowan–Robinson, M., 1990. ApJ, **358**, 60
Loveday, J., Peterson, B.A., Efstathiou, G., & Maddox, S.J. 1992. ApJ, **390**, 79
Maffei, B. 1994. PhD Dissertation, Université Paris VII

Marzke, R.O., Huchra, J.P., & Geller, M.J. 1994. ApJ, **428**, 43
Mather, J.C., *et al.* 1994. ApJ, **420**, 439
Mathis, J.S., Mezger, P.G., & Panagia, N. 1983. A&A, **128**, 212
McGaugh, S.S. 1994. Nat, **367**, 538
Moutarde, F., Alimi, J.-M., Bouchet, F. R., & Pellat, R., 1991. ApJ, **382**, 377
Press, W.H., & Schechter, P.G. 1974. ApJ, **38**, 357
Puget, J.L., Abergel, A., Boulanger, F., Bernard, J.P., Burton, W.B., Désert, F.X., & Hartmann, D. 1995. *preprint*
Rocca–Volmerange, B., & Guiderdoni, B. 1988. A&ASS, **75**, 93
Scalo, J.N. 1986. *Fundam. Cosmic Phys.*, **11**, 1
Schaeffer, R., & Silk, J. 1985. ApJ, **292**, 319
Schaller, G., Schaerer, D., Meynet, G., & Maeder, A. 1992. A&ASS, **96**, 269
Soifer, B.T., & Neugebauer, G. 1991. AJ, **101**, 354
White, S.D.M., & Rees, M.J. 1978. MNRAS, **183**, 341
White, S.D.M., & Frenk, C.S. 1991. ApJ, **379**, 52
Zurek, W.H., Quinn, P.J., & Salmon, J.K. 1988. ApJ, **330**, 519
Zepf, S.E., & Koo, D.C. 1989. ApJ, **337**, 34

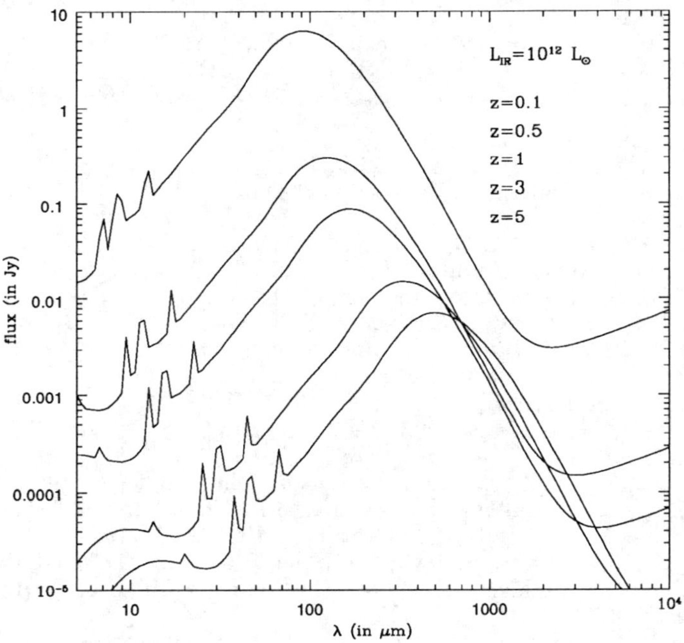

Fig. 1: Observer–frame model spectra of a $10^{12} L_{\odot}$ galaxy at various redshifts. Note that the apparent flux in the submm/mm range is almost insensitive to redshift.

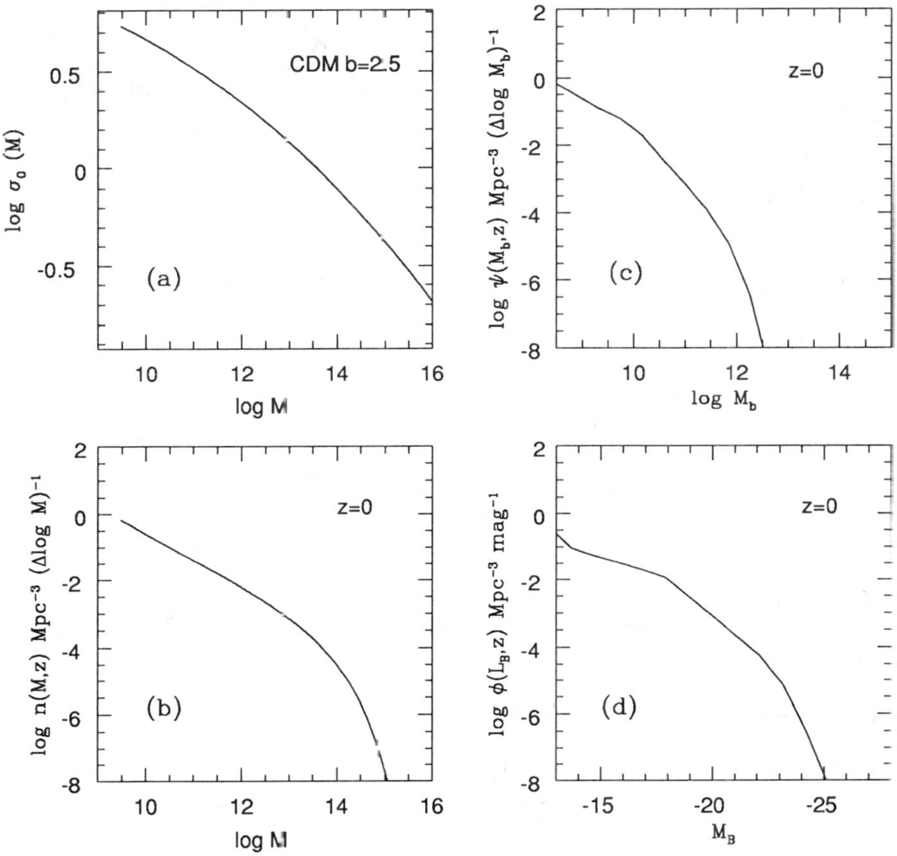

Fig. 2: These four panels illustrate the steps involved in relating the power spectrum of linear fluctuations $P(k)$ to the galaxy luminosity function $\phi(L)$. The curves are computed for $H_0 = 50$ km s^{-1} Mpc^{-1}, $\Omega_0 = 1$ and the CDM model with $b = 2.5$. Panel *a*: variance $\sigma_0(M)$ versus perturbation mass M. See text for the definition of σ_0. Panel *b*: number density $n(M, z)$ of collapsed haloes at redshift $z = 0$ versus halo mass M. Panel *c*: number density $\psi(M_b, z)$ of baryonic cores at redshift $z = 0$ versus core mass M_b. Panel *d*: galaxy luminosity function $\phi(L_B, z)$ at $z = 0$ versus absolute blue magnitude M_B.

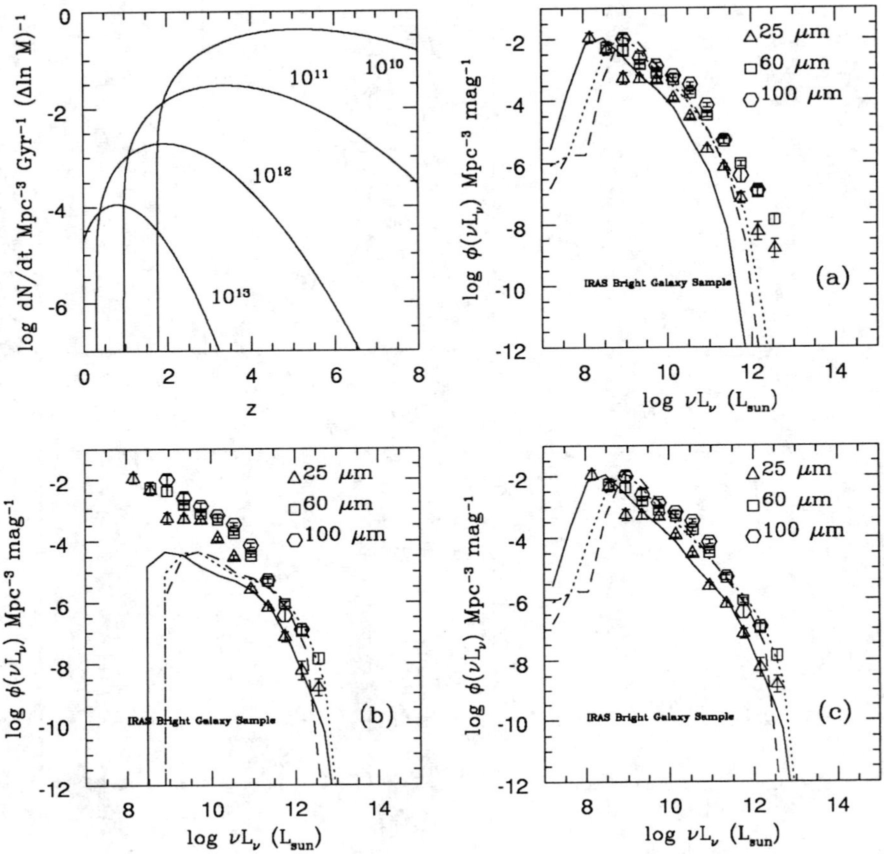

Fig. 3: Upper left–hand Panel: Formation rate of galaxies within dark matter haloes of masses 10^{10}, 10^{11}, 10^{12} and $10^{13} M_{\odot}$, for the CDM power spectrum with $b = 2.5$, $H_0 = 50$ km s^{-1} Mpc^{-1} and $\Omega_0 = 1$.

Fig. 4: Predicted IRAS luminosity functions at 25 μm, 60 μm and 100 μm (solid lines, dashes and dots). Panel *a*: regular star formation in disks involving 90 % of the available cold gas. Panel *b*: initial starbursts in optically–thick clouds involving 10 % of the available cold gas. Panel *c*: resulting luminosity functions for both modes of star formation.

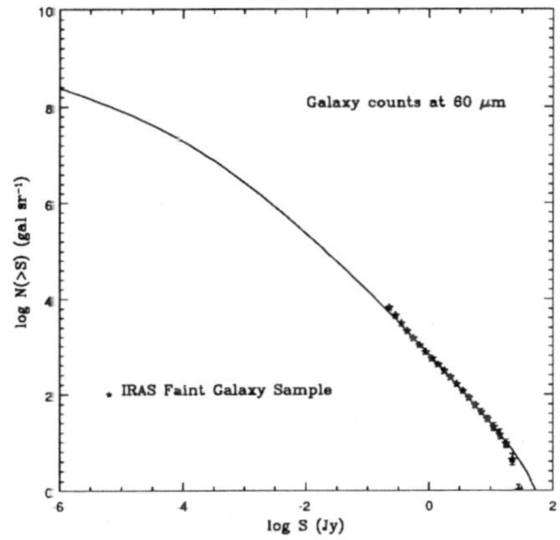

Fig. 5: Predictions for the 60 μm counts, compared to the IRAS faint galaxy counts.

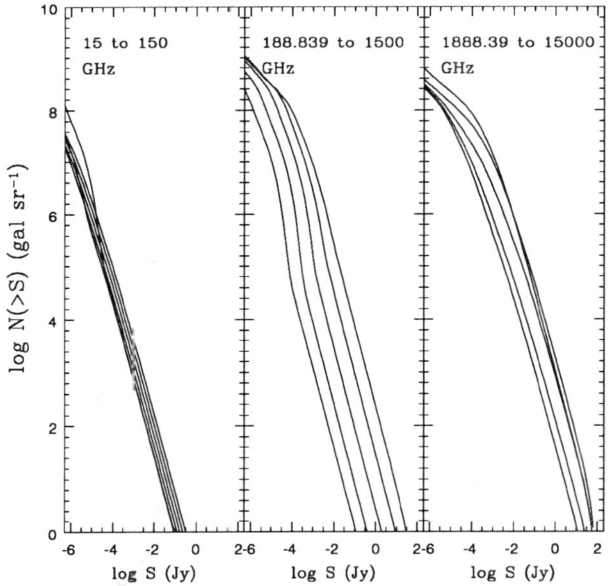

Fig. 6: Predictions of faint galaxy counts at various frequencies. The right–hand Panel shows the typical bend due to the effect of curvature and positive "k–correction". The middle Panel shows the bump due to the "negative k–correction" in the submm/mm range. The left–hand panel gives the counts in the mm and cm range for normal galaxies.

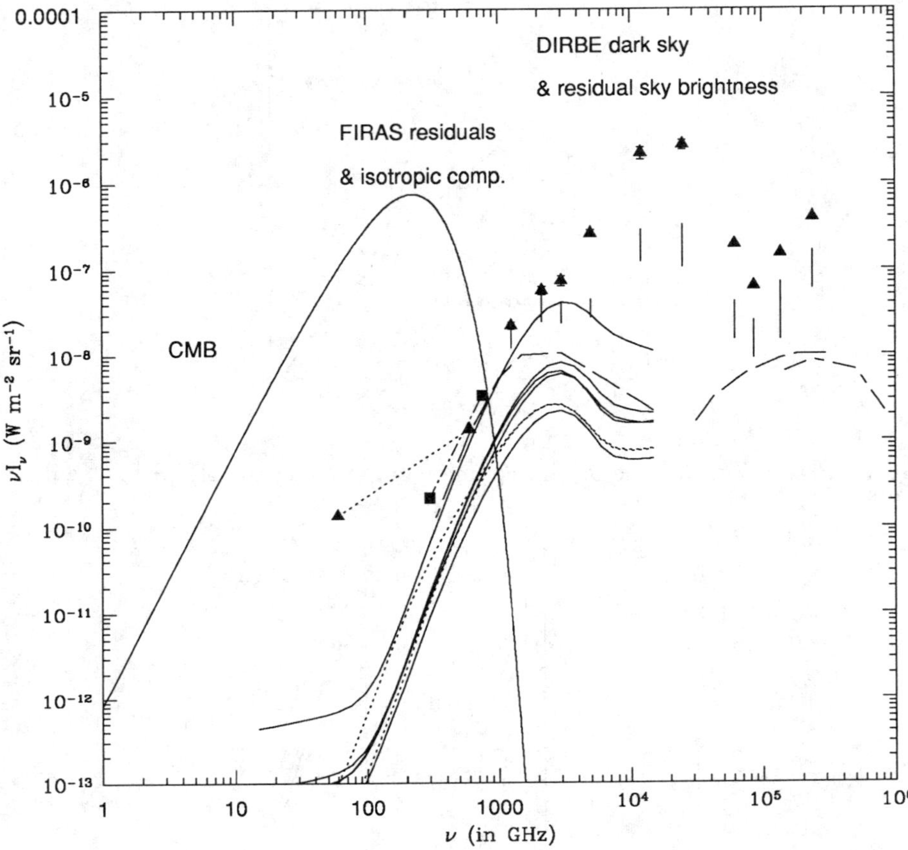

Fig. 7: Predictions of the diffuse background of galaxies (solid lines) within the $\Omega_0 = 1$ CDM universe ($b = 2.5$), compared to the current upper limits for DIRBE and FIRAS, the CMBR and other models. The upper solid line gives the prediction for a typical "strong burst" model (see text) and $\Omega_b = 0.20$ which violates nucleosynthesis constraints and overestimates the current IRAS LF. The lower solid line is for the "minimal model" with $\Omega_b = 0.05$, which fits the IRAS LF. The three other solid lines are for the "strong burst" model with $\Omega_b = 0.05$ and $b = 2.5$ or 1, and an "all burst" model (see text). The DIRBE dark sky and residuals are from Hauser (1995) and the FIRAS residuals are from Mather *et al.* (1994). The squares are the isotropic component extracted by Puget *et al.* (1995) from the FIRAS data. The dashes show a typical model by Franceschini *et al.* (1994), for strong evolution (his case c). The short and long dashes are predictions of the optical background from fits of the faint galaxy counts (Guiderdoni & Rocca–Volmerange 1990). The dotted lines give no–evolution predictions integrated up to redshifts $z_{for} = 3$ and 9.

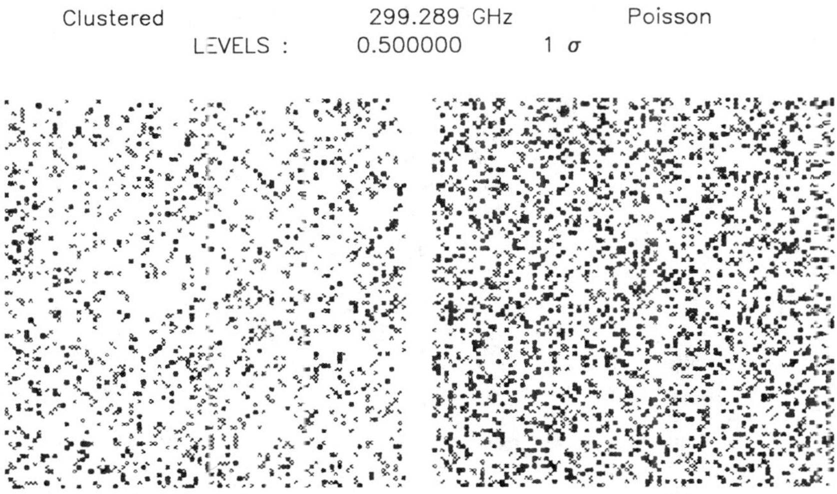

Fig. 8: Maps of unresolved galaxies for $0 \leq z \leq 5$, at 299 GHz, with clustering (left panel) and without clustering (right panel) for the "minimal" model. The spatial resolution is 1.5 arcmin. Two isocontours of brightness are indicated : $\sigma/2$ (gray) and σ (black), where σ is the standard deviation measured on each map.

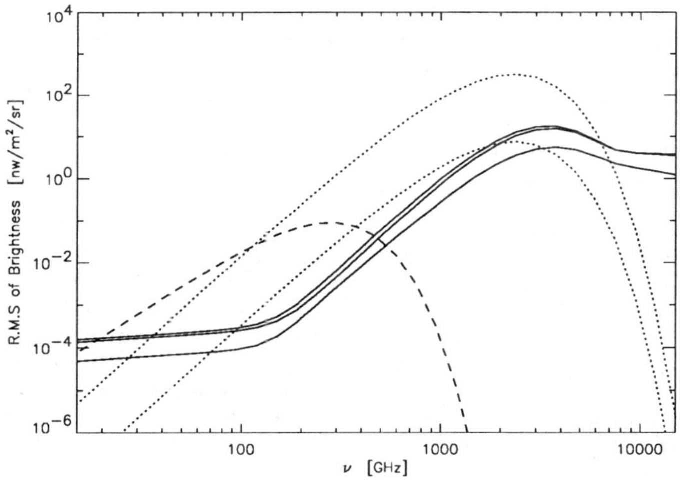

Fig. 9: Contribution to fluctuations for CMBR (dashes), galactic dust (dots) and faint galaxies (solid lines) as a fonction of frequency when the spatial resolution is 1.5 arcmin. For dust, the upper curve corresponds to 50% of the sky and the lower curve to the clearest region as observed by IRAS at 100 μm. For galaxies, the upper curve shows the "strong burst" model with source clustering, and the other curves show the "minimal" model with (middle curve) and without (lower curve) clustering.

DUST IN HIGH REDSHIFT RADIOGALAXIES

Gianfranco De Zotti and Paola Mazzei
Osservatorio Astronomico di Padova
Vicolo dell'Osservatorio 5
I-35122 Padova (Italy)

ABSTRACT

The radio selection has provided, by far, the richest samples of high-z galaxies, most likely of early type, which may therefore yield crucial direct information on the period when the bulk of stars were born. In order to investigate the role of dust in these phases, we have analyzed recent observations of a number of high-z radio galaxies in the framework of a self-consistent scenario for the evolution of early type galaxies. The available data can be fully accounted for by "opaque" models similar to that already used by Mazzei and De Zotti (1994a) to fit the spectrum of the hyperluminous galaxy IRAS F10214 + 4724. However, there is still considerable latitude for models. Good fits can be obtained with galactic ages ranging from 0.05 to 2 Gyr. Crucial constraints may be provided by ground-based sub-mm measurements and by observations with the forthcoming Infrared Space Observatory (ISO).

1. INTRODUCTION

As pointed out by other speakers at this Workshop, there are good reasons to believe that the far-IR emission of disk galaxies, kept, on the average, rather steady thoughout the galaxy lifetime (except, perhaps, during the first Gyr, when bulges were formed), with relatively minor changes in the emission spectrum due to the dust probably being somewhat warmer in the past.

A much more spectacular evolution is expected in the case of spheroidal galaxies which, locally, generally emit in the far-IR less than 1% of their bolometric luminosity (Mazzei & De Zotti 1994b). During their early evolutionary phases, when a substantial metal enriched interstellar medium was present, their far-IR emission could have comprised a large fraction of the bolometric luminosity, which might itself have been an order of magnitude larger than it is today. Altogether, we may expect an evolution of their far-IR luminosity by up to three orders of magnitude, i.e. much stronger than even the evolution of quasars in the optical band. On the other hand, their detailed evolutionary behaviour is highly uncertain, as it depends on the time dependence of poorly know functions, such as the Star Formation Rate (SFR), the Initial Mass Function (IMF), and the dust properties.

If, for example, we make the simple assumption that a constant fraction of metals is locked up in dust grains and that the dust properties did not change much during the galaxy lifetime, we have that the effective optical depth of galaxies to dust absorption is, during early phases, the product of two rapidly varying functions: the rapidly increasing gas metallicity and the rapidly decreasing gas fraction. Relatively minor differences in the time dependence of these two functions, consistent with the available data, yield substantially different histories of the effective optical depth, ranging from cases of galaxies optically thin at all times (so that the dominant fraction of the emission always comes out

© 1996 American Institute of Physics

in the optical) to cases of a prolonged optically thick phase lasting a few Gyr, where most (up to over 90%) of the light comes out in the far-IR. Models of the latter type provide a very good fit to the observed SED of the hyperluminous galaxy IRAS F10214 + 4724 (Mazzei & De Zotti 1994a). Of course, all intermediate cases are also possible. Optical/near-IR photometry of deep samples is not enough to clearly discriminate among the various possibilities (cf. Mazzei & De Zotti 1995).

Of course, we would like to observe directly the spectral energy distribution of high-z galaxies. These, however, proved to be remarkably elusive, a fact that may already hint at the possibility that they are strongly obscured, but other explanations are also possible. So far, the only effective method to find high-z galaxies has been optical identifications of radio sources, which are invariably identified as elliptical galaxies. Of course, the interpretation of the observed SED is complicated by the need of disentangling the starlight from the emission of the active nucleus, and of properly allowing for effects induced by nuclear activity, e.g. star formation triggered by the radio jet as it propagates outwards from the nucleus. Nevertheless, high-z radio galaxies can provide unique information on early phases of galaxy evolution, when intense star formation occurred.

In this paper we exploit the evolutionary population synthesis models for early type galaxies worked out by Mazzei, De Zotti & Xu (1994), taking into account dust absorption and re-emission, to analyze the continuum spectra of the best studied $z > 3$ radio galaxies, namely: 4C41.17, the second highest redshift galaxy known, B2 0902 + 343, the first $z > 3$ galaxy discovered, and 6C1232 + 39. We adopt $H_0 = 50\,\mathrm{km\,s^{-1}\,Mpc^{-1}}$.

2. MODELLING THE SED FROM UV TO FAR–IR

The galaxy is assumed to be a close system, i.e. both galactic winds and inflow of intergalactic gas are neglected. We have adopted Schmidt's (1959) parametrization of the star–formation rate (SFR), $\psi(t)$:

$$\psi(t) = \psi_0 \left(\frac{m_b}{10^{11}\,\mathrm{M_\odot}} \right) f_g^n \;\; \mathrm{M_\odot\,yr^{-1}}, \tag{1}$$

where $f_g = m_{\rm gas}/m_b$ is the fractional mass of gas in the galaxy (m_b is the total mass in baryons), assumed to be, initially, unity. Current analyses favour n = 1–2 (Buat 1992), but $n = 0.5$ has also been advocated (Madore 1977). Successful models for early type galaxies require $\psi_0 \geq 100\ \mathrm{M_\odot\,yr^{-1}}$ for $n \geq 1$ or $\psi_0 \geq 35\ \mathrm{M_\odot\,yr^{-1}}$ for $n = 0.5$.

Also, we have adopted Salpeter's (1955) initial mass function (IMF), with an upper mass limit $m_u = 100\ \mathrm{M_\odot}$ and a lower mass limit, m_l, ranging from 0.01 $\mathrm{M_\odot}$ to 0.5 $\mathrm{M_\odot}$.

Stellar lifetimes are taken into account (i.e. we have not resorted to the instantaneous recycling approximation). The gas is assumed to be well mixed and uniformly distributed. The variations with galactic age, T, of the fractional gas mass $f_g(T)$ and of the gas metallicity $Z_g(T)$ are obtained by numerically solving the standard equation for chemical evolution. The synthetic spectrum of stellar populations as a function of the galactic age has been obtained using the theoretical isochrones given by Bertelli et al. (1990), as extended by Mazzei (1988) up to 100 $\mathrm{M_\odot}$ and to an age of 10^6 yr.

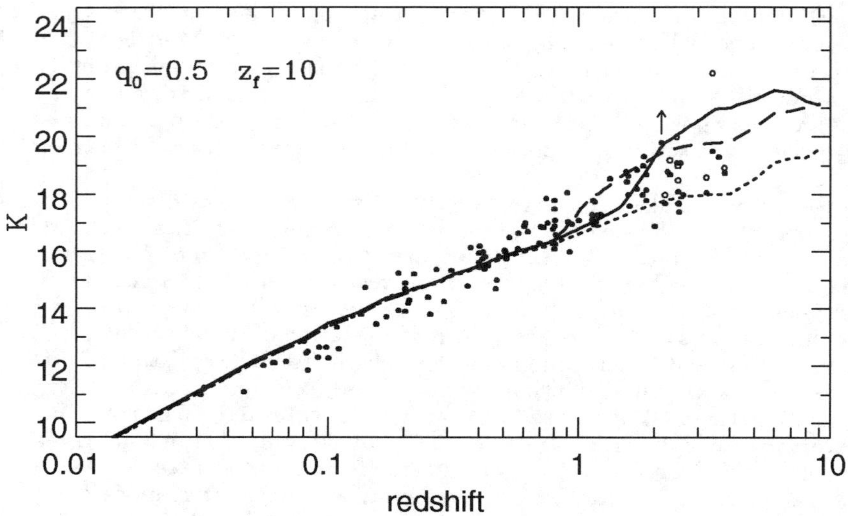

Figure 1: Predicted vs observed K–z diagram of radiogalaxies for a galaxy formation redshift, $z_f = 10$, and of $q_0 = 0.5$. In each panel, the solid line refers to $\psi_0 = 100\,M_\odot\,yr^{-1}$ and $n = 0.5$ (dusty early-type galaxies); the dashed line, to $\psi_0 = 50\,M_\odot\,yr^{-1}$ and $n = 0.5$; the dotted line to $\psi_0 = 100\,M_\odot\,yr^{-1}$ and $n = 1$. The data points are those collected by Eales et al. (1993a; their Fig. 3). The open symbols correspond to line-corrected magnitudes.

The effect of dust has been taken into account based on the following assumptions: the dust to gas ratio is proportional to the gas metallicity $Z_g(T)$; the "standard" grain model (Draine & Lee 1984), incorporating a power law grain size distribution, holds at any time. PAH molecules are included following Puget & Léger (1989).

Corrections for internal extinction have been computed assuming that the elliptical galaxies are spherically symmetric, with a density profile described by King's (1966) formula, and adopting the interstellar extinction curve given by Seaton (1979) for $\lambda \leq 0.37\,\mu m$ and by Rieke & Lebofsky (1985) at longer wavelengths. The optical depth has been normalized requiring that the far-IR to blue luminosity ratio at the present time is equal to the mean value, $L_{FIR}/L_B \simeq 7 \times 10^{-3}$, derived by Mazzei & De Zotti (1994b) for a complete sample of local elliptical galaxies. Its evolution follows directly from that of the gas fraction and of the metallicity.

The diffuse dust emission spectrum is modelled taking into account two components: warm dust, located in regions of high radiation intensity (e.g., in the neighborhood of OB clusters) and cold dust, heated by the general interstellar radiation field. Their relative normalization at the present time is obtained fitting the resulting spectrum to the mean local far-IR colours of ellipticals (Mazzei & De Zotti 1994b). The warm to cold dust emission ratio is assumed to be proportional to the star formation rate; warm dust is therefore dominant in the early evolutionary phases considered here.

As illustrated by Fig. 1, our models satisfactorily account for the increasing spread of K-band magnitudes of radiogalaxies at $z \geq 1$ in terms of the combined effect of evolution of stellar populations and of dust extinction.

3. FITTING THE SEDs OF HIGH-z RADIO GALAXIES

In interpreting the photometric data on radio galaxies it is necessary to take into account that the observed luminosity is a mix of stellar and nuclear emissions. As for the latter, we have adopted the mean continuum spectrum of local AGNs, derived by Granato & Danese (1994). It is also necessary to take into account that low resolution continuum spectra may be distorted by contributions of intense emission lines in the broad-band filters (Eisenhardt & Dickinson 1992; Eales et al. 1993b; Eales & Rawlings 1993).

In general, after subtraction of the emission line contributions to broad-band fluxes, the SED turns out to be consistent with a young starburst, of age in the range 0.05–0.3 Gyr, assuming that the UV continuum is dominated by stellar emission. If, on the other hand, the AGN provides most of the rest-frame emission at 1000Å, stellar populations must be older and galactic ages as large as 1–2 Gyr are derived.

In any case, there is considerable latitude for models, implying, among other things, that little can be inferred from optical/near-IR photometry alone on the amount of dust extinction. Millimeter and sub-mm observations of 4C41.17 and B2 0902+343 imply large masses of dust and the upper limits on 6C1232−39 are still compatible with a similar situation. If so, the total baryonic masses and the bolometric luminosities must be large: $m_b > 10^{12}\,M_\odot$, $L_{bol} > 10^{13}\,L_\odot$.

3.1 4C41.17

The published data on this $z = 3.8$ radio galaxy are shown in Fig. 2. The $2.2\,\mu m$ flux is from the line-free K_S magnitude in a 4" diameter beam measured by Graham et al. (1994). The other oprtical/near-IR points are from Chambers et al. (1990).

This radiogalaxy has been clearly detected at mm and sub-mm wavelengths (Dunlop et al. 1994; Chini & Krügel 1994). As discussed in both papers, it is highly implausible that the observed emission is of non-thermal origin. The present models straightforwardly account for it in terms of dust re-radiation, dominated by warm dust, heated by the hot, young stars.

Panel a) of Fig. 2 illustrates fits by "young" models. The galactic ages range from 0.05 to 0.1 Gyr; thus, 4C41.17 may really be a protogalaxy, in the sense that only a small fraction of stars have yet formed; in fact, our models yield a gas fraction $f_g \simeq 0.9$. The best fit non-thermal contribution (solid line) is $\simeq$ 35%–50% of the continuum emission at $\lambda = 1000$ Å.

The bolometric luminosity is in the range $1.2–3.6 \times 10^{13}\,L_\odot$. Most of the uncertainty follows from the lack of far-IR measurements. Measurements with the forthcoming Infrared Space Observatory (ISO) should be able both to assess the bolometric luminosity and to strongly constrain the dust temperature distribution.

The internal reddening, in the observer's frame, implied by the models is $E(R-K) = 1.0–1.6$ (the largest value corresponding to the highest value of T), consistent with Eales and Rawlings's (1993) estimate [$E(R-K) = 1.6$] based on line ratios. Models imply a very high SFR, ranging from about 5000 up to $\simeq 8000\,M_\odot\,yr^{-1}$, where the lowest value correspond to the highest m_l.

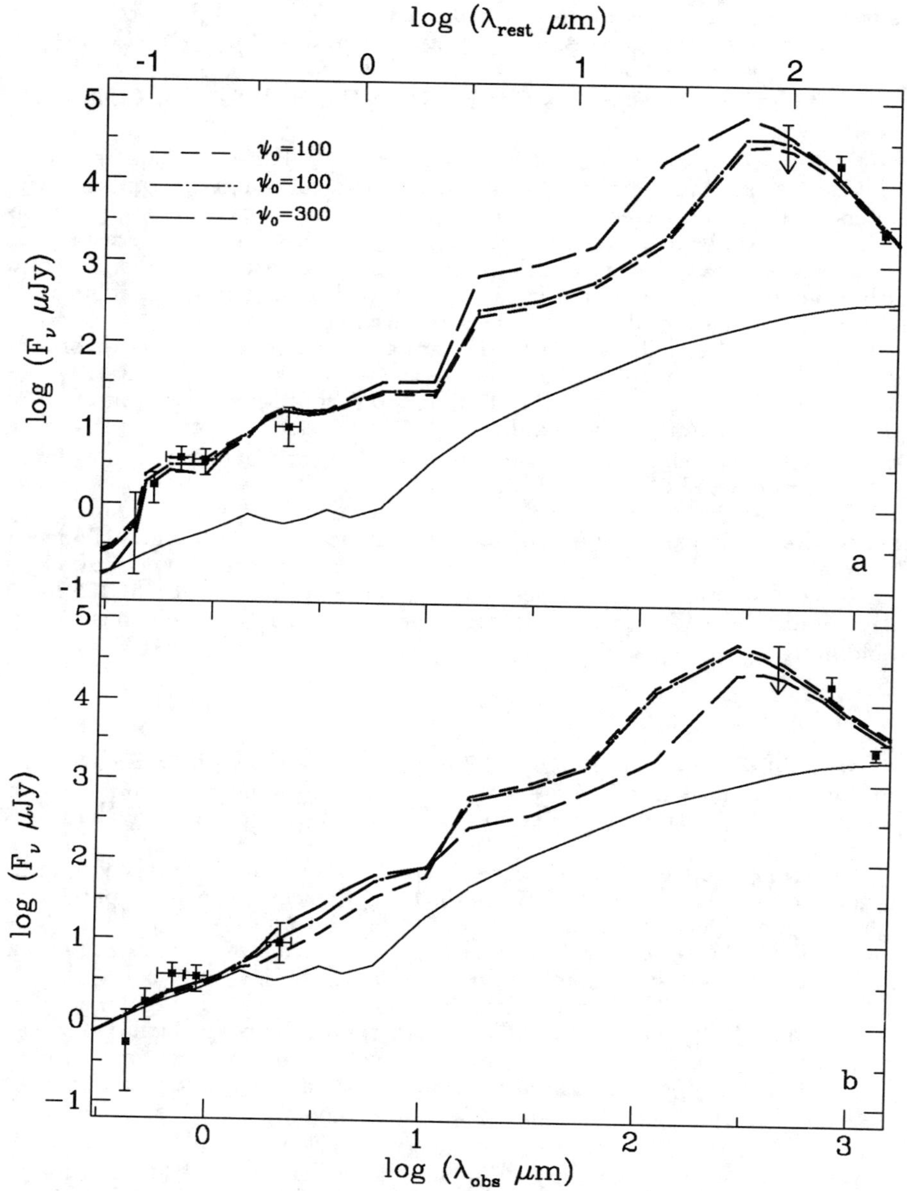

Figure 2: Spectral energy distribution (SED) of $4C41.17$. All models have $n = 0.5$ [cf. eq. (1)]. The solid line shows the adopted non-thermal SED. Models in the upper panel have ages of 0.1 Gyr (short dashed and dot-dashed lines) or of 0.05 Gyr (long dashes). The lower panels shows models with ages of 2 Gyr. The data points are described in the text.

3.2 B2 0902 + 343

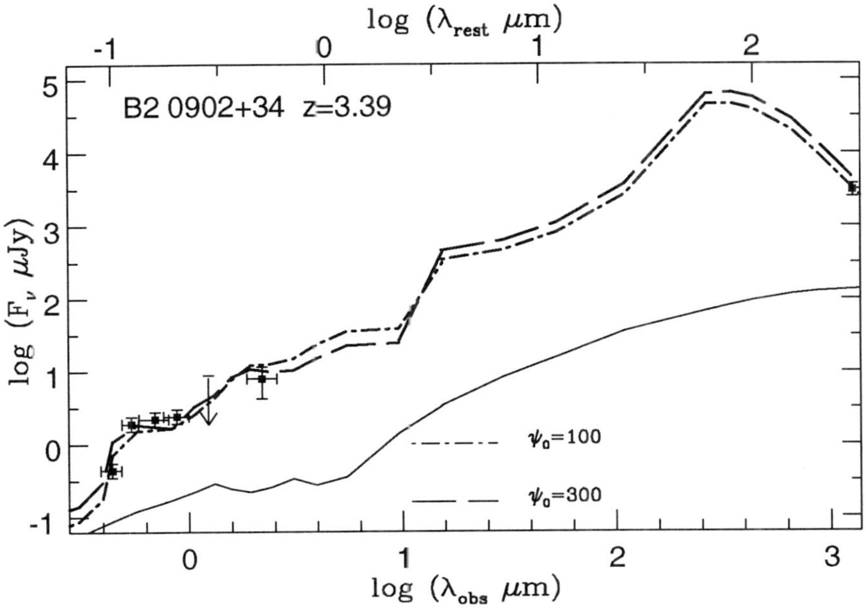

Figure 3: SED of B2 0902 + 343 for ages of 0.3 Gyr (dot-dashed line) and 0.05 Gyr (long dashes). All models have $n = 0.5$. The solid line shows the adopted non-thermal SED. The data points are described in the text.

Shown in Fig. 3 are the optical observations of Lilly (1988), the K-band flux, after subtraction of the line emission, estimated by Eales et al. (1993b; see also Eisenhardt & Dickinson 1992), and the detection at 1.3 mm by Chini & Krügel (1994).

Evidences of strong extinction have been reported by Eales & Rawlings (1993), who estimated $E(R-K) = 2.1$ mag from line ratios, and by Eisenhardt & Dickinson (1992).

*3.3 6C*1232 + 39

Deep optical, infrared, and radio images of this $z = 3.22$ radio-galaxy have been obtained by Eales et al. (1993a). The K-band continuum flux is very uncertain: the upper limit in Fig. 4 is from Eales & Rawlings (1993). Indications of dust obscuration of the center of the galaxy are reported by Eales et al. (1993a). The upper limits at 1.3 mm are from Chini & Krügel (1994).

Because of the lack of a reliable estimate of both the dust extinction and the far-IR emission, models allow the possibility of either small or large (up to $E(R-K) = 1.7$ mag) dust extinction. Fits are obtained for ages in the range 0.05–1 Gyr.

4. CONCLUSIONS

Our main conclusions can be summarized as follows.

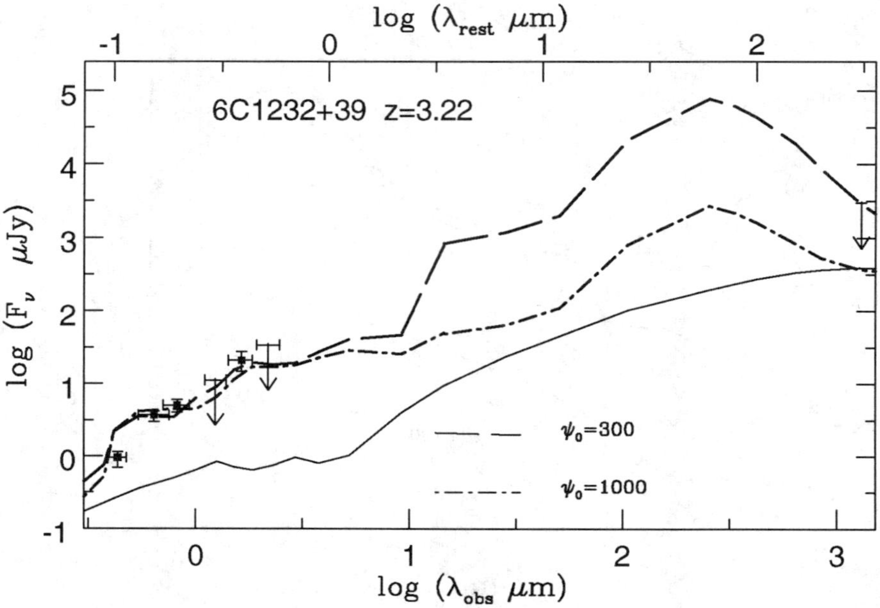

Figure 4: SED of 6C1232 + 39. The model represented by long dashes has $n = 0.5$, age of 0.1 Gyr, and strong dust absorption; the dot-dashed line corresponds to $n = 1$, age of 0.2 Gyr and weak dust absorption. The solid line shows the adopted non-thermal contribution. The data points are described in the text.

The analysis of the observed SEDs of the best studied $z > 3$ radiogalaxies shows that all the ingredients (stellar population synthesis, dust extinction, contribution of the AGN) need to be taken simultaneously into account. Ignoring one of them may severely bias the conclusions.

Optical/near-IR photometry alone does not allow to tell whether or not these galaxies are strongly or weakly obscured by dust. As illustrated by Fig. 4, similarly good fits in the optical/near-IR can be obtained with widely different amounts of dust extinction. Far-IR/sub-mm data are essential to discriminate among the various possibilities.

The couple of high-z radiogalaxies for which sub-mm/mm measurements have been obtained, contain large amounts of dust. Their SED can be accounted for by the general class of models fitting the continuum spectrum of the hyperluminous galaxy IRAS F10214 + 4724 at $z = 2.3$. While this is consistent with the rather general arguments suggesting that dust extinction plays an important role during the early evolution of elliptical galaxies (Franceschini et al. 1994), it is still impossible to draw firm conclusions on the role of dust in these phases. The planned large scale survey with the ISO photometer will no doubt shed light on this problem.

In general, allowance for dust reddening leads to decrease the estimated ages of radiogalaxies (McCarthy 1993), thus alleviating the problem of galaxy ages uncomfortably high in comparison with the age of the universe as well as the difficulty to understand the alignement between optical/infrared continua and the (presumably short lived) radio jets. On the other hand, if there is a substantial non-thermal component contributing in the UV, the stellar component cannot be too blue, implying ages of 1–2 Gyr. Thus, once we allow for the different components, each of which is poorly determined, we end up with a large uncertainty in the ages of radiogalaxies. Far-IR/sub-mm data, assessing the amount of dust absorption, are very helpful in constraining also the galactic ages.

ACKNOWLEDGEMENTS

It is a pleasure to acknowledge useful discussions with L. Danese and A. Franceschini. Work supported in part by ASI.

REFERENCES

Bertelli, P., Betto, R., Bressan, A., Chiosi, C., Nasi, E., & Vallenari, A. 1990, A&AS, 85, 845

Buat, V. 1992, A&A, 264, 444

Chambers, K.C., Miley, G.K., & van Breugel, W. 1990, ApJ, 363, 21

Chini, R., & Krügel, E. 1994, A&A, 288, L33

Draine, B.T., & Lee, H.M. 1984, ApJ, 285, 89

Dunlop, J.S., Hughes, D.H., Rawlings, S., Eales, S.A., & Ward, M.J. 1994, Nat, 370, 347

Eales, S.A., & Rawlings, S. 1993, ApJ, 411, 67

Eales, S.A., Rawlings, S., Dickinson, M., Spinrad, H., Hill, G.J., & Lacy, M. 1993a, ApJ, 409, 578

Eales, S.A., Rawlings, S., Puxley, P., Rocca-Volmerange, B., & Kuntz, K. 1993, Nat, 363, 140

Eisenhardt, P., & Dickinson, M. 1992, ApJ, 399, L47

Franceschini, A., Mazzei, P., De Zotti, G., & Danese, L. 1994, ApJ, 427, 140

Graham, J.R., et al. 1994, ApJ, 420, L5

Granato, G.L., & Danese, L. 1994, MNRAS, 268, 235

King, I. 1966, AJ, 71, 64

Lilly, S.J. 1988, ApJ, 333, 161

Madore, B.F. 1977, MNRAS, 178, 1

Mazzei, P. 1988, Ph.D. thesis, ISAS, Trieste

Mazzei, P., & De Zotti, G. 1994a, MNRAS, 266, L5

Mazzei, P., & De Zotti, G. 1994b, ApJ, 426, 97

Mazzei, P., & De Zotti, G. 1995, MNRAS, submitted

Mazzei, P., De Zotti, G., & Xu, C. 1994, ApJ, 422, 81

Mazzei, P., Xu, C., & De Zotti, G. 1992, ApJ, 256, 45

McCarthy, P.J. 1993, ARA&A, 31, 369

Puget, J.L., & Léger, A. 1989, ARA&A, 27, 161

Rieke, G.H., & Lebofsky, M.J. 1985, ApJ, 288, 618

Salpeter, E.E. 1955, ApJ, 121, 161

Schmidt, M. 1959, ApJ, 129, 243

Seaton, M.J. 1979, MNRAS, 187, 73P

Windhorst, R.A., et al. 1991, ApJ, 380, 362

CONTRIBUTION FROM AGNs TO THE IR BACKGROUND

Gian Luigi Granato, Alberto Franceschini
Osservatorio Astronomico di Padova - Vicolo Osservatorio 5 - I35122 - Italy

and Luigi Danese
SISSA - Strada Costiera 11 - 34014 Trieste - Italy

ABSTRACT

The IR spectra of broad– and narrow–line Active Galactic Nuclei (AGN) are modelled on the assumption of existence of dust tori around the active nuclei. The outcomes of the model are compared with broad IR spectra, visibility of broad lines in the IR, the extension of the IR emission and the X–ray properties of type 1 and 2 AGN. The contribution to the mid–IR background is then estimated properly taking into account local Luminosity Functions, cosmological evolution and spectral properties. Under plausible assumptions, the (dusty) AGN may yield a significant fraction of the extragalactic background in the range 10-50 μm.

1. Introduction

The Cosmic Infrared Background Radiation (CIBR) is usually ascribed to integrated emission of normal galaxies. On the other hand AGN exhibit significant IR emission just in the mid–IR where galaxies are at a minimum. Moreover in the light of unified models, which have been worked out to explain the observed differences between broad– and narrow–line AGN, many 'hidden' (not easily detected by serendipitous surveys in the optical and soft X–ray energy bands) AGN surrounded by dusty tori may exist. The tori absorb a significant fraction of nuclear light at shorter wavelengths and reradiate it in the IR with a peak around 20-40 μm. Hence it is worthwhile to explore the problem of the AGN contribution to CIBR. In the following we present models of IR Spectral Energy Distributions (SED) worked out on the assumption of existence of dust tori around the active nuclei. Then Luminosity Function and evolution of AGN in IR bands are discussed and predictions of the contribution from AGN to CIBR are presented. In particular we estimate the contribution by 'obscured' AGN which may possibly produce a significant fraction of the Hard X–ray Background (HXRB).

2. IR emission in AGNs: observations and models

In models aiming to AGN unification dust is organized in toruslike structures, responsible at the same time for the IR radiation as well as for the apparent differences between broad– and narrow–lined AGN. Strong support to this view comes from spectropolarimetry of a number of Seyfert 2 galaxies, showing that in polarized light they have the same spectra as Seyfert 1's. Less direct evidences are the ionization cones and the ionizing photon deficit.

Various structures have been investigated, depending on equatorial optical depth, covering factor, shape, as well as radial and axial dust density

© 1996 American Institute of Physics

distribution. The models can be broadly divided into two classes. A first one includes models with extremely high optical depths ($A_V \gtrsim 600$) and extremely compact structure (a few pc). In the second family of models tori are more extended (many tens of pc), whereas optical depths cover a large range of values ($10 \lesssim A_V \lesssim 1500$). We refer in the following to the two classes of models as "compact" and "extended" respectively.

Compact models have been worked out by Krolik and coworkers (see Pier and Krolik 1993 and references therein) mainly on the basis of theoretical considerations about the possible dynamical equilibrium and the origin of the structure. However the involved physics is very uncertain.

Extended models have been investigated by Granato and Danese (1994), motivated by observations showing that NGC 1068 and other Seyfert 2 galaxies have mid–IR, CO and HCN emission extending over tens of parsecs (Braatz et al. 1993, Cameron et al. 1993, Tacconi et al. 1994). An efficient computer code has been worked out which solves for the radiative transfer in an axially symmetric circumnuclear dust distribution (Granato and Danese, 1994). The program computes SEDs emitted in various directions as well as visibility maps. In our computations, the inner radius of the cloud has been set by the sublimation of dust when its radiative equilibrium temperature raises above ~ 1500 K, and we used a standard galactic dust composition. Among the many conceivable free parameters of the model, the most important ones are the covering factor f of the dust, the ratio between the outer and the inner radius r_o/r_i, the exponent β of the power law dependence on r of the dust density ($\propto r^{-\beta}$) and the optical thickness of the cloud A_V along an equatorial direction.

We found that to reproduce the shapes of observed IR SEDs of Seyfert 1 galaxies and of NGC 1068 extended and moderately thick tori are required: $r_o/r_i \sim 10^2 - 10^3$, where $r_i[\text{pc}] \sim L_{\text{bol},46}^{1/2}$, and $A_V \sim 10 - 300$. Besides the prediction of the extended emission, the additional merit of the model is that the IR broad band spectra of AGN are fitted by torus emission by varying optical depth and the ratio r_o/r_i with no additional components as those required by compact models. It is worth remembering that when $A_V \gtrsim 20$, even mid–IR emission is self-absorbed in the torus, and therefore significantly anisotropic.

Detection of broad components in hydrogen IR lines in some Seyfert 2s (Goodrich et al. 1994, Ruitz et al. 1994) indicates that, at least in a fraction of Seyfert 2s, the optical depth at around 1.3 μm should not exceed significantly 1, and thus $A_V \sim 10$. On the other hand direct observations of HCN emission of NGC 1068 revealed extended emission with A_V in excess of 70 mag (Jackson et al. 1993). Thus data suggest a broad distribution of optical depths in the tori around AGN.

The photoelectric absorption observed in the hard X–ray band spectra of Seyfert 2 galaxies are explained by assuming a broad distribution of column densities along the lines of sight $N_H \sim 10^{22} \div 10^{25}$. These values, adopting a standard dust–to–gas ratio, translates into $A_V \sim 5 \div 5000$. However very large values of A_V must be taken with caution. In particular it has been suggested that the X–ray absorption may originate in a gas–rich region inner to the dust sublimation radius (*e.g.* Lawrence 1991). Alternatively the fraction of heavy elements condensed in dust grains may be smaller in tori than in ISM of our Galaxy. Soon coming observations with ISO Short Wavelength Spectrograph of broad IR hydrogen lines in Seyfert 2 galaxies compared with hard X–ray observations will help to solve this problem.

In conclusion extended models nicely fit the observed IR broad band spectra and IR maps of AGN. They are also flexible enough to be exploited to predict the contribution to the CIBR under a large variety of assumptions.

3. A mid-IR background dominated by AGNs?

In principle, the required ingredients to estimate the contribution from AGN to the mid–IR background are (i) the distribution of model parameters affecting the spectral properties, in particular A_V; (ii) the distribution of covering factor, which is linked to the relative number of broad– and narrow–line AGN; (iii) their local space density and evolution with z.

As we have seen, at the moment we are able to put only weak constraints on (i) and (ii), particularly if we take into account for the fact that the distribution of model parameters are likely to be functions of both z and L (*e.g.* Lawrence 1991). As for point (iii), IR LLFs are rather well defined but evolution is scarsely known.

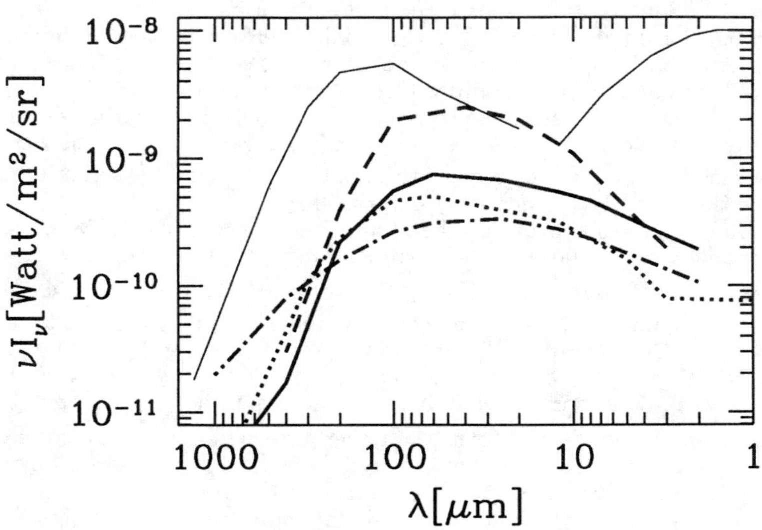

Figure 1. Comparison between the contribution to CIBR from AGN under different assumptions (thick lines) and that from normal galaxies (thin lines, see text).

Nevertheless information is sufficient to infer plausible estimates of upper and lower limits of the AGN contribution to CIBR. We adopt the LLFs computed at 12 μm by Rush, Malkan and Spinoglio (1993). We also assume that all obscured AGN have the same IR spectra as NGC1068, while all "type 1s" have a SED well represented by the mean SED of Seyfert 1s computed by Granato and Danese (1994). Then minimal and maximal prediction (respectively dotted and dashed thick lines in figure 1) differ for the adopted evolution rate and the relative number of obscured objects. Namely we used a luminosity evolution rate $L(z) = L_o(1+z)^\alpha$ with $\alpha = 3.5$, as possibly suggested by the optical counts, for the maximal case and $\alpha = 2.75$, more suitable for evolution in X–ray bands, for the minimal one, both extrapolated up to $z = 4$ with $q_o = 0.05$. In addition, the lower line is computed assuming the same number density for broad and narrow-line AGN, *i.e.* $f = 0.5$, while the maximal estimate refers to $f = 0.8$ (four obscured AGNs for every not obscured one). From Figure 1 it is apparent that in the maximal hypothesis the contribution of dusty AGN in the range 10–50 μm may exceed those of other galaxy populations as estimated by Franceschini et al. (1994) (thin solid line).

Spectral fit to HXRB requires substantial contribution from X–ray absorbed AGN (*e.g.* Franceschini et al. 1993). The distribution of N_H used to reproduce the HXRB (*e.g.* Comastri et al. 1995) can be exploited to infer a distribution of A_V and corresponding spectra of obscured AGNs. For a standard galactic dust to gas ratio, we would have $A_V \simeq 0.5 \times 10^{-21} N_H$. The dot–dashed curve is indeed computed using the distribution of N_H by Comastri et al. and this conversion law. The resulting CIBR is comparable to the previous minimal case. On the other hand, as already pointed out, there is a case for a dust to gas ratio much lower than "standard" in AGN. Assuming a ratio $A_V/N_H \simeq 0.5 \times 10^{-22}$ which is by a factor of 10 lower than in local ISM, the contribution to the CIBR in the mid–IR (solid line) is a significant fraction of the total.

References

Braatz, J.A., Wilson, A.S., Gezari, D.Y., Varosi, F. & Beichman, C.A., 1993, ApJ, 409, L5

Cameron, M., Storey, J.W., Siebenmorgen, R., & Lee, T.J., 1993, ApJ, 419, 136

Comastri, A., Setti, G., Zamorani, G., Hasinger, G., 1995, AA, in press

Franceschini, A., Mazzei, P., De Zotti, G., Danese, L., 1994, ApJ, 427, 140

Franceschini, A., Martin–Mirones, J.M., Danese, L., De Zotti, G., 1993, MNRAS, 264, 35

Jackson, J.M., Paglione, T.A.D., Ishizuki, S. & Rieu, N.Q., 1993, ApJ, 418, L13

Granato, G., & Danese, L., 1994, MNRAS, 268, 235

Lawrence, A., 1991, PASP, 99, 309

Lawrence, A., 1991, MNRAS, 252, 586

Mushotzky, R.F., Done, C., & Pounds, K.A. 1993, ARAA, 31, 717

Pier, E.A., & Krolik, J.H. 1993, ApJ, 418, 673

Rush, B., Malkan, M.A., & Spinoglio, L., 1993, ApJS, 89, 1

Tacconi, L.J., Genzel, R., Blietz M., Cameron, M., Harris, A.I., & Madden, S., 1994, ApJ, 426, L77

RADIATIVE TRANSFER IN CLOUDY GALAXIES: EVOLUTION OF FAR-IR LUMINOSITY

Adolf N. Witt & Karl D. Gordon
The University of Toledo, Toledo, OH 43606 USA

ABSTRACT

The partial absorption of stellar radiation and its re-emission as thermal radiation by interstellar grains in galaxies is the principal process by which galaxies become important emitters of far- infrared and sub-millimeter radiation. The efficiency of conversion, given a fixed amount of dust with set optical properties, depends largely upon the distribution of the dust relative to the illuminating sources. Within the same basic geometry, it is the structure of the dust-carrying ISM which determines ultimately the ratio of observed far-IR to UV/optical luminosities of galaxies.

We report on new multiple scattering radiative transfer calculations for galactic environments with two-phase clumpy media, which are compared with observations of samples of isolated and interacting/merging galaxies. We conclude that in addition to starbursts occurring in the latter sample, changes in the structure of the interstellar media in interacting galaxies contribute to their high far-IR luminosity with a near normal mass of interstellar matter. The changing structure of the ISM in galaxies over time is an important factor in the evolution of far-IR luminosity function of galaxies.

1. Introduction

It is generally accepted that the integrated thermal radiation of dusty galaxies is a major component of the cosmic infrared background. It is probably the dominant component in the wavelength range from 30 to 300μm. Current models (*e.g.* Franceschini *et al.* 1991) predict a background in this spectral region which is still below the upper limits set by COBE (Hauser 1995, this volume) by factors of five to ten. Models which include evolution of galaxies generally consider the changes of the spectral energy distribution and the luminosity of the stellar populations which power dusty galaxies (*e.g.* Guiderdoni & Rocca-Volmerange 1987), but they ignore the possible changes arising from an evolution of the conversion efficiency of UV/optical/near-IR stellar emission into thermal emission by dust in the mid- and far-IR. That such evolution is indeed occurring is suggested by deep IRAS source counts ($F_\nu > 50$ mJy) at 60μm. Such counts (Hacking *et al.* 1987; Franceschini *et al.* 1988; Lonsdale *et al.* 1990; Gregorich *et al.* 1995) show that IRAS galaxies corresponding to K-magnitudes of 14 to 15 ($z \approx 0.04$ to 0.06) are more numerous by factors of two to four than predicted by no-evolution models, whereas deep K-magnitude counts of galaxies (Yoshii & Peterson 1995; Gardner 1995, this volume) show no indication of evolution effects at such bright magnitudes. We, therefore, must examine answers to the following questions:

1. What determines the efficiency of a galaxy as a far-IR radiator?
2. How can this efficiency change as a function of time or as a function of environment?

© 1996 American Institute of Physics

2. What Makes a Galaxy "infrared-active"?

Continuous extinction by interstellar dust is the dominant source of opacity in the interstellar spaces of galaxies in the spectral range $\lambda > 912$ Å where most photospheric emission from stars occurs. The average extinction law for the Galaxy (Fig. 1) appears to be at least representative of such laws for galaxies in general in the sense that the opacity increases toward shorter wavelengths (*e.g.* Brosch *et al.* 1990). Any shift in the evolution of the spectral energy distribution of the dominant stellar sources toward shorter wavelengths will lead to an increased conversion of stellar photons into far-IR emission. Fig. 1 also shows the wavelength-dependence of the dust albedo. Over the important UV-opt. spectral range, only about 50% of photons are absorbed upon a first interaction with dust grains; multiple scattering and thus greater optical depths will be required for complete absorption.

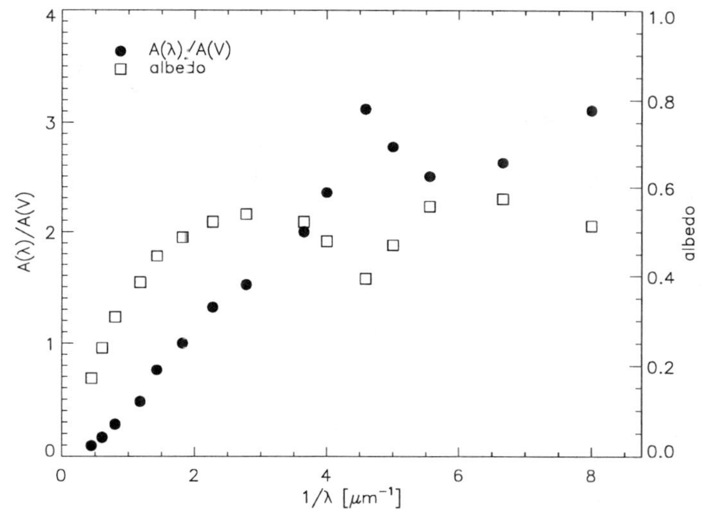

Fig. 1.— The average Galactic extinction curve (filled circles), normalized to V, for $R_V = 3.1$ and the wavelength dependence of the grain albedo (open squares).

Obviously, the total amount of dust present is an important parameter. ISM-poor systems or galaxies where the dust-to-gas ratio is low will generally not be found among IR-luminous objects. However, more important than the total amount of dust is the distribution of the dominant sources, be they stars or active galactic nuclei, in optical depth space. This distribution is determined by the geometry of the dust relative to the sources and, equally important, by the structure of the dusty medium.

3. Radiative Transfer Models

We have conducted radiative transfer calculations for spherical systems containing dust with average Galactic characteristics (Fig. 1). Various geometries as introduced by Witt, Thronson, & Capuano (1992, hereafter WTC) have been employed. Important improvements beyond WTC included the detailed wavelengths dependence of the transfer at wavelengths shortward of the U-band, an extension to visual optical depths of 30 to 40 and most importantly, full radiative transfer in a 2-phase clumpy medium, with multiple scattering. The calculations were done using the Monte-Carlo approach.

We illustrate results for the case of the "dusty" galaxy, representing a uniform mixture of sources and dust. Fig. 2a shows the distribution of sources in optical depth space for a spherical, uniform homogeneous distribution of dust and stars, with a radial optical depth of unity. As expected, the majority of the sources are located near the surface and see optical depths < 1 for most of their photons emitted away from the center of the distribution. A fraction of photons, however, is emitted toward the center region and thus encounters optical depths $1 < \tau < 2$. In Fig. 2b, the identical dust mass has been converted into a randomly-clumpy 2-phase distribution. The filling factor of clumps is 15% by volume, the density ratio of interclump matter to that in clumps is $k_2/k_1 = 0.01$ and the size of clumps is equal to 0.1 in units of the radius of the system. Clumpiness results in a dramatic change of the distribution of sources (left spatially unchanged) in optical depth space. The general consequence is a substantial broadening of the distribution, resulting in a large number of sight lines of very low optical depth near zero, where light escapes through spaces between clouds through the low-density intercloud medium. On the other hand, there now are also numerous lines of sight passing through several clumps, resulting in optical depths $\tau \gg 2$.

The net effect, however, is a substantial reduction of the effective opacity of the medium, for the identical dust mass and the identical basic dust-source geometry.

The consequences of a reduced effective opacity resulting from the structure of the interstellar medium are shown in Fig. 3. We have assumed a spectral energy distribution for the radiating stellar population equal to that of NGC 4449 (WTC 1992, Bruzual 1986), a relatively unreddened starburst galaxy. By embedding this source distribution into different dust distributions, the ratio of the far-IR flux to the residual flux integrated from 0.0912-2.2μm (stellar and scattered light) could be computed. This quantity is equal to about $0.5 \times$ L(IR)/L_B.

The following general conclusions can be drawn:

1. For a given dust mass and basic geometry, a homogeneous dust distribution is always more efficient in converting UV-optical light into far-IR emission than is any clumpy distribution.

2. While conserving mass, more numerous smaller clumps produce a higher effective opacity than fewer larger-sized clumps.

3. The role of the interclump medium is very important. Evaporation of clumps and resultant increases in the interclump medium density greatly enhance the effective opacity of the system.

4. The conversion efficiency of UV-optical light into far-IR emission for a constant geometry and dust mass can be changed by substantial factors by changing the structure of the medium.

We find, in general, the most conversion-efficient geometry is that of a homogeneous spherical shell, followed by uniformly distributed source/dust mixtures.

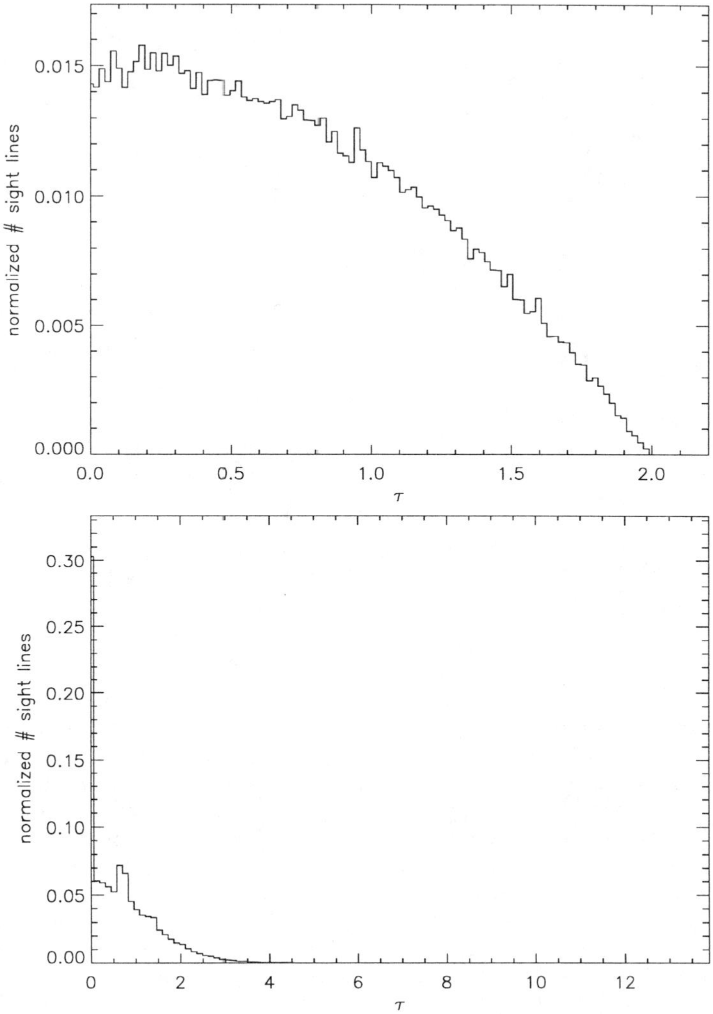

Fig. 2.— (a) top, Distribution of sources in optical depth space for a homogeneous, spherical distribution of dust and sources; (b) bottom, distribution in optical depth space of the same sources for a two-phase clumpy distribution of the identical dust mass.

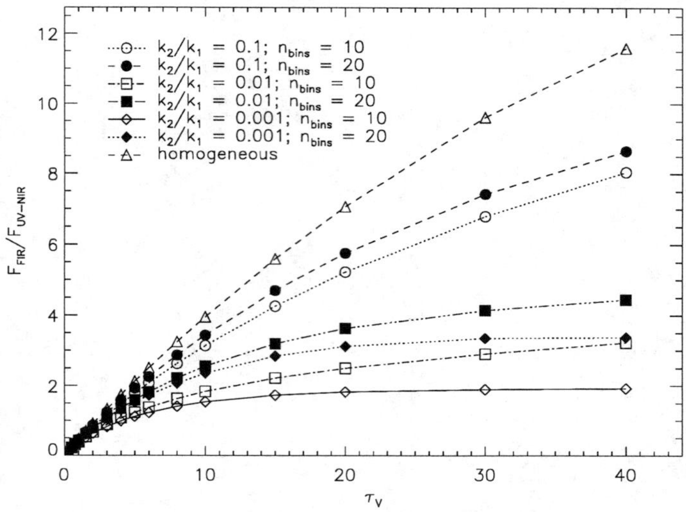

Fig. 3.— The ratio of far-IR flux to residual UV/opt/near-IR flux for extended spherical source distributions of constant-mass dust with varying degrees of clumpiness.

4. Isolated and Interacting Galaxies

In order to compare samples of galaxies with greatly different average ratios of $L(IR)/L_B$, we examine the samples of isolated galaxies and of interacting/merging galaxies defined by Young *et al.* (1986). For these objects, 13 in each group, we obtained values for the IR luminosity, L(IR), the blue luminosity, L_B, and the total mass of interstellar hydrogen, $M(HI + H_2)$, from the compilation of Young *et al.* (1989). We plotted these objects in Fig. 4, showing the ratio $L(IR)/L_B$ as a function of $M(HI + H_2)/L_B$. As noticed by Young *et al.* (1986) these two groups of galaxies have similar ratios of $M(HI + H_2)/L_B$ but differ by nearly an order of magnitude in $L(IR)/L_B$, with the interacting galaxies the more IR-active ones. However, there is significant overlap between the two samples; not all interacting galaxies are exceptionally IR- luminous.

In Fig. 5 we show these two samples in comparison with predictions from our radiative transfer models, with optical depths suitably scaled to match the mass of interstellar hydrogen. The comparison suggests that interacting/merging systems derive their higher IR luminosity from the fact that the dominant sources are embedded in extended, homogeneous dusty media. The two most luminous sources in the sample are Mrk 231 and Arp 220. The less luminous interacting galaxies appear to match clumpy distributions where clumps and sources are nevertheless similarly distributed. The isolated galaxies are well represented by "cloudy" models with filling factors of $(10 \pm 5)\%$, where the clouds are located in the interior and most stars are outside the region occupied by dust. This closely resembles the conditions in typical, undisturbed spiral disks, where most of the interstellar matter is concentrated in a thin central layer, with particular clumping in spiral features.

Young *et al.* (1986) originally suggested that the high IR luminosity of

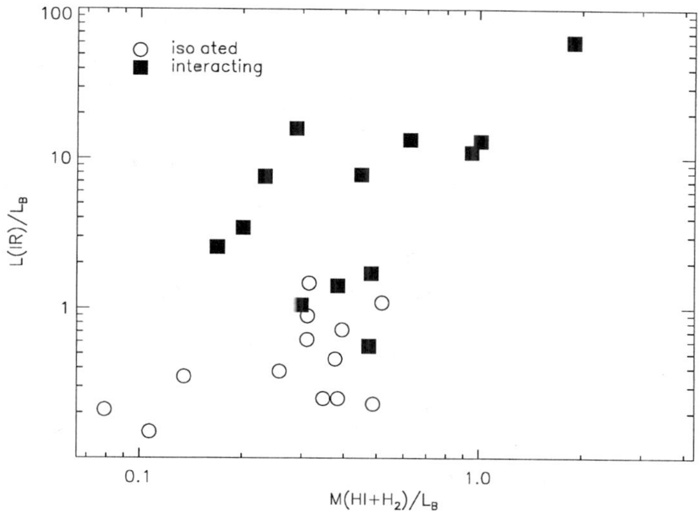

Fig. 4.— Samples of isolated and interacting galaxies of Young *et al.* (1986) show systematically different ratios of $L(IR)/L_B$ for comparable ratios of $M(HI + H_2)/L_B$ (in units of $M_\odot/L_\odot$).

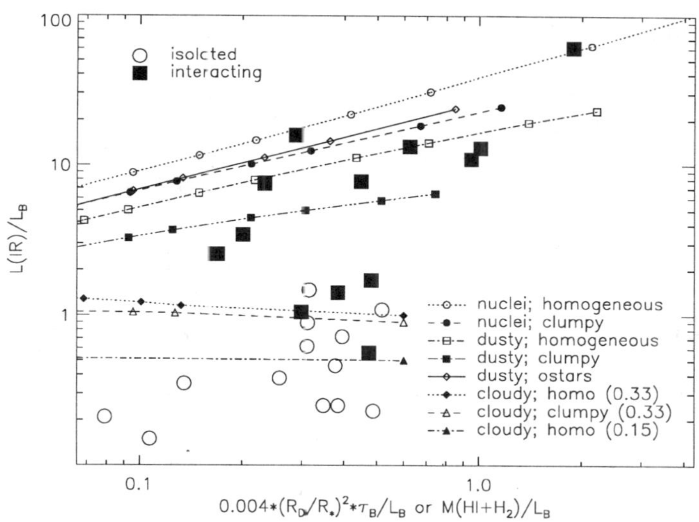

Fig. 5.— Comparison of radiative transfer models with samples of isolated and interacting galaxies (Young *et al.* 1986).

their sample of interacting galaxies resulted from starbursts. These starburst must occur deep in the interior of massive clouds. We have produced one model labeled "O-stars" which is based on a clumpy distribution of the medium but has all sources concentrated at the centers of clumps. In terms of the far-IR luminosity produced, it is very similar to homogeneous dust distributions enveloping extended star distributions, in as much as they have very similar distributions in optical depth space. There is independent evidence that starburst galaxies are distinguished by a more generally distributed interstellar medium characterized by numerous, small clouds (Casoli *et al.* 1991; Aalto *et al.* 1991, Braine & Combes 1993, Carral *et al.* 1994). This suggests that even after the star-forming regions in starbursts break out of their original locations, the general dispersal of the interstellar medium causes conditions where the efficiency of conversion of stellar photons into thermal dust emission remains high for an extended period. We conclude, therefore, that the structure of the interstellar medium, and the evolution of this structure and its distribution with respect to the dominant sources, must be considered in the IR-luminosity evolution of dusty galaxies.

Acknowledgments

This work was supported by NASA Grant NAGW-3168 to The University of Toledo.

References

Aalto, S., Black, J.H., Booth, R.S., & Johansson, L.E.B. 1991, 247, 291

Braine, J. & Combes, F. 1993, A&A, 269, 7

Brosch, N., Almoznino, E., Grosbol, P., & Greenberg, J.M. 1990, A&A, 233, 341

Bruzual, A.G. 1986 in Spectral Evolution of Galaxies, ed. C. Chiosi & A. Renzini (Dorchrecht: Reidel), p. 263

Carral, P., Hollenbach, D.J., Lord, S.D., Colgan, S.W.J., Haas, M.R., Rubin, R.H., & Erickson, E.F. 1994, ApJ, 424, 223

Casoli, F., Dupraz, C., Combes, F. & Kazèz, I. 1991, A&A, 251, 1

Franceschini, A., Mazzei, P., & De Zotti, G. 1991, in "The Early Observable Universe from Diffuse Backgrounds" ed. B. Rocca-Volmerange, J.M. Deharveng, & J. Tran Thanh Van (Editions Frontieres) p. 249

Franceschini, A., Danese, L., De Zotti, G., & Xu, C. 1988, MNRAS, 233, 175

Gardner, J.P. 1995, this volume

Gregorich, D.T., Neugebauer, G., Soifer, B.T., Gunn, J.E., & Herter, T.L. 1995, IRAS Preprint No. 121

Guiderdoni, B. & Rocca-Volmerange B. 1987, A&A, 186, 1

Hauser, M.G. 1995, this volume

Hacking, P., Condon, J.J., & Houck, J.R. 1987, ApJ, 316, L15

Lonsdale, C.J., Hacking, P., & Conrow, T.P. 1990, ApJ, 358, 60

Yoshii, Y., & Peterson, B.A. 1995, ApJ, 444, 15

Young, J.S., *et al.* 1986, ApJ, 311, L17

Young, J.S., Xie, S., Kenney, J.D.P., & Rice, W.L. 1989, ApJS, 70, 699

Witt, A.N., Thronson, H.A., & Capuano, J.M. 1992, ApJ, 393, 611

THE EFFECTS OF DUST AND METALLICITY EVOLUTION ON THE COSMIC INFRARED BACKGROUND

Eli Dwek

Laboratory for Astronomy and Solar Physics
NASA Goddard Space Flight Center
Greenbelt, MD 20771

and

Frank Városi

Hughes STX
4400 Forbes Blvd, Lanham, MD 20706
and
Laboratory for Astronomy and Solar Physics
NASA Goddard Space Flight Center
Greenbelt, MD 20771

ABSTRACT

The contribution of galaxies to the cosmic infrared background (CIB) is a complicated function of their star formation history, the evolution of their metallicity and dust content, and details of the transfer of the stellar energy output through the dusty interstellar medium. We present preliminary results of models examining the effects of dust evolution on the spectral appearance of galaxies, in order to understand their evolving contribution to the CIB.

1. INTRODUCTION

The cosmic infrared background (CIB) contains the cumulative radiative emission from normal and active galaxies, as well as that from more "exotic" sources such as decaying relic particles, massive population III objects, or accreting black holes (e.g. Bond, Carr, Hogan 1991).the latter one including the radiative decay of various relic dark matter particles. Our current best determined upper limits on the CIB (see Hauser 1995; Stecker 1995; Dwek & Slavin 1994) suggests that normal galaxies contribute most of the CIB energy, ruling out some of the more exotic energy sources.

As summarized by several speakers in this workshop (De Zotti & Mazzei 1995, Franceschini et al 1995, Rocca-Volmerange 1995, Rowan-Robinson & Pearson 1995, and Lonsdale 1995), the contribution of galaxies to the CIB is a complicated function of their dynamical and star formation history, their nucle-

© 1996 American Institute of Physics

osynthesis yield, their spectral energy distribution, the evolution of their dust content, and the transfer of the radiative stellar output through the dusty interstellar medium (Witt & Gordon 1995).

This paper presents the results of our preliminary efforts examining the relations between the various physical processes that determine the contribution of normal galaxies to the UV–IR extragalactic background. Our model starts with a simple prescription for the star formation history in a hypothetical galaxy, and follows the evolution of its stellar spectral energy distribution (SED), its metallicity, and its gas and dust content. The last two factors determine the evolution of the galactic opacity, and using a simple radiative transfer model we follow the transfer of the stellar SED through its dusty interstellar medium. We then calculate the cosmological contribution from a population of such galaxies to the CIB, and compare it with current observational constraints. Features of our model that were not considered in previous studies include detailed modeling of the evolution of dust, including its compositional changes. The abundance of the dust cannot indefinitely trace that of the metals as a function of time, supernova remnants destroy grains and return metals locked up in dust back to the gas phase. The gas and dust opacities evolve, and the models calculate the escaping stellar radiation, the fraction of absorbed starlight, the emerging infrared spectrum, and the escaping Lyα emission as a function of time. The models presented here are highly idealized and not designed to match any observational constraints, such as K-band or 60μm galaxy counts, or the current DIRBE upper limits on the CIB. They should therefore be regarded as a preliminary investigation of the couplings between the various physical processes that effect the UV–IR appearances of galaxies, and their contribution to the CIB.

2. MODEL INGREDIENTS

2.1 Evolution of Dust and Metallicity

The evolution of the mass $m(A)$ of a given element A in the interstellar medium (IMS) of a galaxy can be symbolically written as:

$$\frac{dm(A)}{dt} = -\left\{\begin{matrix} stellar \\ birth \end{matrix}\right\} + \left\{\begin{matrix} stellar \\ production \end{matrix}\right\} + \{infall\} \tag{1}$$

where the first term represents the depletion of the element from the interstellar gas by star formation, the second term its enrichment by stellar nucleosynthesis, and the third term its enrichment due to infall. We assume here a primordial composition for the infalling gas. The evolution of the mass of the same element A that is locked up in dust can be written in a similar fashion as:

$$\frac{dm_d(A)}{dt} = -\left\{\begin{matrix} stellar \\ birth \end{matrix}\right\} + \delta_A \times \left\{\begin{matrix} stellar \\ production \end{matrix}\right\} + \{infall\} - \left\{\begin{matrix} supernova \\ destruction \end{matrix}\right\} \tag{2}$$

where the first three terms describe the identical processes as in eq.(1), and the fourth represents the destruction of grains by supernova remnants (SNRs). The parameter δ_A represents the condensation efficiency of the element A in the stellar ejecta (stellar winds or supernovae ejecta).

We follow the evolution of metallicity by using a simple one-zone chemical evolution mode, taking into account the finite stellar lifetime in calculating the return of the stellar ejecta back to the interstellar medium (ISM). The model is characterized by a constant gas consumption rate due to star formation, given by $\dot{M}(SFR) = 0.5\ M_{\odot} yr^{-1}$, a Miller-Scalo initial mass function (Scalo 1986), and a constant infall rate equal to $\dot{M}(SFR)/2$. The galactic age was taken to be $t_0 = 14\ Gyr$. The nucleosynthesis yield of stars with masses $1M_{\odot} \leq M \leq 8M_{\odot}$ was taken from Renzini & Voli (1981), and for stars with masses $12 \leq$M(M$_{\odot}$) ≤ 40 from Weaver & Woosley (1993), in both cases the yields were selected for stellar metallicities of $Z = Z_{\odot}$. SNIa yields were taken from Thielemann, Nomoto, & Hashimoto (1993), and the ratio of the SNIa to SNII rate was taken to be 0.15 (Thieleman, Nomoto, & Hashimoto 1991).

The composition of the newly−formed dust was determined by the C/O ratio in the stellar ejecta. We assumed that graphite formed when this ratio exceeded unity, and silicates formed when $C/O < 1$. In the former case, all the carbon in the ejecta was assumed to be locked up in dust, i.e. $\delta_C = 1$, with $\delta_A = 0$ for all other elements A. In the latter case, we assumed that all the Mg, Si, S, Ca, Fe, and a mass fraction equal to $16 \times \{m(Mg)/24 + m(Si)/28 + m(Fe)/56\}$ of the O in the ejecta is locked up in dust, and took $\delta_C = 0$. This limits the production of graphite (or carbon) dust to stars with masses between ≈ 2 - 9 M$_{\odot}$, and silicate−type dust to the remaining stars ($M \approx\ 1-2,\ 9-40\ M_{\odot}$).

Figure 1 depicts the essential results of this simple model. It illustrates that the chemical evolution model used here satisfies the $G-$dwarf problem (e.g. review by Rana 1990), and reproduces the abundance of metals in the solar neighborhood. Figure 1c depicts the evolution of dust and metallicity as a function of time. An important aspect of the model is that *the abundance of the dust does not follow that of the metallicity, but rather levels off when the dust production rate is balanced by its destruction rate.* This departure from the trend of the metallicity evolution is caused by the presence of the additional fourth term in eq. (2), representing the grain destruction by SNRs. To reproduce the observed dust-to-gas mass ratios of about 0.003, and 0.004, for respectively, graphite and silicate dust, we had to significantly reduce the SNR destruction rate from its nominal value. Grain lifetimes in our model are $\approx 1\ Gyr$, about ten times their nominal values. Clearly, more realistic models are needed to reproduce the observed interstellar depletions in a way that is consistent with our current understanding of grain destruction processes in the interstellar medium.

Another important feature of the model is that the onset of the formation of graphite grains lags behind that of the silicates. This results from the fact that graphite grains are predominantly produced in stellar winds from low-mass stars, whereas silicates are primarily formed in SN ejecta, which are "instantaneously" recycled back into the ISM. The lag time is about $3 \times 10^7\ yrs$, a sizeable fraction of the duration of a starburst. This effect may therefore have important consequences for the dust opacity and spectral appearance of starburst regions when their line of sight traverses predominantly through the newly-formed dust.

2.2 Evolution of the Stellar Energy distribution and Dust Opacity

The stellar energy distribution (SED) of the model galaxy was followed using standard population synthesis models (e.g.Rocca-Volmerange 1992; Charlot &Bruzual 1991). Evolutionary tracks for for stars with initial solar metallicity abundances, Z=0.02, and metal poor stars with Z=0.001 were taken from

Schaler et al. (1992). Stellar spectra were taken from the library of synthetic spectra generated by Kurucz (1992).

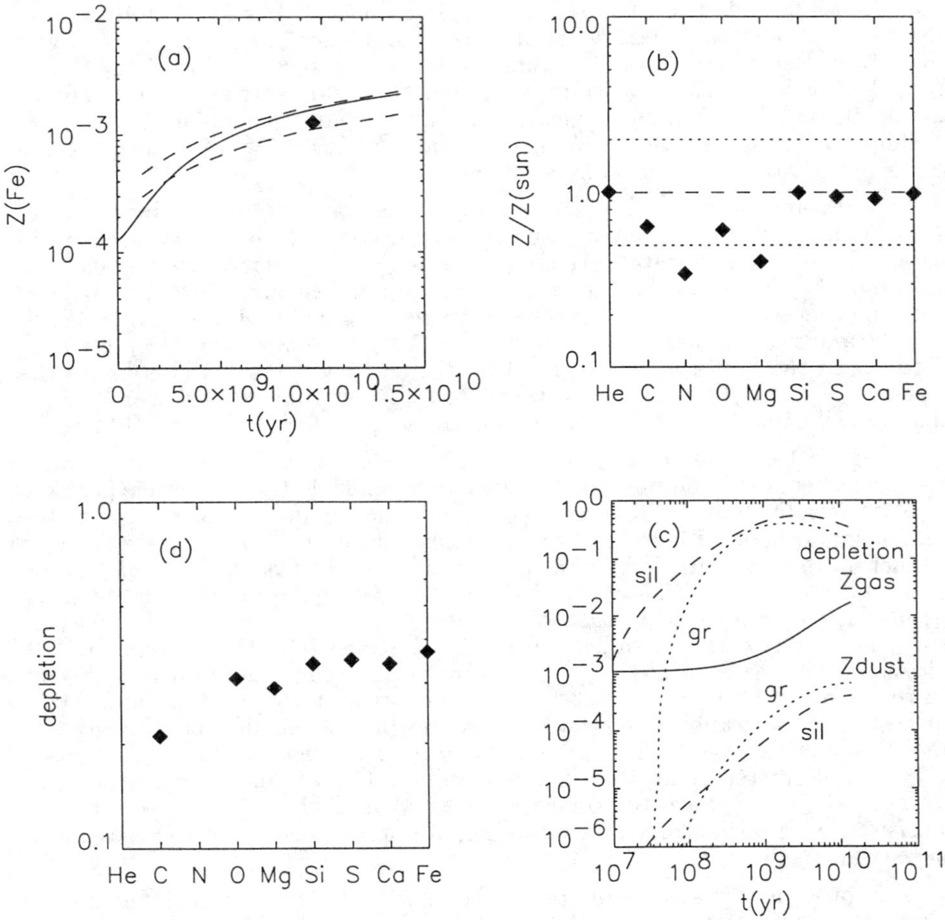

Figure 1: Selected results from the simplified chemical evolution model: (a) the evolution of the Fe abundance as a function of time. The dashed lines show the constraints of its evolutionary history as it is imposed by the solar neigborhood G−dwarfs (e.g. Rana 1990), and the filled diamond represents its solar value (Anders & Grevesse 1989); (b) total abundances (gas + dust) of select elements at time $t = 9.5\ Gyr$, normalized to their solar abundances. The horizontal dotted lines indicate the values where the normalized abundances are a factor of two below and above their solar value; (c) the evolution of the total metallicity Z, and Z_d, the dust-to-gas mass ratio of graphite and silicate grains, and Δ, the fraction of carbon and silicate elements locked up in dust; (d) the depletion Δ_A of the various refractory elements A in interstellar dust, defined as $\Delta_A \equiv m_{dust}(A)/m_{tot}(A)$.

To calculate the emerging stellar spectrum we simply characterized a galaxy as a spherical or a disk-like homogeneous distribution of sources (stars) and absorbers (gas and dust). A galaxy opacity is then completely characterized by its optical dimensions: $\tau_R(\lambda_0)$, its optical radius at some reference wavelength λ_0, and for a disk-shape galaxy by an additional parameter $\tau_z(\lambda_0)$, its opacity in the vertical direction. We chose the spherical and disk galaxies to have equal volumes, and a thickenss-to-diameter ratio of 0.02 for the disk. For the spherical galaxy the photon escape probability is given by the analytical expression given for the pure absorption case by Osterbrock (1989) and generalized by Lucy et al. (1989) to include the effect of scattering:

$$p(\tau, \omega) = p(\tau, 0)[1 - \omega + \omega\ p(\tau, 0)]^{-1} \tag{3a}$$

where

$$p(\tau, 0) = \frac{3}{4\tau}[1 - \frac{1}{2\tau^2} + (\frac{1}{\tau} + \frac{1}{2\tau^2})\ e^{-2\tau}], \tag{3b}$$

τ is the radial optical depth of the galaxy at some wavelength, and ω the dust albedo at the same wavelength.

We confirmed the validity of eqs. (3) with Monte-Carlo simulations which are compared to the analytical formulae in Figure 2.

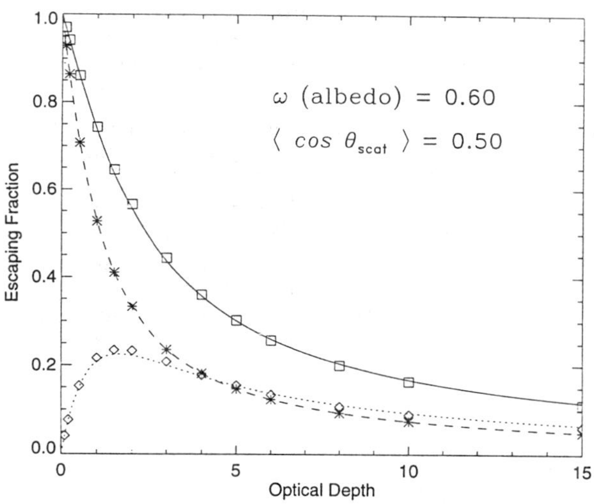

Figure 2: The escape probability from a shperical and homogenous galaxy as a function of radial optical depth. Calculations were performed for a grain albedo of 0.6, and a scattering asymetry parameter, $< \cos(\theta) >$, equal to 0.5. Monte-Carlo simulations, represented by diamonds, squares, and asterixes, are compared to analytical expressions [computed using eqs. (3)] represented by the various curves. The dashed curve is the direct escape probability, that is, the probability that a photon escapes the nebula without undergoing any scattering; the dotted curve represents the additional escape probability via multiple scattering events (the MC simulations followed up to 17 scattering events); and the solid curve represents the total escape probability.

The Figure shows that the the numerical simulations are in reasonable agreement with the analytical expression, with deviations less than ± 6 %. Comparison for other values of the albedo, showed that in general the relative error was always $< 10\%$. We then used Monte-Carlo simulations to calculate the escape probability from a homogeneous disk. The results of these simulations showed that the angle-averaged escape probability from a disk can be approximated by that of a sphere with an effective optical radius given in terms of the disk optical axes by:

$$\tau_R(eff) = \frac{3\tau_z \tau_R}{2\tau_z + \tau_R} , \tag{4}$$

which in our case gives $\tau_R(eff) = 0.183 \times \tau_R(spherical\ galaxy)$. The effect of the disk geometry is thus to reduce the overall opacity of the galaxy, and the amount of UV-visual radiation that is transformed to IR emission. The optical dimensions of a galaxy evolve as a function of time because of its evolving gas and dust content. Figure 3 depicts the evolution of the escape probability as a function of wavelength for two cases. The first case (shown in Fig. 3a), includes only the effects of dust opacity, whereas the second case (Fig 3b) includes the effect of gas absorption as well. The figure shows that the gas effectively absorbs all the radiation shortwards of the Lyman continuum. This is an obvious overestimate of the absorbtivity of the gas, since the gas in real galaxies is clumpy, enabling a fraction of the Lyc photons to escape. We assume that a fraction equal to 0.68 of the absorbed ionizing photons are isotropically re-emitted as Lyα photons. Assuming that the Lyα emitters are homogeneously distributed throughout the galaxy, we calculate the Lyα-photon escape probability, and their contribution to the heating of the dust.

Figure 4 shows the evolution of the energy output from the galaxy. L_{tot} and L_{esc} represent, respectively, the total and escaping stellar luminosity. L_{grf} represents the luminosity absorbed by the graphite grains, and L_{sil} that absorbed by the silicates. L_{gas} is the stellar *Lyc* luminosity that is absorbed by the gas. Energy conservation requires that $L_{tot} - L_{esc} = L_{grf} + L_{sil} + L_{gas}$. The figure shows that initially a large fraction of the stellar radiation escapes the galaxy since the dust abundance, and hence the galactic opacity, is small. As more dust is created, the optical depth of the galaxy increases, resulting in a decrease in the escaping luminosity. The luminosity absorbed by the gas decreases with time, reflecting the increased softening of the SED with increasing stellar metallicity.

2.3 Evolution of the Dust Temperature and the Infrared Luminosity

The energy absorbed by the dust is reemitted as infrared radiation. The dust temperature T_d for a given dust composition is given by the energy balance equation:

$$L_{abs} = 4\mathcal{M}_d \int_0^\infty \pi < \kappa(\nu) > B_\nu(T_d)\, d\nu \tag{5}$$

where L_{abs} is the stellar luminosity absorbed by the given grain specie, M_d its total mass, and $< \kappa(\nu) >$ is the mass-weighted average of $\kappa(\nu) \equiv 3Q(a,\nu)/4\rho a$ over a $\sim a^{-3.5}$ grain size distribution, where ρ is the mass density of the dust, a its radius, and $Q(a,\nu)$ its absorption efficiency.

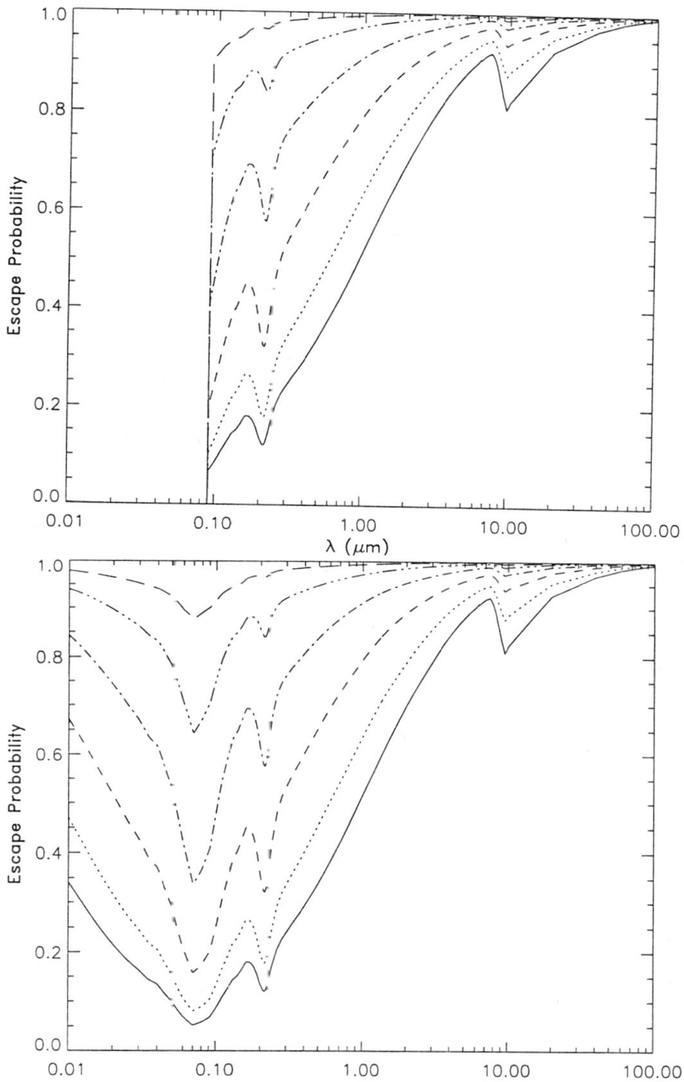

Figure 3: The escape probability from a spherical galaxy is shown as a function of wavelength for several epochs (0.1, 0.24, 0.55, 1.2, 2.8, and 6,2 *Gyr*, respectively from the top) for two cases: (a) when only dust opacity is included in the radiative transfer; (b) with gas absorption included. In the second case, almost all the radiation shorwards of the Lyman continuum ($\lambda = 912$ $\mathring{A}$) is absorbed by the gas. In both cases the escape probability does not change much between ~ 6 and 14 *Gyr*.

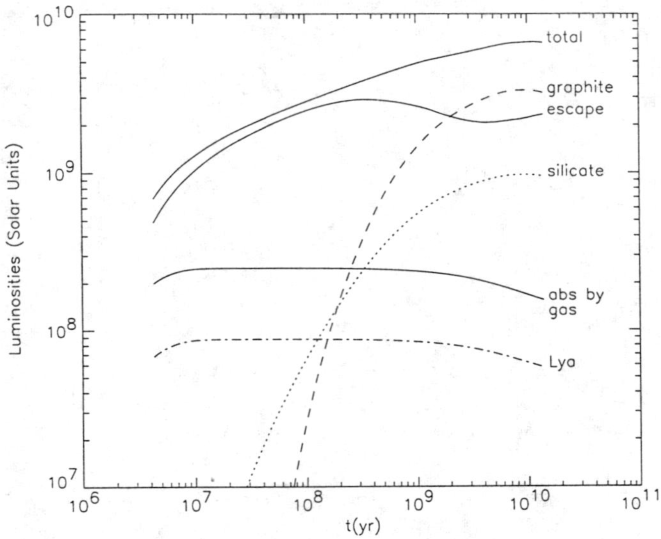

Figure 4: The evolution of the nebular energetics. See text for details.

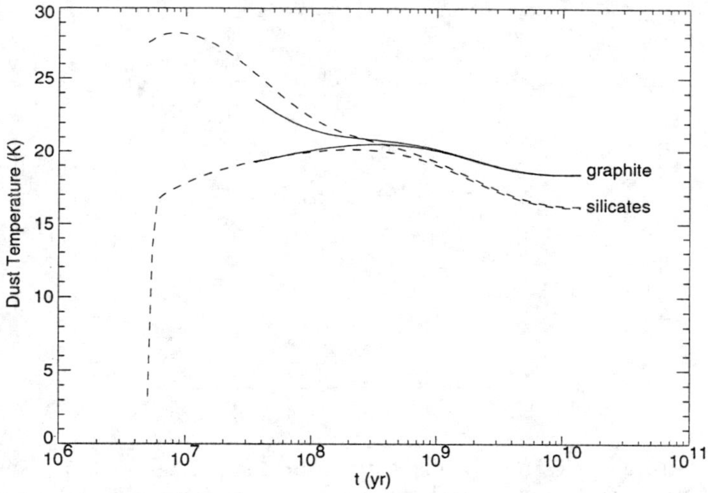

Figure 5: The evolution of the temperature (given by eq. (4)) of silicate and graphite grains, (represented by dotted and solid lines, respectively) for a spherical galaxy. For each dust composition, the lower curve represents the dust temperature resulting from the direct capture of stellar radiation, and the upper one the temperature attained with the additional capture of Lyα photons. Graphite does not form until after $\sim 10^7$ yrs.

Figure 5 shows the evolution of the graphite and silicate dust temperature as a function of time. The figure shows that Lyα photons have a significant effect on dust temperature during the early stages of galactic evolution. This is the result of the fact that Lyα is resonantly scattered by the gas, so that most most of them get eventually absorbed by even a little amount of dust.

It is obvious that the derived bimodial dust temperature distribution is unrealistic. Our simple model neglected details of the distribution of the dust around the ionizing stars, and the effects of the stochastic heating of very small dust particles. To create a more realistic (but nevertheless still simplistic) dust temperature distribution, we will adopt a power law distribution in grain temperatures: $P(T_d)\ dT_d \sim T_d^{-\alpha}\ dT_d$, where the maximim temperature was chosen to be 1500 K, and its minimum value was determined from the condition that the total IR emission integrated over the dust temperature distribution equal that of the single temperature case. The adopted value of α is equal to the value one would derive for a spherical distribution of dust particles decreasing in number density as $\sim\ r^{-0.8}$ from a central heating source. For this simple geometry we get that:

$$
\begin{aligned}
\alpha(graphite) &= 7.6 \quad for\ T < 70\ K \\
&= 6.5 \quad for\ T \geq 70\ K \\
\alpha(silicate) &= 7.6 \quad for\ T < 150\ K \\
&= 5.4 \quad for\ T \geq 150\ K
\end{aligned} \tag{6}
$$

The resulting temperature distribution is similar to that of a population of interstellar dust grains of which a fraction is undergoing temperature fluctuations.

3. THE INTEGRATED CONTRIBUTION TO THE CIB

Having obtained the evolution of the UV$-$IR flux distribution of one galaxy, we can estimate the contribution of a population of such galaxies to the CIB. To do so, we normalize their current number density $n_G(0)$ so that $n_G(0) \times L_B = \mathcal{L}_B$, where $\mathcal{L}_B$ is the observed luminosity density of field galaxies given by $\mathcal{L}_B = 0.44 \times 10^8 L_\odot Mpc^{-3}$ (in units of $L_\odot = 3.826 \times 10^{33} erg\ s^{-1}$ for $H_o = 45 km\ s^{-1}\ Mpc^{-1}$; Efstathiou, Ellis, Peterson 1988), and L_B is the observed blue luminosity of a galaxy. From our model calculations $L_B = \{2 \times 10^8 L_\odot, 5 \times 10^8 L_\odot\}$ for, respectively, a spherical and a disk$-$shaped galaxy, giving a correspondent number density of $n_G(0) = \{0.5 Mpc^{-3}, 0.2 Mpc^{-3}\}$, for these galaxies. The number density of galaxies in a comoving volume element dV_c is given by $n_G(z) = n_G(0)\ (1+z)^{3+\gamma}$, where γ is a parameter taking into account the possibility of evolution in their number density. Their total contribution to the specific intensity per sr^{-1},I(ν), of the CIB is given by:

$$I(\nu) = \frac{c}{H_o}\int_0^{z_*} \frac{n_G(z)\ S(\nu,t_G)\ dz}{[\Omega_o(1+z)^5 + \Omega_\Lambda(1+z)^2 - (1+z)^4(\Omega_o + \Omega_\Lambda - 1)]^{1/2}} \tag{4}$$

where $S(\nu,t_G)$ is the galaxy's SED at age $t_G = t(z) - t(z_*)$, where $t(z)$ is the age of the universe at redshift z, z_* is the redshift at the onset of galaxy formation, and all other symbols have their usual meaning (e.g. Carroll, Press, & Turner 1992).

The results of our calculations are shown in Figures (6) and (7), for spherical and disk shaped galaxies, respectively. Each figure shows the cumulative contribution of the galaxies to the CIB for different values of γ, the density evolution parameter. For each value of γ we plot two curves: the first curve is the cumulative contribution of of the galaxies if they were dust-free, and the second curve represents their cumulative output taking dust absorption and the resulting IR emission into account. The total energy output in both cases is identical. Comparison of figures 6 and 7 shows that the effect of dust on the redistribution of the stellar energy output is greater for spherical galaxies. This is not surprising, since disk galaxies are on the average more transparent than spherical ones of the same volume. The figures also show that the model falls below the lower limit on the CIB obtained from IRAS and K-band galaxy counts, unless one invokes evolution with $\gamma \geq 2$. The failure of the $\gamma = 0$ models stems mainly from the constant star formation rate adopted here. An exponentially increasing rate of star formation in the past would increase the cumulative contribution of galaxies to the CIB. More realistic models are planned for future studies.

ACKNOWLEDGEMENTS

This research was supported by the NASA Astrophysics Theory Program ATP 188-44-23.

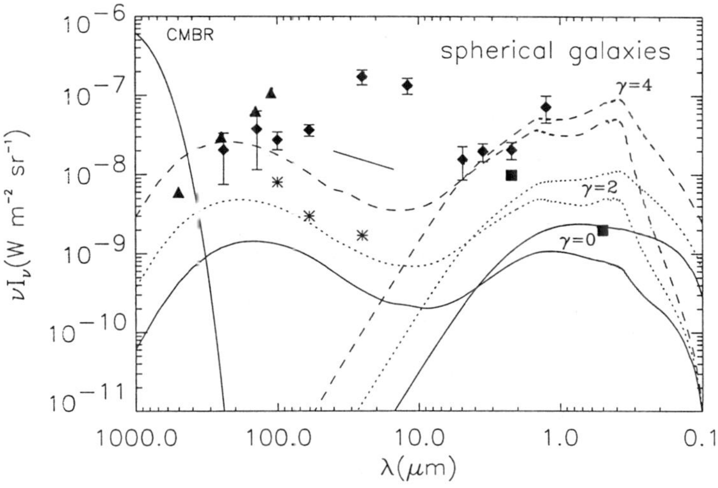

Figure 6 - Model calculations for the contribution of spherical "galaxies" to the CIB are plotted for various values of the density evolution parameter γ. See text for more detailed explanation. Model results are compared with current states of measurements: the range of DIRBE residuals (diamonds) and the FIRAS dark sky limit (triangles; see summary by Hauser 1995); the lower limits derived from IRAS galaxy counts (asterixes, Gregorich et al.1995) and from K-band counts (squares, Cowie et al. 1990); and the upper limit derived by Dwek & Slavin (1995) from γ−ray observations of Mrk 421 (solid line).

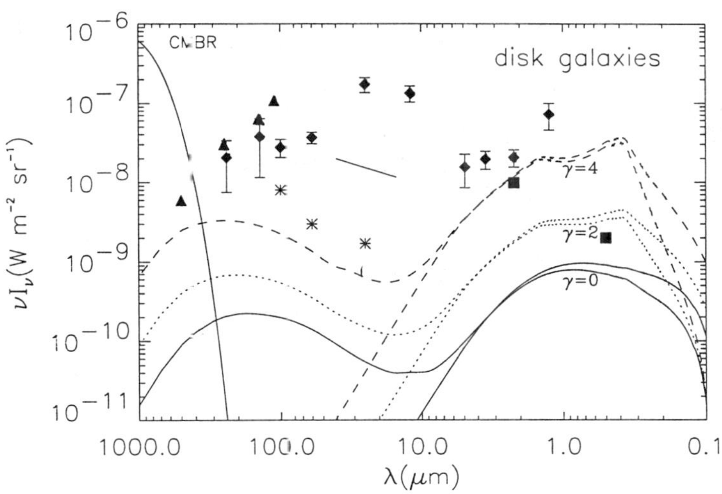

Figure 7 - Same as Figure 6 for disk−shaped "galaxies".

REFERENCES

Anders, E., Grevesse, N. 1989, Geochim. Cosmochim. Acta, 53, 197
Bond, J. R., Carr, B. J., & Hogan, C. J. 1991, ApJ, 367, 420
Charlot, S., & Bruzual, G. A. 1991, ApJ, 367, 126
Carroll, S. M., Press, W. H., & Turner, E. L. 1992, ARA&A, 30, 499
Cowie, L. L., Gardner, J. P., Lilly, S. J., & McLean, I. 1990, ApJ, 360, L1
De Zotti, G., & Mazzei, P. 1995, this Workshop
Dwek, E., & Slavin, J. 1994, ApJ, 436, 696
Efstathiou, G., Ellis, R. S., & Peterson, B. A. 1988, MNRAS, 232, 431
Franceschini, A. et al. 1995, this Workshop
Gregorich, D. T. et al. 1995, AJ, AJ, 110, 259
Hauser, M. G. 1995, this Workshop
Kurucz, 1992, in The Stellar Populations of Galaxies, eds. B. Barbuy, & A. Renzini (Dordrecht: Kluwer), 225
Lonsdale, C. 1995, this Workshop
Lucy, L. B., Danziger, I. J., Gouiffes, C., Bouchet, P. 1989, in Supernovae, ed. S. E. Woosley (New–York: Springer–Verlag), 82
Osterbrock, D. E. 1989, Astrophysics of Gaseous Nebulae and Active Galactic Nuclei (Mill Valley, CA: Univ. Science Books), Appendix 2
Rana, N. C. 1991, ARA&A, 29, 129
Renzini, A., & Voli, M. 1981, A&A, 94, 175
Rocca-Volmerange, B. 1992, in The Stellar Population of Galaxies,eds. B. Barbuy, & A. Renzini (Dordrecht: Kluwer),357
Rocca-Volmerange, B., & Fioc, M. 1995, this Workshop
Rowan-Robinson, M.,& Pearson, C. 1995, this Workshop
Scalo, J. M. 1986, Fund. Cosmic Phys. 11, 1
Schaller, G., Schaerer, D., Meynet, G., & Maeder, A. 1992, A&AS, 96, 269
Stecker, F. W. 1995, this Workshop
Thielemann, F.,-K., Nomoto, K., & Hashimoto. 1993, in Origin and Evolution of the Elements, ed. N. Prantzos et al. (Cambridge Univ. Press), 297
Thielemann, F.,-K., Nomoto, K., &Hashimoto, M. 1991, in "Supernovae", eds. J. Audouze et al. (Elsevier Science Publishers), 46
Timmes, F. X., Woosley, S. E., & Weaver, T. A. 1995, ApJS, 98, 617
Weaver, T. A., & Woosley, S. E. 1993, Phys. Rept., 227, 65
Witt, A., & Gordon, K. D. 1995, this Workshop

THE PHYSICAL STATE OF THE INTERGALACTIC MEDIUM OR CAN WE MEASURE Y?

James G. Bartlett, Domingos Barbosa, Alain Blanchard
Observatoire de Strasbourg, U.L.P., 11, Rue de l'Université, 67000 Strasbourg, FRANCE
Unité associée au CNRS
bartlett@astro.u-strasbg.fr

ABSTRACT

We present an argument for a *lower limit* to the Compton-y parameter describing spectral distortions of the cosmic microwave background (CMB). The absence of a detectable Gunn-Peterson signal in the spectra of high redshift quasars demands a high ionization state of the intergalactic medium (IGM). Given an ionizing flux at the lower end of the range indicated by the proximity effect, an IGM representing a significant fraction of the nucleosynthesis-predicted baryon density must be heated by sources other than the photon flux to a temperature $\gtrsim$ few $\times 10^5\,K$. Such a gas at the redshift of the highest observed quasars, $z \sim 5$, will produce a $y \gtrsim 10^{-6}$. This lower limit on y rises if the Universe is open, if there is a cosmological constant, or if one adopts an IGM with a density larger than the prediction of standard Big Bang nucleosynthesis.

1. Introduction

Arriving from high redshift, the photons of the cosmic microwave background (CMB) carry information on the state of the material through which they have traveled, in particular the intergalactic medium (IGM). Any interaction at low redshifts between the IGM and the photons leads to a characteristic spectral distortion of the CMB commonly referred to as the *Compton y-distortion* (Zel'dovich & Sunyaev 1969), due to its physical origin in Compton scattering between electrons and photons and the fact that it is described by a single number called the y parameter (defined below). The deviation from a pure blackbody appears in the Wien tail of the CMB, in the far infrared at wavelengths from about 1 mm to 100 microns. Observations of this background thus provide constraints on the density and temperature of the ionized component of the IGM; an upper limit on y places *upper* limits on the temperature and density. The current limit from FIRAS is $y < 2.5 \times 10^{-5}$ (Mather et al. 1994).

Studies of quasar spectra also provide information on the IGM: the absence of a detectable Lyα absorption trough indicates that the IGM contains extremely little neutral Hydrogen (Gunn & Peterson 1965). This so-called Gunn–Peterson (GP) test places stringent limits on the quantity of HI found in the IGM since a redshift of about 5 and forces us to conclude that either their is no uniformly distributed IGM or that it is very highly ionized. For an assumed total (nonzero) IGM density, the GP test thus provides a *lower* limit to the temperature of the ionized component.

As these two types of observations constrain the IGM in the temperature-density plane in *opposite* directions, one may hope to draw interesting conclusions concerning the physical state of the IGM by combining the two types of studies. This is what we shall explore in this contribution by asking the question "given the GP limits, what is the corresponding lower bound to y?"

© 1996 American Institute of Physics

2. The Observables

The two observational quantities under consideration are the optical depth to Lyα absorption and the Compton parameter y. For an observed frequency ω_0, the former may be written as an integral along the photon path, $\omega(z) = \omega_0(1+z)$, up to the redshift z_{qso} of the quasar under observation (we write our equations in units for which $c = k = h/2\pi = 1$):

$$\tau[\Omega_{igm}, T, J_{24}; H_0, \Omega, \Lambda] = \int_0^{z_{qso}} dz |\frac{dt}{dz}| n_{HI}(z) \sigma_\alpha[\omega(z)], \tag{1a}$$

where $\sigma_\alpha(\omega)$ is the absorption cross-section; n_{HI} is the density of neutral Hydrogen; Ω_{igm} is the mass density of the IGM relative to the critical density – $n_{igm} = \Omega_{igm}(3H_0^2/8\pi G)/m_p$, m_p being the proton mass (in our simple calculation we consider only Hydrogen in the IGM); T is the temperature of the IGM and J_{24} is the photon flux at the Hydrogen ionization threshold in units of 10^{-24} W/m^2 Hz ster (in more sane units of cgs, and not the PC units enforced during the conference, this corresponds to the familiar J_{21}!). Studies of the proximity effect around quasars suggest that $J_{24} \sim 0.3 - 3$ (Bajtlik et al. 1988), although estimates of the quasar contribution to the ionizing background favor values at the low end of (in fact below) this range (Madau 1992). The function $|\frac{dt}{dz}|$ contains all of the cosmological parameters: Ω, the total mass density of the universe; H_0, the Hubble constant (hereafter $H_0 = 100h$ km/s/Mpc); and Λ, a possible cosmological constant. If we restrict ourselves for a moment to the case of a critical universe ($\Omega = 1$), this expression reduces to

$$\tau = 4.4 \times 10^5 (1 - \chi)(\Omega_{igm} h^2) h^{-1} (1 + z)^{1.5}, \tag{1b}$$

in which appears the *ionized* H fraction $\chi(\Omega_{igm}, T, J_{24})$. This ionized fraction is defined as n_p/n_{igm}, where n_p is the density of free protons. We have explicitly written the dependence on IGM density in terms of $\Omega_{igm} h^2$ for comparison with the predictions of Big Bang nucleosynthesis theory. This theory constrains the number $\Omega_{bbn} h^2 \sim 0.013$ (Walker et al. 1991). We see immediately that for any non-negligible quantity of intergalactic gas, χ must be very close to unity to avoid producing an observable amount of absorption, *i.e.* the medium must be highly ionized. This in turn implies, as we will quantify in the following section, a high temperature, particularly for the low fluxes J_{24} suggested by integration of quasar emission. For our purposes, the most useful limits on τ have been given by Giallongo et al. (1994): $\tau < 0.05$ at $z = 4.3$.

The Compton parameter is defined as the integral along the line-of-sight of the Compton optical depth times an effective energy transfer coefficient T/m_e, where m_e is the electron mass:

$$y[\Omega_{igm}, T; \Omega, H_0, \Lambda] = \int_0^{z_{qso}} dz |\frac{dt}{dz}| \left(\frac{T}{m_e}\right) \chi n_{igm} \sigma_T, \tag{2a}$$

which for a critical universe becomes (σ_T being the Thompson cross-section)

$$y = 10^{-6} T_4 (\Omega_{igm} h^2) h^{-1} \left(\frac{1+z}{6}\right)^{1.5}. \tag{2b}$$

In the second expression we have defined T_4 as $T/10^4$ K and assumed that $\chi = 1$. Notice also that the numerical value is referred to a redshift of 5.

From these formulae, we easily see the physics helping us to constrain the IGM. A limit on τ eliminates high density, *low* temperature regions of the IGM temperature-density plane while limits on y eliminate high density, *high* temperature regions of this phase space (figures 1 and 2). Thus we "squeeze" the IGM into the left-most portions of the diagram, towards low total IGM density. The redshift dependence in equations (1) and (2) provides motivation to search for the GP effect at the highest possible redshifts, although one must remember that the increasing Lyα forest line density makes this a more and more difficult proposition (beyond the primary difficulty of finding such high redshift quasars). All of this assumes that we know the dependence of χ on the density and temperature of the IGM, or, in other words, that we know the ionization mechanism. In the following section we consider ionization by collisions and radiation.

A remark on the dependence of the constraints on the cosmological model, which enters only through the expansion time dt/dz – All models for which $\Omega < 1$, with or without a cosmological constant, result in tighter constraints on the IGM: The slower deceleration in these scenarios implies a longer expansion time at any given redshift z (dt/dz at any given z increases relative to the critical case) and therefore larger optical depths to *both* Lyα absorption and Compton scattering. Lowering the Hubble constant for a fixed, physical IGM density, $\Omega_{\rm igm}h^2$, has the same effect. In both case the reason is the same – the Universe is older.

3. Constraints

To proceed and actually constrain the IGM, we must model the ionization physics of the gas, in order to find the functional form of $\chi(\Omega_{\rm igm}, T, J_{24})$. Assuming ionization equilibrium maintained by a flux of ionizing photons and by collisions, we write

$$\alpha n_{\rm p}^2 = \Gamma_{\rm pi} n_{\rm HI} + \Gamma_{\rm c} n_{\rm p} n_{\rm HI}, \tag{3}$$

where α is the recombination rate; $\Gamma_{\rm pi}$, the photoionization rate, dependent on J_{24}; and $\Gamma_{\rm c}$, the collisional ionization rate. Recall that we consider only Hydrogen in our simple calculation. The assumption of ionization equilibrium eliminates any dependence on the state of the medium prior to the redshift under consideration. Although the recombination rate is smaller than the expansion rate at redshifts less than about 5, this assumption seems reasonable because even the small quantity of gas capable of recombining in one expansion time would quickly violate the GP limits if there were no compensating source of ionization.

To get a feel for the numbers, consider the case of pure photoionization in a critical Universe with $h = 1/2$. Putting $\Gamma_{\rm c} = 0$, $\Gamma_{\rm pi} \sim 4 \times 10^{-12} J_{24}$ s^{-1} and $\alpha \sim 4 \times 10^{-13} T_4^{-0.7}$ cm^3 s^{-1} (Peebles 1993) in equation (3), one finds $(1-\chi)/\chi^2$, and, with the approximation that $\chi \approx 1$, one then arrives (Eq. 1b) at $\tau = 2 \times 10^3 (\Omega_{\rm igm} h^2)^2 h^{-1} T_4^{-0.7} J_{24}^{-1} [(1+z)/6]^{4.5}$. Thus, in order to satisfy $\tau < 0.05$ at a redshift of 5, an IGM with a density $\sim \Omega_{\rm bbn} = 0.013/h^2$ subjected to an ionizing flux $J_{24} \sim 1$ must have a temperature of $T_4 \gtrsim 50$. This same gas would then produce a $y \gtrsim 10^{-6}$. Results of a more general and careful calculation, taking into account both photoionization and ionization by collisions, are shown in the figures.

4. Conclusion

The goal of our discussion is to show the nature of the constraints imposed on the IGM by combining the GP test with limits on the y parameter. The quantitative results are given in the figures for the current GP and y limits, as well as for an eventual limit of $y < 10^{-6}$. Spectral observations of the CMB with the latter sensitivity would either provide the evidence for or eliminate the possibility of hiding a large quantity of baryons in the IGM. This would be an important result for some cosmological scenarios, such as the PIB model (Peebles 1987a; 1987b), and of special interest in light of the cluster "baryon crisis" (White et al. 1993). At a level of $y \sim 10^{-6}$, one is at the threshold of eliminating an IGM containing most of the baryons predicted by standard nucleosynthesis, particularly for low density Universes (figure 2). Such a conclusion would have important implications for the nature of the dark matter in galactic halos. There is the additional hope of improving the GP limits with high resolution spectra of quasars at ever larger redshifts. A combined effort of this kind would permit us to draw interesting and important conclusions on the history of the baryonic content of the Universe during galaxy formation.

References

Bajtlik, S., Duncan, R.C., & Ostriker, J.P. 1988, ApJ, 327, 570
Giallongo, E., D'Odorico, S., Fontana, A. et al. 1994, ApJ, 425, L1
Gunn, J.E., & Peterson, B.A. 1965, ApJ, 142, 1633
Madau, P. 1992, ApJ, 389, L1
Mather, J.C. et al. 1994, ApJ, 420, 439
Peebles, P.J.E. 1993, Principles of Physical Cosmology, Princeton University Press (Princeton:New Jersey)
Peebles, P.J.E. 1987a, ApJ, 315, L73
Peebles, P.J.E 1987b, Nature, 327, 210
Walker, T.P., Steigman, G., Kang, H-S, Schramm, D.M., & Olive, K.A. 1991, ApJ, 376, 51
White, S.D.M., Navarro, J.F., Evrard, A.E., Frenk, C.S. 1993, Nature, 366, 429
Zel'dovich, Ya. B., & Sunyaev, R.A. 1969, Ap&SS, 4, 301
vskip 3 truecm

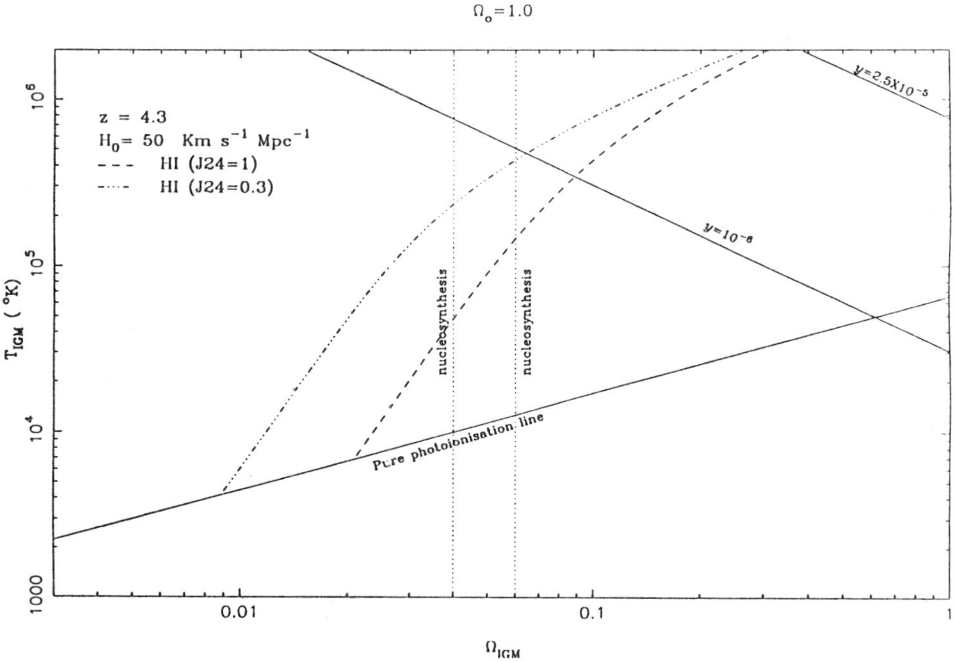

Figure 1 The constraints in the temperature-mass plane of the IGM for a flat universe with a Hubble constant of 50 km/s/Mpc. The Giallongo et al. (1994) GP limit ($\tau < 0.05$ at $z = 4.3$) eliminates the space below and to the right of the dashed curve. The calculation accounts for ionization by both collisions and an ionizing flux of $J_{24} = 1$. The change in slope along this curve at the higher temperatures signals the increasing importance of collisional ionization. The dotted-dashed curve shows the corresponding boundary for a flux $J_{24} = 0.3$; this flux is more in line with estimates of the quasar contribution to the ionizing background (Madau 1992). The constraints imposed by observations of the CMB are drawn for the actual FIRAS limit and for a limit of $y < 10^{-6}$. These constraints eliminate the space above and to the right of the respective curves. Big bang nucleosynthesis bounds the baryon density to the range indicated by the vertical dotted lines. The straight line labeled *Pure photoionisation line* gives the minimum temperature possible for the medium at a given τ (here $= 0.05$), corresponding to the case where the gas is only heated by the ionizing photons; it is parameterized by J_{24}. For this reason, we see the dashed and dotted-dashed curves terminate at two different points along this line.

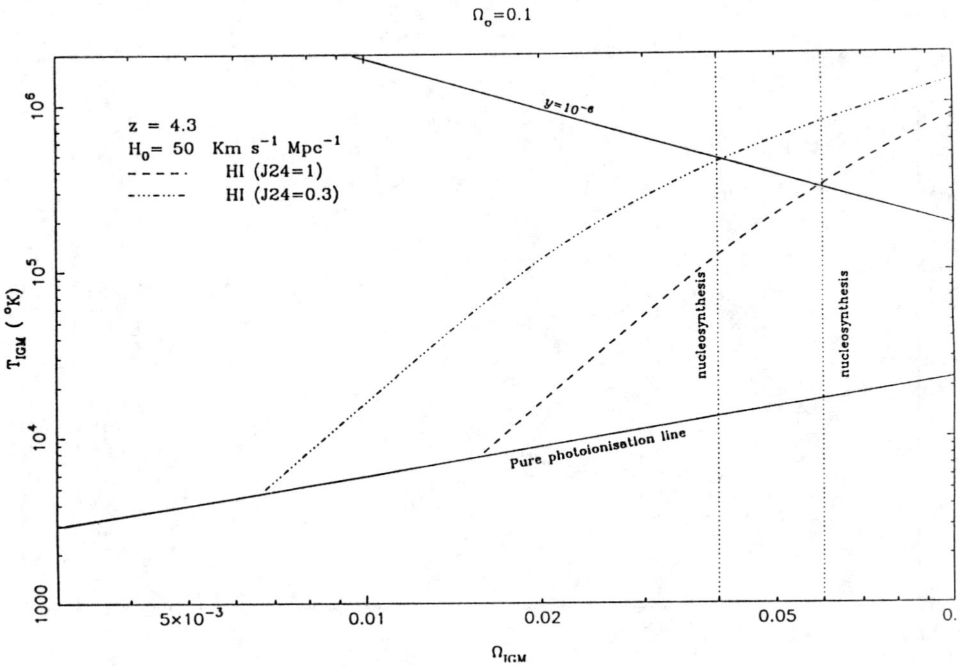

Figure 2 Same as figure 1 for $\Omega = 0.1$. Note the change of scale along the abscissa.

THE MM/SUB-MM FOREGROUNDS
AND FUTURE CMB SPACE MISSIONS

François R. Bouchet
Institut d'Astrophysique de Paris, CNRS
98 bis Boulevard Arago, F-75014 Paris France

Richard Gispert & Jean-Loup Puget
Institut d'Astrophysique Spatiale
Bât. 211, CNRS-Université Paris XI, F-91405 Orsay Cedex, France

ABSTRACT

Following COBE, several second generation satellites dedicated to mapping the anisotropies of the Cosmic Microwave Background (CMB) at high resolution and sensitivity are currently under study by space agencies. In particular COBRAS/SAMBA is in a phase A study at ESA, with a decision expected mid 1996, for a launch around 2003. Many crucial scientific questions will be solved if one succeeds to know with high precision the sub-degree angular scales of the anisotropies. The extent to which that can be achieved depends on our ability to separate the CMB fluctuations from those due to foregrounds.

After briefly reviewing the scientific potential of such high precision experiments, we present below a modeling of the galactic foregrounds which suggests that the CMB anisotropies will by far dominate the other sources, in a large fraction of the sky, at least at some frequencies. Additionally, this modeling shows that the scientifically important small scales can be reached more easily at relatively high frequencies (past the black-body peak), while lower frequencies are more appropriate for degree scale anisotropies. We proceed by describing the current COBRAS/SAMBA concept. Finally, we describe detailed simulations of 10 by 10 degree patches of the millimeter and sub-millimeter sky, and the first results obtained concerning the accuracy of the foregrounds separation. In particular, we present recovered maps of the Sunyaev-Zeldovich effect which suggest that much more than 1,000 clusters will be detected.

1. SMALL SCALE ANISOTROPIES OF THE CMB

A central problem in cosmology is the building and testing of a full and detailed theory for the formation of (large-scale) structures in the Universe. It is widely believed that the observed structures today grew by gravitational instability out of very small density perturbations. Such perturbations should have left imprints as small temperature anisotropies in the cosmic microwave background (CMB) radiation.

The COBE satellite has opened a new era by reporting the first detection of such temperature anisotropies (Smoot et al. 1992). However, COBE was limited by poor sensitivity, and restricted to angular scales greater than 7 degrees, much more extended than the precursors of any of the structures observed in the Universe today. As a result, these measurements are compatible with a wide range of cosmological theories. On the other hand, high precision measure-

© 1996 American Institute of Physics

ments at the sub-degree scale will be extremely discriminating, allowing strong constraints on large scale structure formation theories, the ionization history, the cosmological parameters (such as the Hubble and cosmological constants, or the type and amount of dark matter), and very early Universe particle theories. These goals could be met by producing nearly full-sky maps of the background anisotropies with a sensitivity of $\frac{\Delta T}{T} \sim 10^{-6}$ on scales from 10 arc minutes to tens of degrees.

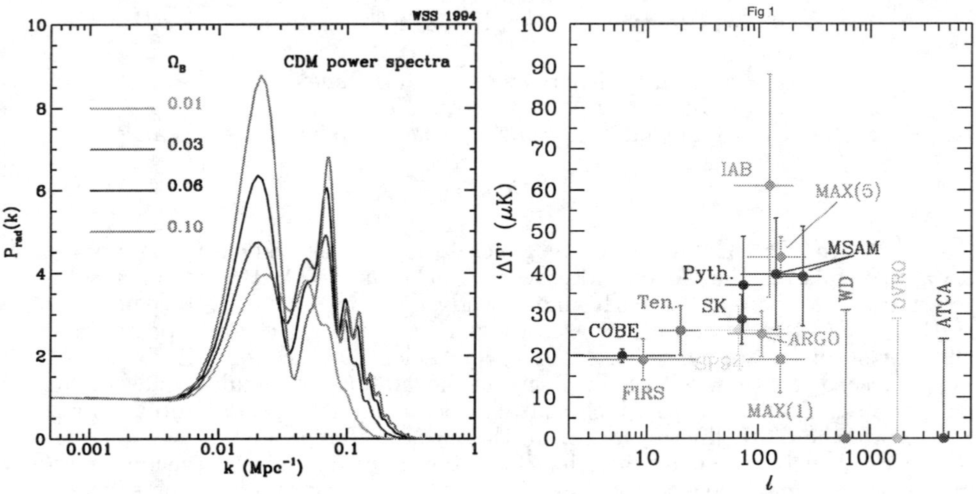

Figure 1: (a) Variation with the baryonic content of the Universe of the power spectrum for a CDM model (b) Current state of the CMB measurements. Both plots are courtesy of the Berkeley CMB group (in particular Hu, Sugiyama & Silk).

More precisely, the *statistics of the background anisotropies*, if Gaussian, would favor inflationary models, whereas non-Gaussian fluctuations would rather favor models in which irregularities are generated by primordial (often topological) defects such as strings, monopoles and textures. The *shape of the primordial fluctuation spectrum*, according to most theories of the origin of the fluctuations in the Universe corresponds to potential fluctuations $\delta\phi$ which should be independent of the scale of the irregularities (λ). Even a few percent deviation from this prediction would have extremely important consequences for the inflationary models. These models also predict a specific *ratio of tensor* (generated by gravitational waves) *vs. scalar anisotropies* depending notably on the scale of the fluctuations. The best way to learn about the *re-ionization history of the universe* would be through the small-scale anisotropies which may be partly erased if the inter-galactic medium was re-ionized at high redshift. These small scale anisotropies depend also on the *baryon density* (see fig.1.a), the *nature of the dark matter*, and the *geometry of the Universe* (and of course on the *initial spectrum of irregularities*).

Observations of the structure of the CMB on scales down to 10′ will resolve structures comparable in scale to clusters of galaxies ($\sim 10h^{-1}$Mpc, where h is Hubble's constant in units of 100 km s^{-1}Mpc^{-1}), and so allow a much more direct link between observations of *galaxy clustering*, *galaxy peculiar velocities* and *temperature anisotropies*. Additionally, the detection of at least a thousand rich clusters by the *Sunyaev-Zeldovich effect* (caused by the frequency change of CMB photons scattered by hot electrons in the gaseous atmospheres of rich clusters of galaxies), in combination with X-ray observations would constrain the evolution of rich clusters of galaxies and provide another handle on H_0. Finally, the Doppler effect on the same clusters should provide a unique measurement of the peculiar velocity of clusters with optical depth larger than 10^{-2} and the velocity dispersion of the weaker ones.

2. MODEL OF THE GALACTIC FOREGROUNDS

The final accuracy of the measurements of the cosmological anisotropies will rely on the ability to separate these fluctuations from those coming from the galactic background and from unresolved extra galactic sources. It is thus crucial to know the expected contribution of these foregrounds, in order to predict the expected fraction of the sky for which no "cleaning" would be necessary. Here we describe our model for the galactic emission to do this.

2.1 Dust emission

The spectrum of the dust emission in the wavelength range of interest has been measured by the FIRAS instrument aboard COBE with a 7 degree beam (Wright et al. 1991, Reach et al. 1995). Several balloon experiments have detected the dust emission at high galactic latitude in a few photometric bands with angular resolution from 30 arc minutes to 1 degree (Fischer et al. 1995, Page et al. 1990, De Bernardis 1992, Meinhold and Lubin 1991, Cheng et al. 1994...). The average spectrum at high latitude exhibits a cold component and can be modeled with a dust at 21.3 K, with an emissivity ν^{α} of spectral index $\alpha = 1.28$. This corresponds to the region B4 with $37 < b < 53$ and $-53 < b < -37$ in the PhD thesis of Barnes. It is also compatible with the analysis of Reach et al. 1995, who find $T = 22.2$, $\alpha = 1.12$ between 30 and 60 degrees, and $T = 20.7$, $\alpha = 1.36$ between -30 and -60 degrees. We used the DIRBE data at $240\mu m$ as a spatial template[2], and the average spectrum above to predict the long wavelength behavior of the dust.

This assumed spectral behaviour is a *worst* case from the point of view of the analysis of CMB anisotropies, because it assumes that the "cold component" is all interstellar and is present wherever dust emission is detected at $240\mu m$. Indeed, the decomposition in two components with two temperatures (and $\alpha = 2$) in the paper by Reach et al. (1995) shows that the properties of this cold component are variable; the ratio of the optical depth of the cold to warm component increases at high latitudes. It is likely that most of this cold component is in fact instrumental or extragalactic (Désert, this conference, Puget et al. 1995). In the second case it is also a foreground to the CMB, but with quite different properties in term of inhomogeneities. The interstellar cold component is then concentrated in the higher column density lines of sight of

[2] We used the $240um$ DIRBE map instead of the much less noisy $100\mu m$ one, because it should be a better tracer of the dust column density, while the $100\mu m$ emission has a stronger dependence on the dust temperature. This means that we will be over pessimistic in estimating the dust contribution to the foregrounds variance.

the ISM, the more diffuse phase having an emissivity law with $\alpha = 2$. Using $\alpha = 2$ instead of $\alpha = 1.28$ would reduce the contribution at long wavelength by a significant factor, about 2.7 at 1mm, and 5.8 at 3mm.

We verified that the power spectrum of the DIRBE fluctuations is decreasing approximately like the 3rd power of the spatial frequency, ℓ, in agreement with the determination of Gautier et al. (1992) who used instead the $100\mu m$ IRAS data down to a resolution of 4 arc minutes.

2.2 Free-free emission

The ionized gas in the interstellar medium can be traced in several ways, e.g. through its H_α emission (Reynolds 1992), the N-II fine structure line at $205\mu m$ seen in the FIRAS data, or pulsar dispersion measures.

The N-II measurements lead to a somewhat higher emission measure than the H_α ones, the latter observations being at present biased toward low column density regions. Taking this into account, Bennet et al. (1992) deduce an average brightness of the free-free emission at the galactic pole of 7 μK at 53 GHz. The power spectrum of the very limited available data from Reynolds appears very similar to the ℓ^{-3} one of the dust. Furthermore, this H_α data shows a part correlated with the H-I clouds traced by their 21 cm emission line.

We make the hypothesis that all the ionized Hydrogen of the ISM is associated with dust, as suggested by the observations of individual ionized regions. We can thus use the DIRBE 240 micron map as a spatial template for the dust *and* the free-free emissions correlated with H-I. In addition, we model the part uncorrelated with H-I by using the same spatial template, but North/South inverted (in galactic coordinates); this second, uncorrelated, component accounts for 5% of the total dust and 50% of the free-free emission. The spectrum of the free-free emission is modeled as $I_\nu \propto \nu^{-0.16}$, and normalized to give on average 7 μmK at 53 GHz.

2.3 Synchrotron emission

We use a spectral index of $\alpha = -0.7$ to extrapolate at higher frequency the spatial template provided by the full sky map at 408 MHz of Haslam et al. (1970). Due to the beam size of the observations their poor spatial range, and the presence of heavy stripping, the detemination of the spatial power spectrum is quite uncertain. We found however that the power spectrum of that map at high latitude is also similar to the dust one, i.e. it varies roughly $\propto \ell^{-3}$.

2.4 Characterization of the fluctuations

By using the model outlined above, one can generate at any desired frequency full sky maps with (for instance) a 1 degree FWHM resolution for the dust, free-free, and synchrotron[3] emissions, and for their sum. We found convenient to express all the emissions in equivalent thermodynamical temperature fluctuations.

In order to obtain *local* estimates of the level of fluctuations at interme-

[3] Given the amount of stripping present in the 408 MHz maps, we are in fact overestimating the synchrotron fluctuations, i.e. we are overly pessimistic.

diate angular scales of the various components, we compute the mean, $< T >$, and the *rms* fluctuation, σ_T, of the maps over square patches with 3 degrees on a side (i.e. containing about 10 beams of 1 degree FWHM). Figure 2 displays the distribution function of σ_T and $< T >$ for the "Dust+Free-Free" and the "Synchrotron" components. Clearly, the local level of fluctuations is well correlated with the mean. At low levels, σ_T is roughly proportional to $< T >^{1.15}$ for the "Dust+Free-Free" component, while σ_T varies only as $< T >^{0.5}$ for the "Synchrotron" component. Since the often dominating (according to frequency) "Dust+Free-Free" contribution to the variance is increasing faster than the mean value, in particular at high mean fluxes, the variance map is in fact more contrasted than the map of mean values, as was found by Gautier et al.

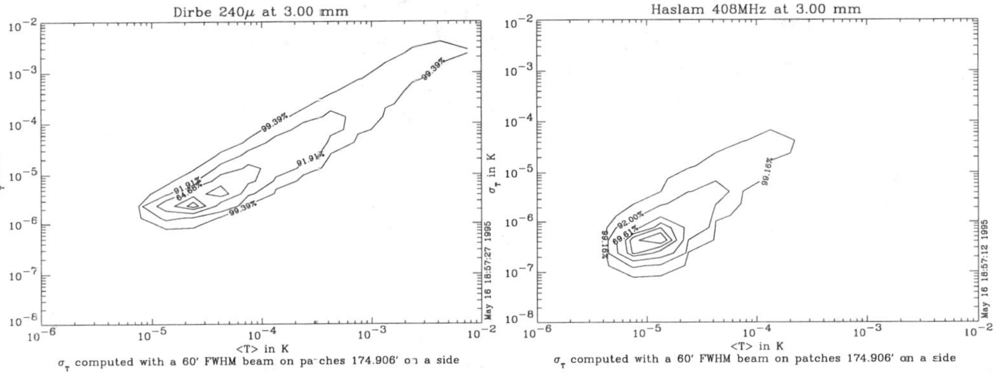

Figure 2: Contributions to temperatures distribution at 3mm. (a) "Dust+Free-Free". (b) "Synchrotron".

This figure 2 also shows that, at 3mm, 70% of the patches for the synchrotron have a mean temperature less than 2×10^{-5}, and fluctuations σ_T about 20 times smaller. For the dust, 65% of the patches have a mean temperature less than 2×10^{-4}, and fluctuations σ_T about 10 times smaller (in the range of angular scales probed).

If we sum the galactic contributions, we can thus generate full-sky maps of the fluctuation levels in different regions, at any frequency. Figure 3 shows an example of such a map at 3mm. Clearly, large regions of the sky are quite "clean" around the degree scale[4]. Since it was found that the overall shape of the power spectrum is similar in different regions, we are in effect measuring the local variations of the normalization of the angular power-spectrum. The map in figure 3 provides direct graphical evidence that a large fraction of the sky will have quite low levels of pollution at the angular scales probed.

[4] Note that our estimator of the local level of the fluctuations amounts to restrict the contribution to the variance to a range of angular scales ℓ (indexing the order of the spherical harmonics Y_{lm}), with $\delta\ell$ comparable to ℓ.

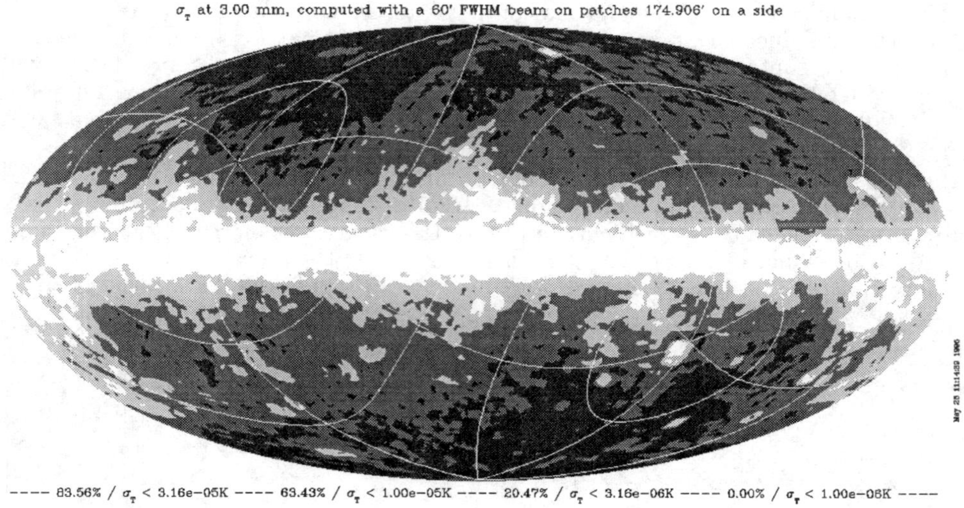

Figure 3: Map at 3mm of the temperature fluctuations due to all galactic foregrounds. The fluctuation levels, denoted by 4 different grey half-tones, are obtained by computing the local variance over square patches with 3 degrees on a side (see text).

More quantitatively, the figure 4 gives the fraction of the sky for which the *rms* fluctuations of the foregrounds (integrated over a restricted range in ℓ) and of their total contribution are less than a given value. In the 1 to 3 degree integration range, the *rms* total for the best half of the sky is $1. \times 10^{-5}$ at 3mm and 3×10^{-5} at 1.3mm. For a CDM case, the corresponding CMB fluctuations would be of order $3. \times 10^{-5}$ (this rough value was obtained by applying the local variance estimation above to a small 10 par 10 degree realisation of a COBE normalised CDM model; it should thus be a lower limit). Note that while at 1.3 mm the fluctuations come only from the dust, at 3mm the free-free part is substantial.

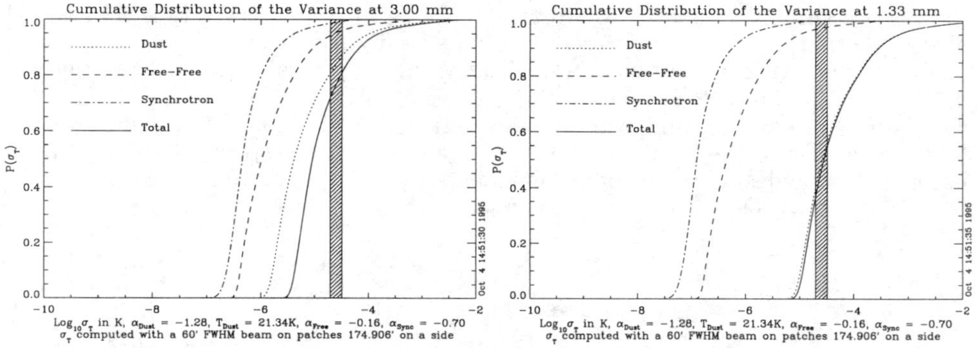

Figure 4: Foreground fluctuations in 3 degree patches for a 1 degree FWHM Gaussian beam. (a) at 3mm. (b) at 1mm. The hatched region shows the

expected range of fluctuations for a COBE normalised CDM model.

We can furthermore use the knowledge of the slope of the power spectra of the components to estimate the results of such measurements on smaller angular scales. Indeed, measurements with a 6 arc minute beam over 18 arc minute patches should be $10^{-1/2}$ smaller than the previous ones on 10 times larger scales. In that case, half of the sky is with *rms* fluctuations smaller than 10^{-6} at 3mm, and $3. \times 10^{-6}$ at 1.3mm. CMB fluctuations (in the CDM case) would still be of order $3. \times 10^{-5}$. At 3mm, the contribution to the fluctuations from synchrotron and dust emission are comparable.

2.5 Consequences

This shows that the scientifically important small scales (structure of the "Doppler" peaks past the first one) can be reached more easily at high frequency. Indeed, by working at low frequency, the foreground is dominated by the free-free component that cannot be taken out because it is measurable only at longer wavelengths where the resolution is not as good. On the contrary, the dust emission can be measured with a better angular resolution at higher frequency. Additionnaly, at high angular resolution, the contribution from extragalactic point sources increases. The model of Toffalati et al. (1995) suggests that they start to overcome the galactic contributions at wavelengths longer than about 2mm and angular scales smaller than about half a degree.

A removal of the dust contribution at the 20% level, which is relatively easy (thanks to the steepness of the dust emission spectrum), brings the dust contribution below the free free one, even at 3 mm. Thus, even if the lowest total contribution from foregrounds is around 3mm, the best window for measuring fluctuations at high angular scale is likely to be closer to the minimum of the Sunyaev-Zeldovich effect, around 1mm.

As far as the extragalactic background due to early galaxies is concerned, it is possible (and even likely!) that these CIBR fluctuations dominate over those coming from the galactic dust distribution in the lowest column density cirrus regions (see Guiderdoni et al., this conference). This would be important, both for the measurement of the small scales primordial fluctuations and because of the own cosmological interest of the CIBR. The only way to detect/separate this component is to have enough channels at high frequency, because the spectrum of high redshift galaxies is rising with frequency. Nevertheless this spectrum has to be flatter than the spectrum of the local dust because the $300\mu m$ to 1mm range is closer to the maximum of the redshifted dust emission. The statistics of the fluctuations will be an important tool which requires the coverage of a fairly large fraction of the sky. From that point of view the steep rise of the curves in figures 4 and 5 means that the "good sky" is a large fraction of the total.

3. MISSION CONCEPT

Measuring the cosmological background anisotropies with an accuracy better than a part in a million, on scales down to about ten arc minutes, is a very ambitious goal. Figure 1 illustrates how far from this goal are present ground and balloon based observations, after more than one decade of effort. One can anticipate that, at best within 10 years, this type of experiments will have located the first Doppler peak and measured it's amplitude with a 10-30% accuracy. The main limitations of these experiments come form the short duration of exposures available, the atmospheric effects, and the difficulties in controlling foregrounds contamination and systematics. On the other hand, although difficult, the goal expressed at he beginning of this section is achievable in the coming decade with a dedicated satellite experiment using the best detectors available today.

Indeed, the European Space Agency is studying a mission (COBRAS-SAMBA) which is aimed at the measurements described above. This mission will be up for selection in the spring of 1996 to become the third medium-size mission of the Horizon 2000 plan. This concept results from the work of the ESA Science Team[6].

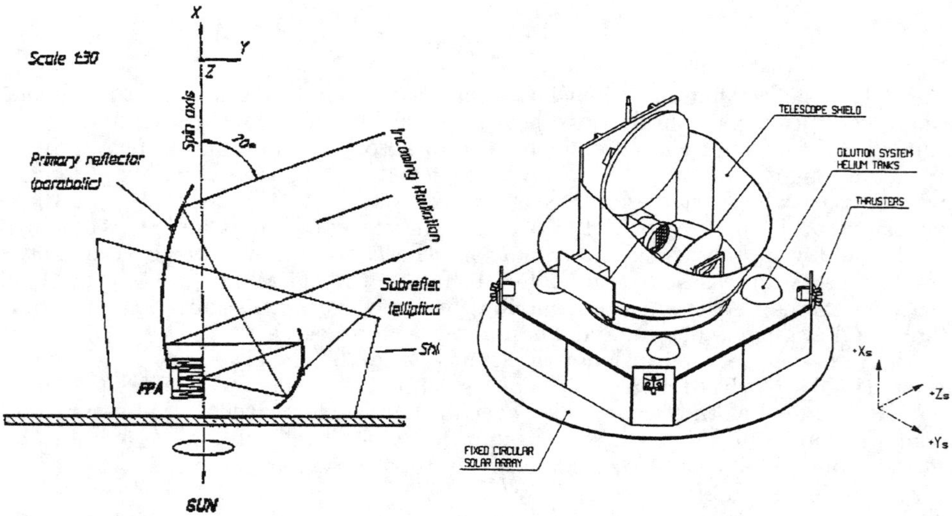

Figure 5:(a) Cross-sectional sketch of the model payload. Shown are the largest flared shield, the telescope and the focal plane assembly. (b) 3D view of the model payload.

The payload will be passively cooled; preliminary thermal models indicate that the focal plane assembly will reach an average temperature of $\sim$100 K, while the telescope optical surfaces will stabilize at $\sim$120 K. These temperatures, in addition to the absence of the atmospheric noise and heat load due to the chosen orbit (around L2, see below), insure that the conditions for a low and stable background are met.

An off-axis 1.5 meter gregorian telescope provides the necessary angular resolution and collecting area to meet the objectives of the mission. The separation of the different components of the foreground emissions requires a broad frequency coverage. The proposed mission will incorporate nine frequency bands

[6] The members which carried the study were : M. Bersanelli, C. Cesarsky, L. Danese, G. Efstathiou, M. Griffin, J.-M. Lamarre, M. Mandolesi, H.U. Norgaard-Nielsen, O. Pace, E. Pagana, J.-L. Puget, J. Tauber and S. Volonté. Requests for copies of the relevant report can be obtained from S. Volonté, ESA HQ, 8-10 rue Mario Nikis, PARIS CEDEX 15, FRANCE.

provided by five arrays of bolometers at high frequency and four arrays of passively cooled tuned radio detectors at low frequency.

The 56 bolometers require cooling to ~0.1 K, which will be achieved with an active system similar to the one foreseen for the ESA cornerstone mission FIRST. It combines active coolers reaching 4 K with a dilution refrigeration system working at zero gravity, both being developed and space-qualified in Europe. The refrigeration system will include two pressurized tanks of ^{3}He and ^{4}He, giving an operational lifetime of at least two years.

On the low frequency side, HEMT (High Electron Mobility Transistors) amplifiers give the required sensitivity to detect temperature anisotropies of the cosmological background on the low-frequency side. The frequency and angular coverage will provide good overlap with the COBE-DMR maps.

The frequency band at which the foreground emission is the lowest (~130 GHz) will be covered by both the tuned radio receivers and the bolometers. The fact that the tuned radio detectors are passively cooled insures that the low frequency channels can be operated for a duration limited only by the spacecraft consumables.

Center Frequency	31.5	53	90	125	150	217	353	545	857
Detector Type	HEMT radiometer arrays				Bolometer arrays				
Det. Temperature	~100 K				0.1-0.15 K				
Cooling needed	Passive				Cryocooler+Dilution system				
Number of Detectors$^♣$	4	14	26	12	8	12	12	12	12
Angular Resolution	30	18	12	10	10.5	7.5	4.5	4.5	4.5
Bandwidth ($\frac{\Delta\nu}{\nu}$)	0.15	0.15	0.15	0.15	0.35	0.35	0.35	0.35	0.35
$\frac{\Delta T}{T}$ Sensitivity$^\diamondsuit$	6.5	6.7	8.7	25.5	2.1	2.2	13.2	99.3	$> 10^3$
$\frac{\Delta T}{T}$ Sensitivity$^\heartsuit$	2.2	2.2	3.0	8.4	0.7	0.7	4.4	33	7600

♣ For the HEMTS, each dual–polarization horn supports two receivers (detectors).

$\diamondsuit$ 1σ sensitivity for a 2 years mission, over 90% of the sky.

$\heartsuit$ 1σ sensitivity for a 2 years mission, over 2% of the sky with deep integration.

Table I: Payload Characteristics. Frequencies are in GHz, angular resolutions in arc minute, $\Delta T/T$ Sensitivity in 10^{-6} units.

COBRAS-SAMBA will be placed into a halo orbit around the L2 Lagrangian point of the Earth-Sun system. This location is much more favorable than a low-earth orbit, or even the L5 Earth-Moon orbit in terms of observing efficiency, since the Earth and Moon would always be located behind the telescope, and stray light would not be a significant issue. The L2 orbit is also very favorable from the point of view of passive cooling and thermal stability.

The requirement of coverage of a large fraction of the sky is obtained by offsetting the optical axis of the telescope by 70° with respect to the spin axis of the satellite directed along the line joining the satellite and the Sun. In this nominal position, the optical axis scans a circle of diameter 140°, centered on a point in the ecliptic plane in the anti-Sun direction. In one year, the observed circle sweeps the whole ecliptic. To increase the sky coverage, the spacecraft rotation axis will be moved away from the ecliptic plane by up to ±15°.

The coverage of the sky is not uniform in terms of integration time per pixel. However, the motion of the spin axis can be chosen in such a way as to give deeper integrations in chosen parts of the sky. The regions of lowest galactic emission are of particular interest for the anisotropy measurement, and many of them will be observed with long observation time during the mission. Table I shows the estimated average sensitivities per pixel at the end of the two year baseline mission, for each frequency channel, assuming an observing strategy as sketched above. The pixel size for each channel is also given in Table I.

As this brief summary of the mission concept shows, a high precision determination of the small scale anisotropies of the CMB is now within reach.

4. SKY SIMULATIONS AND ANALYSES

In order to better assess the level to which can separate the various components, the RUMBA[7] group has undertaken the construction of fake millimeter and sub-millimeter, 10 by 10 degrees maps, as observed by virtual instruments, e.g. the COBRA/SAMBA mission, using theoretical modeling and data extrapolations. Our goal is to create maps as realistic as possible of the relevant physical contributions which may contribute to the detected signals. This astrophysical modeling is followed by simulations of the measurement process itself by a given instrumental configuration. Finally, we have implemented a number of ways of analyzing these fake data, with particular emphasis on the recovery of maps of the Sunyaev-Zeldovich effect.

4.1 Sky Simulations

An early report on our map simulations may be found in Bouchet et al. (1994). For the primary $\frac{\Delta T}{T}$ and the effects from clusters, (thermal and Doppler) we have created realizations corresponding to specific theoretical models (*e.g.* standard CDM). For the galactic emission we use spectral extrapolations of far-infrared data (e.g. IRAS maps and catalogs) and of radio data at 408 MHz, as explained in the previous section. We have implemented a number of hypotheses on the spectral behavior of these components, to test for the sensitivity to our physical assumptions (e.g. 2 component dust models, uncorrelated spatial templates for the dust and free-free component, etc..). For the unresolved component of the emission of galaxies, we are currently modeling the fluctuations by Monte-Carlo simulations, see Guiderdoni et al. in this volume.

All these processes are stored as elementary maps (presently $12.5° \times 12.5°$, with 500^2 pixels, each of size 1.5 arc minutes). At this stage we assume simplified instrumental configurations with top-hat spectral filters (i.e. unit transmission across the entire band), Gaussian lobes and a white noise detector noise, although these assumptions can be easily modified. The final product is a series of 10 by 10 degree[8] map, one per frequency channel, with the appropriate resolution and noise level, and a series of "process" maps to be recovered.

4.1 Maps Analyses

We shall make the hypothesis that the flux $f(\mathbf{r}, \nu)$ measured in each di-

---

[7] This group currently involves N. Aghanim, J.R. Bond, F.R. Bouchet, A. De Luca, R. Gispert, B. Guiderdoni, E. Hivon, J.-L. Puget... The acronym RUMBA stands for Research Upon Microwave Background Anisotropy.

[8] We do not retain 1.25 degree on each side to avoid edge effect after convolution with an instrumental beam

rection $\mathbf{r}$ and every frequency ν is a superposition of n_p processes (physical components) whose contributions can be factorized as a spatial template $x_j(\mathbf{r})$ times a spectral coefficient $s_j(\nu)$

$$f(\mathbf{r}, \nu) = \sum_{j=1}^{n_p} s_j(\nu) x_j(\mathbf{r}).$$

This is unlikely to be realistic for the unresolved contribution from galaxies, since at different frequencies different redshift range dominate, and thus different sources. But this contribution should be rather weak (see e.g the models in Toffolati et al. 1995), and we leave the analysis of this complication to further work. Additionnaly, note that the assumption above does not restrict the analysis to components with a spatially constant spectral behavior. Indeed, these variations are expected to be small, and can thus be linearised too[9].

If there are n_c channels, each with a spectral transmission $t_i(\nu)$, the observation in each direction can be arranged as an n_c component vector $y(\mathbf{r})$ given by

$$y(\mathbf{r}) = Px(\mathbf{r}) + B(\mathbf{r}), \tag{1}$$

where x is an n_p elements vector, B is a vector whose n_c elements are the noise contributions in each channel, and the $n_c \times n_p$ design matrix P describes the "projection" of each component in each channel

$$P_{ij} = \int_0^\infty t_i(\nu) s_j(\nu) d\nu.$$

If all the observations are at the same angular resolution, formula (1) applies also to the smoothed quantities. On the other hand, it should be advantageous to be diffraction limited in each channel. It is then more convenient to work in Fourier space (or with Spherical harmonics for larger maps) where convolutions reduce to products. Then

$$y(\mathbf{k}) = A(\mathbf{k}) x(\mathbf{k}) + B(\mathbf{k}),$$

with $A_{ij}(k) = P_{ij}\ w_i(\mathbf{k})$, if w_i is the angular response of channel i. Every pixel in Fourier space is then independent and it is natural to define a figure of merit

$$\chi^2 = |y - Ax|^2$$

for each mode $\mathbf{k}$. The matrix inversion algorithm by singular value decomposition is precisely designed to solve the problem of finding the vector x that minimizes this figure of merit.

We found that this method works very well provided that one does not attempt to recover modes varying too rapidly on the scale of the largest selected

[9] For instance, in the case of a varying spectral index whose contribution could be modeled by $\nu^{\alpha(\mathbf{r})} x_j(\mathbf{r})$, we would decompose it as $\nu^{\bar{\alpha}}[1 + (\alpha(\mathbf{r}) - \bar{\alpha}) \ln \nu] x_j(\mathbf{r})$. We would thus have two spatial templates to recover, $x_j(\mathbf{r})$ and $(\alpha(\mathbf{r}) - \bar{\alpha}) x_j(\mathbf{r})$ with different spectral behaviors, $\propto \nu^{\bar{\alpha}}$ and $\propto \nu^{\bar{\alpha}} \ln \nu$ respectively.

beam. Typically, we restrict the range of k so that the coarser resolution beam weight $w(k)$ is no less than 0.1 (this selects about 500 modes). By using the seven COBRAS/SAMBA channels with highest frequency and resolution (FWHM $\leq$ 12', $\lambda < 3$mm), we find that one can recover each *individual* mode of the CMB power spectrum, with an error at the 1% level, in a "clean" region of the sky, till $k \sim 1000$. Averaging over the best 25 % of the sky would reduce the uncertainty on the pwer spectrum by a factor of 10. Note that this inversion was done by assuming we make no error in the spectral behavior of the physical processes fitted (dust temperature and all spectral indices) . Still, in these "inversion" attempts, we used separate dust and free-free spectral inversion models, i.e. we did not use the 2 spectral models[10] really used to generate the "observations". Thus our results are representative of the type of inversion error introduced by uncertain spectral models. Still, the correlation coefficient of the initial and recovered CMB maps is larger than 0.995.

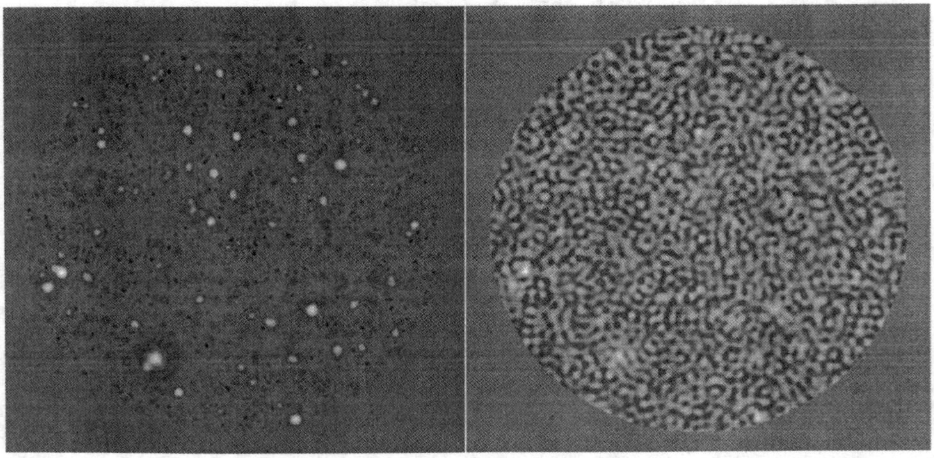

Figure 6: (a) Initial map of y-values, convolved to the resolution of the recovered one. (b) Recovered map by a simple Singular Value Decomposition in Fourier space.

Figure 6 shows an example of initial and recovered maps of the Compton y parameter of the Sunyaev-Zeldovich effect of clusters, when one uses the simple Singular Value Decomposition method. The 25 highest peaks are easily identified in a map of 9 degree diameter (at least in a color version), i.e one would correctly identifies about 8000 clusters over the best half the sky. Of course, this is somewhat optimistic, since this neglects many sources of error (e.g. we did not include in the simulation a background of resolved and unresolved galaxies, we assumed we knew the spectral behavior of the galactic foregrounds, the noise was taken to be an isotropic white noise, etc...) On the other hand, the method is far from optimal, in particular for extracting localized emissions. Assuming that at least 1000 clusters will be detected by a mission like COBRAS/SAMBA is thus a quite conservative statement.

[10] Let us recall that in the model used to generate the "observations", the 2 spatial templates have very different ratios of dust to free-free emission.

Let us now introduce another method called Wiener filtering. Any *linear* recovery procedure may be written as an $n_p \times n_c$ matrix W which applied to the $y(\mathbf{k})$ vector yields the $x(\mathbf{k})$ vector, i.e.

$$W(\mathbf{k})y(\mathbf{k}) = x(\mathbf{k})$$

We required above that the method best reproduce the observational data set. Another possibility is to require a statistically minimal error in the recovery, when the method is applied to an ensemble of data sets with identical statistical properties. More precisely, we want that the difference between the initial and recovered processes be of minimal variance, i.e. we search for the Wiener filter matrix W such that

$$\varepsilon_i^2(\mathbf{k}) \equiv < |Wy - x|_i^2 > = < |W(Ax + B_j) - x|^2 > = < |W_{ij}(A_{jl} - \delta_{il})x_l + W_{ij}B_j|^2 >$$

be minimal for every process i and all mode k, with brackets denoting the ensemble averaging, δ_{il} the identity matrix, and repeating indices implying summation. If we denote by P_l the power spectrum of process l, $P_l(k) = < x_l(\mathbf{k})x_l^\star(-\mathbf{k}) >$, and by $Q_j(\mathbf{k})$ that of the noise in channel j, $Q_j(\mathbf{k}) = < B_j(\mathbf{k})B_j^\star(-\mathbf{k}) >$, the statistical independence of the noise and of the physical components implies[11] that

$$\varepsilon_i^2(\mathbf{k}) = [W_{ij}(A_{jl} - \delta_{il})]^2\, P_l(\mathbf{k}) + W_{ij}^2 Q_j.$$

Requiring that the derivative of ε^2 versus all W_{ij} be zero yields

$$W_{im}(A_{ml}A_{jl}P_l + \delta_{mj}Q_j) = A_{ji}P_i.$$

Thus the Wiener matrix is given by

$$W = S\widetilde{A}^T(\widetilde{A}\widetilde{A}^T + N)^{-1}, \tag{2}$$

where we have defined for convenience the matrix $S, \widetilde{A}$, and N by

$$S_{ij} = \delta_K(i-j)\sqrt{P_j}, \quad \widetilde{A_{ij}} = A_{ij}\sqrt{P_j}, \quad \text{and} \quad N_{ij} = \delta_K(i-j)\sqrt{Q_j},$$

with δ_K standing for Kronecker's Symbol. Note that the meaning of the Wiener matrix definition (2) is made more transparent by treating the noise contributions on an equal footing with the processes. By considering the noises has n_c additionnal processes each contributing in a unique channel, we need to extend the A matrix, with the additionnal $\widetilde{A_{ij}} \propto \delta_K(i-j)\sqrt{Q_j}$. One then has simply $W = S\widetilde{A}^{-1}$.

Thus, provided one knows the power spectra, one gets an optimal linear recovery, with the input data weighted according to a generalization of the usual signal-to-noise ratios in Fourier space (obtained from $\sqrt{P_j}$ and $\sqrt{Q_j}$). One can thus think of the Wiener filtering as a "polishing" stage of a first inversion

[11] Note that we have further assumed statistically isotropic processes, so that $P(\mathbf{k}) = P(k)$

yielding an estimate of the power spectra. Note though that the "typical amplitude" $\sqrt{P_j}$ might be a rather poor indicator for a non-Gaussian process like the Sunyaev-Zeldovich effect from clusters. Other methods for that particular case might then be more appropriate. An obvious approach would be to use wavelet transforms to first "clean" the observations from localised sources, before using the Wiener filtering. When we applied this method in the same conditions that the direct inversion above, we found an improvement on the error of individual modes of the CMB by a factor of 10 (in this case, the inversion is done on nearly 14000 modes of different module, and takes 12 times longer).

In any case, it is clear that a second generation of CMB mission would allow a much better knowledge of the foregrounds than it is currently possible, and a very high precision determination of the CMB anisotropies.

ACKNOWLEDGMENTS

We thank the RUMBA group participants for permitting us to show before publications a report on the early results of their activity.

REFERENCES

Barnes, W.J., 1994, PhD thesis at the University of Chicago

Bennet, C.L. et al., 1992, *Astrophys. J.* **397**, L7

Bouchet, F.R., R. Gispert, N. Aghanim, J.R. Bond, A. De Luca, E. Hivon, & B. Maffei, 1994, Proceedings of the meeting "Far Infrared and Submillimeter Space Missions in the Next Decade", Paris, France, Eds. M. Sauvage, Space Science Review,

Boulanger, F., et al., this conference

Cheng, E.S. et al., 1994, *Astrophys. J.* **422**, L37

De Bernardis, P., Melchiorri, F., Melchiorri, B. and Vittorio, N., 1992, *Astrophys. J. Lett.* **396**, L57

Désert, F.-X., 1995, this conference

Fischer M. L., Clapp, A., Devlin, M., Gunderssen, J. O., Lubin, P. M., Meinhold, P. R., Richards, P. L. and Smoot, G. F., 1995, *Astrophys. J.* **444**, 226

Gautier, T.N., Boulanger F., Perault M., & Puget J.-L. 1992, *Astron. J.* **344**, 35

Guiderdoni, B. et al., 1995, this conference

Haslam C.G.T., Quigley M.J.S., Salter C.J., 1970, *MNRAS* **147**, 405

Meinhold, P. and Lubin P.1991, *Astrophys. J.* **370**, L11

Page L.A., Cheng E.S., & Meyer, S.S., 1990, *Astrophys. J.* **355**, L1

Puget J.-L., Abergel A., Bernard, J.-P., Boulanger F., Burton W.B., Désert F.-X. and Hartmann D, 1995, Astron & Astrophys. Lett., *Submitted.*

Reach W., et al., 1995, ApJ, in press.

Reynolds, 1992, *Astrophys. J.* **392**, L35

Smoot, G.F. et al. 1992, *Astrophys. J.* **396**, ,L1

Toffolatti et al. 1995, *Astro. Letters and Comm.*, **32**, 125.

Wright E.L. et al. 1991, *Astrophys. J.* **381**, 200

FUTURE DIRECTIONS

EGBIRT and DESIRE: MEASURING THE CIBR AT 3 AU

John C. Mather
Code 685, NASA Goddard Space Flight Center
Greenbelt, MD 20771

Charles A. Beichman
IPAC/Caltech, M.S. 100-22
770 So. Wilson Ave., Pasadena, CA 91125

ABSTRACT

Why can't we find the first galaxies? We have not found either the stork or the cabbage leaf. A space mission with a small infrared telescope taken to 3 AU from the Sun could address these questions by measuring the accumulated radiation produced by all such objects.

The Cosmic Infrared Background Radiation (CIBR) is the main repository for the energy liberated in early generations of galaxies and clusters of galaxies, but it has never been detected. Its measurement would be a quantum leap for cosmology, but Galactic stars and dust and interplanetary dust have set the limits so far. A new instrument, observing 3 AU from the Sun, could reduce the effects of the foregrounds by two orders of magnitude. Studies of two versions of such an instrument were already proposed to NASA, the GSFC (Goddard Space Flight Center) Extragalactic Infrared Background Telescope (EGBIRT) and the JPL (Jet Propulsion Laboratory) Deep Space Infrared Experiment (DESIRE). Both involve very small telescopes with array detectors, cooled to reduce detector dark currents and telescope emission. The visible and near IR regions are scientifically very promising. As of this writing the DESIRE team is proposing a MidEx (mid-sized Explorer) mission and now includes some members of the EGBIRT team.

A small deep space mission could easily measure the predicted levels of the CIBR at a price far less than that of COBE. The technology is here and the time is ripe for overcoming this major obstacle to understanding our origins.

1. INTRODUCTION - INTERPRETATION OF THE CIBR

It is probably impossible to observe the formation of the first generation of galaxies and other luminous objects directly, but they *can* be observed indirectly with a deep space mission to measure the cosmic infrared background radiation (CIBR). We still do not even know whether galaxies were made top-down, from the initial evolution of large scale structure, or bottom-up, from the combination of small objects in nonlinear clustering processes, so observations are very important. The CIBR is the main accumulator for luminosity of all known categories of objects, including the unresolvable high redshift precursors of galaxies.

A measurement of the CIBR would be immediately interpretable. The first step is a comparison with extrapolations of observed point source counts. We already have theoretical upper limits on the CIBR based on the present metal abundances and the theory of stars. Close agreement would show that the early universe was "simple," in the sense that what we already see is all we get. We

© 1996 American Institute of Physics

would conclude that there was no bright burst of initial energy star formation, no significant release of gravitational energy in cluster formation, quasars and black holes, no decay of primordial elementary particles, and no significant dust obscuration in the distant objects or in intergalactic space.

A disagreement would constitute a discovery of a previously unknown and energetically very significant population of extragalactic objects, or an unexpected evolution of known types. We can not overstate the significance of such an event. Each alternative explanation makes a specific prediction of both the ordinary spectral intensity of the CIBR and the spatial fluctuation power spectrum. For instance, the decay of primordial elementary particles with a specified mass, decay mode, and half life produces a very simple and predictable spectrum, and if the decay occurs before the clustering of matter, there is no associated spatial fluctuation. For another example, if there was a burst of star formation at a particular epoch, and there was no significant obscuring dust, then we can determine the redshift of the event from the wavelength at which the CIBR has peak intensity. If the measured CIBR exceeds the predictions from the measured metal abundances, but has the spatial fluctuations expected from quasars and galaxies, we would conclude that there was significant gravitational energy release, or that the helium and metals produced in the sources were trapped within them and not expelled back into space. If there was significant absorption by dust, then we would see that the CIBR has a very reddened spectrum, with a reradiated thermal spectrum at much longer wavelength than otherwise expected.

There are many hypotheses with many theoretically free parameters, but the proposed missions would produce a rich data base of spatial and spectral information in thousands of fields. The confrontation promises to be very interesting, even if a measured excess could not be immediately interpreted.

2. ZODIACAL LIGHT - IMPROVEMENTS AT 3 AU

The zodiacal light at 1 AU from the Sun is much brighter than the predicted CIBR. As a result the CIBR has not yet been detected, even after the careful analysis done by the COBE DIRBE team (Hauser 1994). The zodiacal light comes from two processes, scattering of sunlight by interplanetary dust grains, and reradiation of the solar energy absorbed by the same grains. There is a narrow window around 3 μm, but even there the DIRBE residuals are about 20% of the foreground. The grains themselves come from several sources, including the collision and continued abrasion of asteroids, the decay of comets, and interstellar grains penetrating the heliosphere. The comets themselves may have more than one source, although the recent Hubble discovery of numerous bodies in the Kuiper belt beginning at Uranus suggests that the main source may now be known. It is not yet known whether the Oort cloud at 10^4 AU is also important.

The dust grains from asteroids can be studied by a variety of means. Some grains fall to Earth and survive long enough to be captured in the upper atmosphere. They are sharply concentrated in the ecliptic plane, so space probes can measure them in situ by counting their impacts and determining their velocities, and by observing their scattered sunlight as a function of distance from the Sun. Larger grains can be studied as meteoroids. Some of the asteroid families are sufficiently dominant as sources that their progeny can be followed as they evolve into the zodiacal rings found by IRAS. Close to Earth, some of them come into orbital resonance with the Earth, where they have been observed by the COBE (Reach *et al.* 1995). The resonance phenomenon depends on the size of the grains, so that the details of the resonance should be interpretable in terms of the particle size spectrum. The orbits of the asteroidal grains have been modeled in supercomputers, and are well understood.

Although the asteroid source is now reasonably well characterized, it is not the dominant source. About 2/3 of the dust grains come from comets (Dermott 1995). With long, eccentric orbits, their dynamics are determined as much by planetary perturbations as by the Poynting-Robertson drag that draws in the asteroidal grains. The grain size distribution has been observed by COBE for Comet Austin (Lisse 1994). Individual comets may become short-period after encounters with Jupiter or Saturn, so that they decay much more rapidly, and individual objects could become significant time-dependent sources for the cometary grains. In short, the comet situation is complex and there is no particular guarantee of symmetry of the cloud.

Nevertheless, it is known from observations that the total density of the dust diminishes rapidly with distance from the Sun. If all the grains were in circular orbits, and the source were steady, the density would be proportional to $1/r$ from the Poynting-Robertson drag force. Observations confirm that this is not far wrong near the Earth. The COBE DIRBE measurements have been studied very carefully, with models that include the a power law radial dependence of density and temperature, various forms of dependence on distance from the symmetry plane, and various effects (warp, tilt, and offset) of the planetary perturbations on the symmetry plane. One recent DIRBE model yielded a radial exponent of -1.35 for the density and -0.42 for the temperature of the dust, comparable to results from other missions, but different model assumptions produce different values for these exponents.

The DIRBE models and data are not capable of determining the range of validity of such a power law because the modeling problem is fundamentally underdetermined. For example, a uniform shell of dust at 5 AU would not be distinguishable from a cosmic infrared background. However, in situ measurements by the Pioneer 10 Imaging Photopolarimeter (Hanner *et al.* 1974) showed a drop of the visible wavelength zodiacal light by a factor of 100 at 3 AU, beyond which it was undetectable. This drop is much greater than the factor of 9 expected from a $1/r$ density law going to $r = \infty$, confirming that the effective source of most of the grains is within 3 AU of the Sun, and showing that 3 AU is far enough to make an enormous difference to a measurement of the CIBR.

There are also observations of interstellar dust grains penetrating the Solar System. Small grains should be kept out by the solar wind, which is sufficiently magnetized to drag small electrically charged grains. Larger interstellar grain impacts have been observed by the Ulysses (Grün *et al.* 1995). They should adopt a distinctly anisotropic distribution around the Sun because of the solar wind and the focussing effects of the Solar gravitational field. These have not yet been observed photometrically, so the interstellar grains are thought to be a negligible fraction of the zodiacal dust at 1 AU. However, they may well be the dominant part outside the orbit of Jupiter. The extrapolation is uncertain because the Ulysses instrument measured only large grains.

Based on these considerations and on the COBE DIRBE models, extrapolations to 3 AU have been. The largest improvements are found at the visible and near IR wavelengths. The $1/r^2$ dependence of the Solar flux combines reduces the scattered light rapidly, and the modest cooling of the dust (by a factor of order $1/3^{0.42}$) cuts off the Wien tail of the thermal emission. At longer IR wavelengths the improvement from the cooling of the dust is much smaller, and the main uncertainty is in the dust density. In the long wavelength Rayleigh-Jeans region, the $1/r^2$ dependence of the Solar flux is replaced by the $1/r^{0.42}$ dependence of the temperature, so the Hanner *et al.* reduction factor of 100 at visible wavelengths should become a factor of $100/3^{1.58} = 18$ in the far IR.

3. GALACTIC FOREGROUND LIMITATIONS

The galactic foreground comes from stars and interstellar material such as dust, dustlike macromolecules like polycyclic aromatic hydrocarbons, and ordinary atoms, ions, and molecules. Except for the very faintest, stars can be recognized individually with a large enough telescope to gain sensitivity and angular resolution. At a resolution of 1 arcminute at visible wavelengths, the residual unresolved stellar foreground is of order 1% of the total stellar flux. This statement depends on the scale height of the Galaxy for each type of star, and on the distribution of star types. At 2.2 μm, there are about 300 stars per deg^2 brighter than K=14 (Kleinmann *et al.* 1994), with a power law distribution $N \sim f^{-0.7}$. The distribution flattens at fainter levels (Bahcall and Soneira 1980). The integrated brightness of this distribution at 2.2 μm is 82 nWm^{-2}sr^{-1} , so a 1% level is quite interesting compared to predicted extragalactic fluxes. The telescope diffraction limited beam area is proportional to λ^2, but the stellar brightness in the Rayleigh-Jeans limit has $\lambda I_\lambda \sim \lambda^{-3}$. The residual starlight can be made less than 1 nWm^{-2}sr^{-1} over a wide wavelength range.

Much less is known about the near IR properties of the Galactic dust clouds at high latitude. Small dust grains and macromolecules can be heated to high temperatures by individual cosmic rays or UV photons, so that they can emit or fluoresce at wavelengths as short as a few microns. There are specific diagnostic features (tracers) from the PAH's at 3.3 and 7.6 μm which can be used to map the dust distribution. On the other hand, the PAH emission need not be strictly proportional to the continuum emission of the same clouds. The complexity of the clouds and the lack of clear markers limits our ability to model and compensate for their emissions. Initial model fits reported by Hauser (1994) suggest that the cirrus produces only a few percent of the total high latitude sky brightness even at the DIRBE 3.5 μm channel. We estimate that a 10% accuracy of the dust model should be achievable, so the net near IR effect would be under 1% of the total high latitude brightness. The prime areas for viewing are regions identified as free of cirrus clouds, based on the IRAS 12 and 100 μm maps, the COBE DIRBE maps, ISO images, and ground-based surveys of CO and H I.

4. INSTRUMENT DESIGN

The general features of the necessary instrument are easily established. An aperture of the order of 10 cm is needed to achieve the required angular resolution and adequate signal to noise ratio. The diffraction limit of a 10 cm aperture at 5 μm is 12 arcsec. The telescope and detectors must be cooled to suppress the thermal emission and the dark current. A single Indium Antimonide (InSb) 256 x 256 pixel array can have excellent efficiency and low dark current and cover the entire wavelength range from 0.5 to 5 μm. These detectors require cooling to 30 K for good enough performance. For 5 to 35 μm, a 128 x 128 pixel

Si:As array can also work well, but requires a temperature of 5.5 to 6 K for low dark current. Longer wavelength coverage requires liquid helium temperatures and Ge:Ga detector arrays. These are significantly more difficult and expensive, although with enough effort they could certainly be provided.

This degree of radiative cooling is a straightforward matter for a deep space mission, where the main source of heat is a parallel beam of light from the Sun. There are nearly 4π steradians of dark sky for the radiators, and only two baffles are needed between the 300 K spacecraft and the radiatively cooled telescope and a 30 K cryostat shell. The Si:As cooling requirement can be achieved by a small solid hydrogen cooler. The DESIRE cryostat is designed for a lifetime of 20 months with only 2 kg of hydrogen.

The telescope should be an off-axis section of a Gregorian telescope, like the COBE DIRBE and the ZIP telescopes, for low sidelobes. The off axis design avoids scattering from the secondary supports, and the Gregorian version allows the use of Lyot stops to trim off light scattered at the edges of the primary and first field stop. For simplicity the DESIRE mission is proposed with two identical telescopes, one for the InSb detectors with pure radiative cooling, and one for the Si:As detectors with their hydrogen cryostat.

Wavelengths would be determined by fixed filters in several sections in front of the arrays. The DESIRE would provide of order 10 bands from 0.5 to 30 μm wavelength, with some bands tuned specifically to the PAH emissions.

Scale factor calibration is established from observations of known stars. The zero flux reference would be a dark shutter activated by command, as well as a dust cover over the telescope that is activated just once after launch.

Exposures of the order of 300 seconds are required to achieve adequate signal to noise ratios with a small telescope. Longer exposures are not useful because of cosmic ray bombardment of the detectors, and multiple exposures must be made and compared to remove the cosmic ray hits from the data.

Because of the limited telemetry rate, significant data compression is required to enable the mission to survey large parts of the sky. The DESIRE mission can map 1500 deg^2 in its lifetime.

5. SPACECRAFT DESIGN

The required spacecraft must be very small to fit the definition of a MidEx mission. A payload of order 200 kg can be launched by the MedLite rocket, if a final stage kick motor is used. The payload needs very large solar arrays because the sun is so faint at 3 AU. Radioisotope thermal power generators are too expensive and controversial for this application. A fixed high gain antenna is used for communication with the Earth, so for extended periods of time the spacecraft must have its antenna axis pointed towards the Earth. The duty cycle for communications increases in proportion to the square of the distance of the probe from Earth, and the JPL Deep Space Net 34 or 70 m antennas are required for reasonable performance.

The attitude control system must be small to fit the various budgets. The DESIRE concept proposes a 3 axis control system using two momentum wheels and cold gas thrusters. The scientific instrument exposures contain sufficient numbers of known stars to provide their own precise pointing information, so the main attitude control system need only provide approximate control.

The mission trajectory is a simple ellipse from Earth to 3 AU, and the trip to aphelion takes 17 months. A small boost can be achieved by a lunar flyby on the way, but this requires a precisely timed launch, and is not required.

6. DATA PROCESSING AND INTERPRETATION

Data reduction from the mission starts with decompression, quality checking, and calibration of the individual exposures. The multiple overlapping exposures will be compared, and statistically robust averages that eliminate cosmic rays and other variable events will be made. Since part of the calibration is derived from these observations, it is an iterative process. These basic images are the raw data for the mission, and would be made public promptly.

Interpretation begins with fitting the time, spectral, and spatial dependences to models of the zodiacal light and the Galactic foregrounds. Experience with the IRAS and COBE DIRBE shows that this step is long and difficult, but the instrument design and location are so superior to those of previous missions that even the raw data should have immediate cosmological significance.

ACKNOWLEDGEMENTS

We thank the members of the EGBIRT and DESIRE mission concept proposal teams for generating much of the material used in this article. The EGBIRT study team was led by J. Mather and included R. Avant, M. Hauser, T. Kelsall, S. H. Moseley, G.B. Berriman, G. Toller, S. Burdick, W. Cobb, T. Murdock, C. Lisse, W. Reach, E. Tedesco, H. Thronson, E. Wright, J. Pipher, and W. Forrest. The original DESIRE study team was led by C. Beichman and included J.-P. Bernard, R. Cutri, G. Helou, S. Phinney, F. Boulanger, T. Gautier, T. Herter, and M. Sykes.

REFERENCES

Bahcall, J., & Soneira, R. 1980, ApJ(S), 44, 73

Dermott, S., 1995, this conference

Grün, E. *et al.* 1995, Adv. Sp. Sci., in press

Hanner, M. *et al.* 1974, JGR, 79, 3671

Hauser, M. 1994. To be publ, in Proc. IAU Symposium 168, "Examining the Big Bang and Diffuse Background Radiations," Kluwer

Kleinmann, S. *et al.* 1994, in "Science With Astronomical Arrays," ed. N. Epchtein, A. Omont, & B. Burton, p. 11

Lisse, C.M. *et al.* 1994, ApJ, 432, L71

Reach, W. *et al.* 1995, Nature, 374, 321

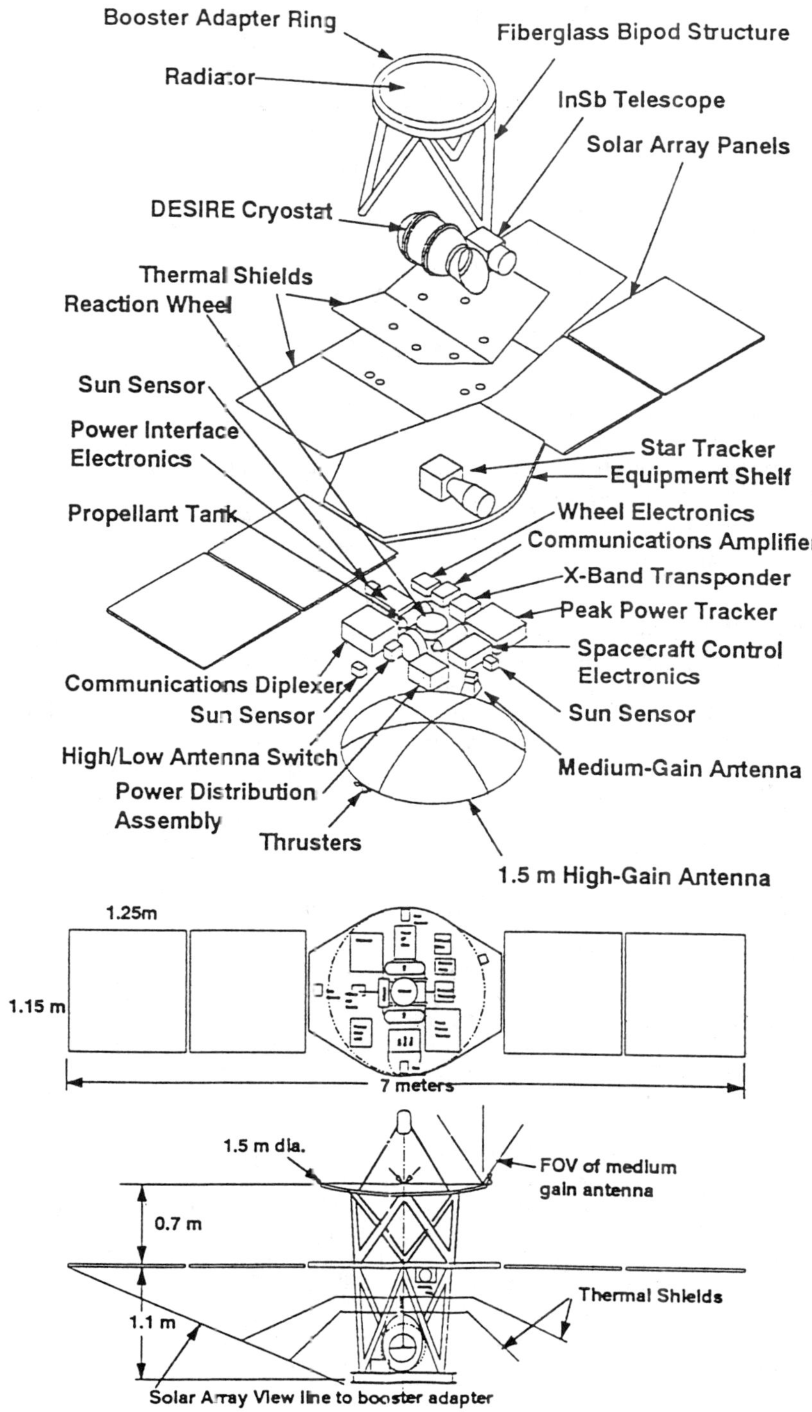
Booster Adapter Ring
Fiberglass Bipod Structure
Radiator
InSb Telescope
Solar Array Panels
DESIRE Cryostat
Thermal Shields
Reaction Wheel
Sun Sensor
Power Interface
Electronics
Star Tracker
Equipment Shelf
Propellant Tank
Wheel Electronics
Communications Amplifier
X-Band Transponder
Peak Power Tracker
Spacecraft Control
Electronics
Communications Diplexer
Sun Sensor
Sun Sensor
High/Low Antenna Switch
Medium-Gain Antenna
Power Distribution
Assembly
Thrusters
1.5 m High-Gain Antenna
1.25m
1.15 m
7 meters
1.5 m dia.
FOV of medium
gain antenna
0.7 m
Thermal Shields
1.1 m
Solar Array View line to booster adapter

COSMIC INFRARED BACKGROUND SCIENCE FROM SIRTF

Edward L. Wright
UCLA Physics & Astronomy, Los Angeles CA 90095-1562
wright@astro.ucla.edu

Michael Werner
Jet Propulsion Laboratory, California Institute of Technology, Pasadena CA

George Rieke
Steward Observatory, Tucson AZ

ABSTRACT

The Space InfraRed Telescope Facility (SIRTF), a cryogenic meter-class infrared telescope in orbit around the Sun at a distance of 1.01 au, will be able to measure both discrete sources and the diffuse background in the infrared. Studies done with SIRTF will help to separate the contributions of different sources for the infrared intensities measured by DIRBE.

1. Introduction

The infrared intensity away from the galactic and ecliptic planes is not dominated by a single component. Several sources make comparable contributions to the total intensity: scattering and thermal emission from interplanetary dust; galactic stars; scattering and thermal emission from interstellar dust; emission from interstellar molecules, ions and atoms; emission from external galaxies; and possibly an intergalactic source. The Space InfraRed Telescope Facility (SIRTF) will have both very high sensitivity to diffuse sources, with a well determined zero level, and the angular resolution necessary to separate discrete sources from diffuse backgrounds.

2. Description of SIRTF

SIRTF is planned by NASA to be an infrared observatory in space. It will cover the 3-180 μm region of the spectrum where atmospheric absorption and emission make sensitive ground-based measurements very difficult. Like IRAS, COBE and ISO, SIRTF will be cryogenically cooled, but unlike these earlier mission, SIRTF will take advantage of the large format infrared arrays now available. These large arrays will give SIRTF an unprecedented speed when performing areal surveys. While point sources will be detectable if they rise above the background noise in the few pixel area covered by the point spread function, very sensitive measurements of diffuse emission can be made by summing large numbers of pixels after removing the bright sources. The zero level of these diffuse flux measurements will be well-determined, because each SIRTF imaging instrument contains either a shutter or a scan mirror that allows the detectors to look either at the sky or at a cold dark reference.

The mass and cost of SIRTF have been minimized by adopting a "warm launch" configuration. Rather than enclosing the entire telescope and instrument package by a toroidal helium tank, as was done on IRAS and ISO, the

© 1996 American Institute of Physics

helium tank for SIRTF will cool only the instruments at launch. Once in space and kept shaded from the Sun, the telescope will rapidly cool to ≈ 50 K by radiative cooling alone. After this precool, the helium vapor leaving the instrument dewar will be used to further cool the optics to ≈ 6 K. Figure 1 shows the warm launch configuration. Further details of the SIRTF design are given by Werner and Fansen (1995).

An ideal environment for radiative cooling is provided by launching SIRTF into a trailing heliocentric orbit. This orbit has a semi-major axis of 1.01 au. As a result, SIRTF will lag behind the Earth with a separation increasing by ≈ 0.1 au/yr. Therefore the Earth and the Moon will be negligible sources of heat, leaving only the Sun to be blocked out by shields. A large radiating area exposed to cold space is thus available, and the outer shell of the telescope dewar will stabilize at ≈ 60 K. Similar thermal advantages can be obtained in a high circular orbit around the Earth (100 Mm altitude to avoid the outer Van Allen belt), but the Δv necessary to reach the high circular orbit is 1 km/sec higher than that needed for the heliocentric orbit. The other alternative considered is the halo orbit around the L2 Lagrange point in the Earth-Sun system. This orbit requires station-keeping because the L2 point is only metastable. At the end of the mission SIRTF will be 0.3 au from the Earth. A high-gain antenna is needed to transmit data to the ground. This antenna will be fixed to the back end of the spacecraft, and data will be dumped by pointing the telescope away from the Earth. Deep surveys of points on or near the ecliptic can be executed during the data dumps.

3. SIRTF Studies of Interplanetary Dust

SIRTF will be very far away from all interfering infrared sources (except for the Sun). Thus SIRTF observations of the interplanetary dust (IPD) will not be affected by off-axis response to the Earth limb or the Moon. The Sunshade required by the thermal design guarantees that SIRTF will also not be affected by the Sun. Furthermore, SIRTF will be able to cover an annular region of the sky, centered on the Sun, that is almost as large as the DIRBE scan path. In principle then, SIRTF could reproduce the DIRBE observations of the IPD. But in practice no future instrument is likely to match the 2×10^8 10-color photometric observations made by DIRBE. The most likely way to correct SIRTF measurements for zodiacal emission is thus to observe a field at a wide range of solar elongations and measure the brightness change. One would then use a geometric model of the zodiacal dust cloud based on DIRBE data to estimate the ratio of total zodiacal flux to the change in zodiacal flux. This procedure would cross-calibrate the SIRTF and DIRBE measurements of the IPD.

But there is one aspect of the interplanetary dust cloud that was not observed by IRAS or DIRBE, and will not be observed by ISO. SIRTF will provide a unique opportunity to study the circumsolar ring of dust predicted by Dermott *et al.* (1994). This is a collection of dust particles whose orbits are stabilized against the Poynting-Robertson (P-R) effect by a net torque provided by perturbations due to the Earth. To get a net torque, particles must be in a resonant orbit, making for example 5 orbits while the Earth makes 6. But resonance is necessary and not sufficient: the orbits must also be somewhat eccentric and one of the perihelia must occur just slightly behind the Earth in its orbit. The resulting forward tug compensates for the net drag from the P-R effect. The requirement that all of the trapped particles have a perihelion just behind the Earth produces a clump of dust. As a result, the trailing edge of the IRAS (Reach 1992) or COBE (Reach *et al.* 1995) scan circle ($\beta = 0$, $\lambda = \lambda_\odot + 90°$) is 3% brighter than the leading edge.

SIRTF in its trailing heliocentric orbit will fly right through this clump of dust. This will provide stereo views of the clump, giving a much better understanding of its structure. Figure 2 shows SIRTF orbit superimposed on the IPD ring and clump predicted by Dermott in a coordinate system rotating with the Earth.

4. Near IR Background

Near infrared wavelengths provide an interesting opportunity to observe the extragalactic background light (EBL) in the 3.5 μm window in the zodiacal light. The DIRBE measurements in the near infrared have significant interfering signals from the zodiacal light and from nearby stars. At 3.5 μm, the zodiacal light seen by DIRBE at the ecliptic poles is 64 kJy/sr, evenly split between scattered sunlight and thermal emission. This is quite similar to the flux from galactic stars, which contributes 65 csc $|b|$ kJy/sr. The residual after removing a zodiacal light model and extrapolating to csc $|b| = 0$ is O(10 kJy/sr). While this residual has the right order-of-magnitude to be an extragalactic background, a much better study of the foregrounds will be necessary before one can be certain. The situation at 2.2 μm is somewhat worse, with the zodiacal light being $\approx$ 80 kJy/sr, and with a larger flux from stars. Thus the precision of determining the EBL will be considerably better at 3.5 μm than at 2.2 μm.

Observations from SIRTF will eliminate the interference from galactic s-tars, since the 1 μJy sensitivity of SIRTF corresponds to a very low luminosity red dwarf at a distance of several kiloparsecs, which is well outside the disk of the Milky Way. At 2.2 μm the resultant turnover in star counts has been seen by Glazebrook etal (1994), with 800 stars with $K < 15$ per square degree but only 1700 per square degree with $K < 17$. For Euclidean source counts following $N \propto S^{-1.5}$, the ratio of counts would have been 16:1 instead of 2:1. Since the resolution of SIRTF will be 2″, less than 0.1% of the pixels will contain galactic stars. For $K > 16$, galaxies are more numerous than stars, and for the deepest exposures, the overlapping images of millions of galaxies per square degree will make it impossible to distinguish individual sources. The extragalactic background light is the sum of the fluxes from these overlapping galaxies plus a hypothetical diffuse component.

At 2.2 μm, ground-based surveys by Cowie *et al.* (1991) and Djorgovski *et al.* (1995) have probably reached the turnover in the counts where the contribution of faint objects to the total intensity starts to converge. The peak in $dI/dm = 0.4 \ln 10\, S^2 dN/dS$ kJy/sr/mag appears to occur at $K \approx 20$. The total intensity contributed by galaxies with $K < 23.5$ is 3 kJy/sr.

In order to determine the contribution of galaxies to the 3.5 μm extragalactic background light, the K band counts will have to be repeated in the L band. This will be impossible for any uncooled telescope, since the thermal background from room temperature optics will be millions of times higher than the zodiacal light.

Thus SIRTF will be needed to study the EBL at 3.5 μm. One possible sequence of observations is:

1) obtain deep exposures of the same field at different solar elongations,

2) blank out the parts of the field with stars,
3) use the intensity change with elongation to scale a physical model of the zodiacal light, and then subtract the total zodiacal light,
4) count galaxies *vs.* magnitude to determine their contribution to the intensity, $I = \int S(dN/dS)dS$,
5) compare this to the mean of the field to see if there is a diffuse residual.

This "diffuse" background might include light from numerous faint galaxies at high redshift. If many dwarf galaxies merge to form the current population of galaxies, then the number counts of very faint galaxies will be quite high, as the blue number counts are. Merging has been suggested as a possible explanation of the large number of faint blue galaxies, and very deep number counts at 3.5 μm, where the redshifted peak emission from stars will occur, will be a valuable tool to distinguish among models for the evolution of the galaxy population.

5. Far IR Background

The far infrared number counts of galaxies that exist to date have been made using the IRAS satellite. The very deep survey of Hacking & Houck (1987) at the North Ecliptic Pole gives a steep $N(S) \propto S^{-1.78}$ at their flux limit of 50 mJy at 60 μm. The galaxy density at this flux limit is 20 per square degree. If the Hacking & Houck power law fit to their counts continued to fainter objects, the infrared background to 60 μm would diverge to infinity. Thus it is very important to extend these source count measurements to lower flux limits. SIRTF will easily be able to measure sources down to very low flux limits at 70 μm, and will thus be limited by confusion and not detector noise. The very high sensitivity and the perfectly stable beam shape of SIRTF will allow the use of super-resolution to extend source counting to fainter sources. Using an final FWHM after super-resolution of $0.5\,\lambda/D$ as the SIRTF beam size, and taking 0.1 sources per beam as the confusion limit, SIRTF will be able to count galaxies at 70 μm to a density of $10^{4.4}$ galaxies per square degree. While the $N(S)$ law at these source densities is not known, all reasonable models and comparison to the K and B band counts suggest that this density will occur on the faint side of the peak in S^2dN/dS, so a SIRTF ultra-deep, super-resolved far infrared survey will be able to determine the contribution of galaxies to the far infrared diffuse background.

At longer wavelengths ≈ 160 μm, the beam size of SIRTF even after super-resolution will restrict the depth to which number counts can be measured. Even so, multi-color deep SIRTF surveys will provide an excellent observational basis for models of the evolving population of infrared luminous galaxies.

6. Conclusion

The Space InfraRed Telescope Facility will provide the ability to determine the contribution of discrete sources to the infrared background that DIRBE lacked. SIRTF will also have a well defined zero level and excellent sensitivity to diffuse background, and can thus duplicate some of the DIRBE measurements. This ability will be especially valuable in the study of the circumsolar ring at dust at 1 au from the Sun, and especially the clump of dust trailing behind the Earth, which SIRTF will fly through, proving "stereo" imagery for the first time.

References

Cowie, L. L., Gardner, J. P., Lilly, S. J. & McLean, I. 1990, ApJL, 360, L1-L4

Dermott, S. F., Jayaraman, S., Xu, Y. L., Gustafson, B. A. S. & Liou, J. C. 1994, Nature, 369, 719-723

Djorgovski, S., Soifer, B. T., Pahre, M. A., Larkin, J. E., Smith, J. D., Neugebauer, G., Smail, I., Matthews, K., Hogg, D. W., Blandford, R. D., Cohen, J., Harrison, W. & Nelson, J. 1995, ApJL, 438, L13-L16

Glazebrook, K., Peacock, J. A., Collins, C. A. & Miller, L. 1994, MNRAS, 266, 65-91

Hacking P. & Houck, J. R. 1987, ApJS, 63, 311

Reach, W. 1991, ApJ, 369, 529-543

Reach, W., Franz, B., Weiland, J., Hauser, M., Kelsall, T., Wright, E., Rawley, G., Stemwedel, S. & Spiesman, W. 1995, Nature, 374, 521-523

Werner, M. & Fansen, J. 1995, to be published in SPIE Proceedings

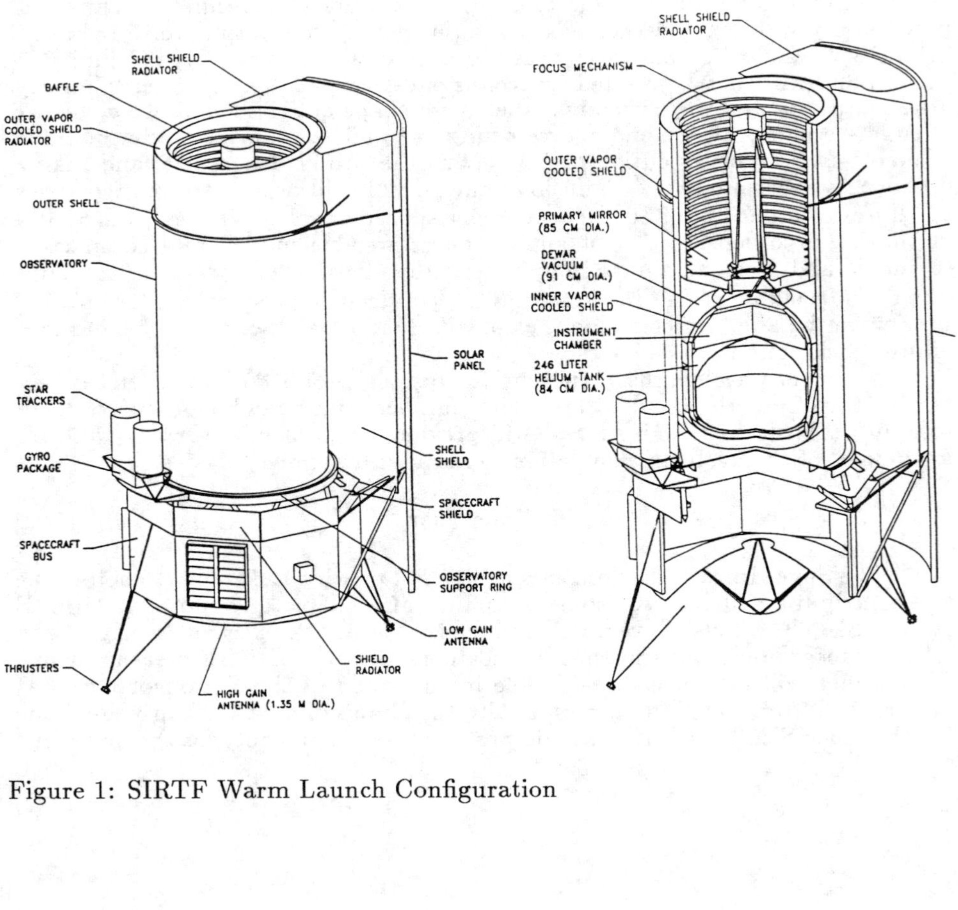

Figure 1: SIRTF Warm Launch Configuration

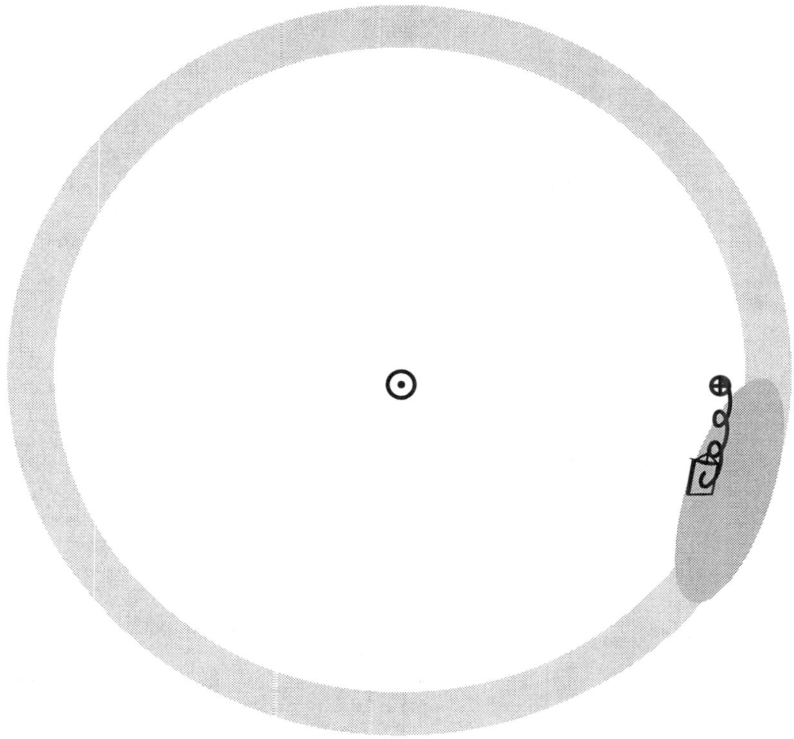

Figure 2: Circumsolar ring and clump from Dermott *et al.* (1994) compared to the SIRTF trailing heliocentric orbit.

POSTER CONTRIBUTIONS

AN OVERVIEW OF THE *DIRBE* AND *FIRAS* DATA PRODUCTS

David Leisawitz
Code 631
NASA Goddard Space Flight Center
Greenbelt, MD 20771

ABSTRACT

Those *COBE* data products that are relevant to "Unveiling the Cosmic Infrared Background" are described. The products are available, in most cases via network transfer, from the National Space Science Data Center (NSSDC). Guest Investigators approved under NASA's Astrophysics Data Program receive support from the *COBE* Guest Investigator Program at the Goddard Space Flight Center.

1. INTRODUCTION

Two *COBE*[1] instruments – the Diffuse Infrared Background Experiment (*DIRBE*) and the Far Infrared Absolute Spectrophotometer (*FIRAS*) – yielded important new information about infrared radiation from sources in the solar system, the Galaxy, and the early Universe. The *DIRBE* in particular was designed to enable the discrimination of the various emission components in order to detect or put interesting limits on the cosmic infrared background.

Boggess *et al.* (1992) summarized the characteristics of the *COBE* instruments and their performance in flight. Detailed information about the *COBE* data products, including an Explanatory Supplement for each of the *COBE* instruments, is provided on the World Wide Web at

http://www.gsfc.nasa.gov/astro/cobe/cobe_home.html.

Many of the products can be obtained by anonymous ftp from the /cobe directory at *nssdca.gsfc.nasa.gov.*

2. *DIRBE* DATA PRODUCTS

The *DIRBE* data products give the sky brightness as observed at 1.25, 2.2, 3.5, 4.9, 12, 25, 60, 100, 140, and 240 μm, plus Stokes Q and U parameters for the 1.25, 2.2, and 3.5 μm bands. Both zodiacal and Galactic emission components are present. Future plans call for the production of a zodiacal light–subtracted set of sky maps. Calibrated time–ordered data are available in addition to those products described below and in Table 1.

---

[1] The NASA Goddard Space Flight Center is responsible for the design, development, and operation of the Cosmic Background Explorer (*COBE*) and preparation of the *COBE* Project Data Sets. Scientific guidance is provided by the *COBE* Science Working Group.

© 1996 American Institute of Physics

Table 1. *DIRBE* Data Products

Product	Description	Number of files	MB/file	Format	Distribution medium[a]
Weekly Sky Maps	Weekly–averaged intensities for 10 bands (1.25, 2.2, 3.5, 4.9, 12, 25, 60, 100, 140, and 240 μm) and Stokes Q and U parameters for 1.25, 2.2, and 3.5 μm bands[b]; 1 file per week	41	28	FITS BINTABLE	c, t
Annual Average Sky Maps	Average of intensities for all 10 bands over the entire cryogenic mission; 1 file per band	10	11	FITS BINTABLE	n, c, t
$\varepsilon = 90°$ **Sky Maps**	Intensities for all 10 bands for the special case of Solar elongation $\varepsilon = 90°$; 1 file per band	10	13	FITS BINTABLE	n, c, t
Galactic Plane Maps	Galactic Plane subset of the $\varepsilon = 90°$ Sky Maps; limited to a 6 month interval	1	6	FITS BINTABLE	n, c, t
Beam Profiles	Effective two–dimensional shape of *DIRBE* beam for each detector	16	0.5	FITS IMAGE	n, c, t
Color Corrections	Color correction factors needed in case source spectrum is not $\nu I_\nu =$ constant	1	0.04	ASCII	n, c, t
System Responses	Normalized system spectral response functions for each wavelength band	1	0.06	ASCII	n, c, t

[a] Data available from the NSSDC on tape (t), CD–ROM (c), or over the network (n).
[b] See Berriman *et al.* (this volume) for description of polarimetry data.

2.1 *Weekly Sky Maps*

Each celestial direction was observed by the *DIRBE* along multiple sight lines through the interplanetary dust (IPD) cloud. A set of 41 Weekly Sky Maps covering the period of optimized cryogenic operation (1989 December 11 to 1990 September 21) offers an unprecedented view of the IPD cloud. These maps provide weekly–averaged intensity values for each photometric band and $0.^\circ3 \times 0.^\circ3$ pixel, plus near–infrared polarimetric data. Each map covers approximately half of the sky.

2.2 *Annual Average Sky Maps*

The Annual Average Sky Maps provide a single, ten–month averaged intensity value per pixel for each of the 10 *DIRBE* bands. Improved sensitivity to faint emission generally results from the coaddition of repeated observations. However, since these maps average over the variable zodiacal light signal, they are useful primarily at wavelengths at which the IPD signature is weak, particularly 140 and 240 μm. The Annual Average Sky Maps also provide information on the depth of sky coverage over the mission.

2.3 *Solar Elongation* $= 90^\circ$ *Sky Maps*

The only fixed solar elongation angle ε at which the entire sky can be mapped is 90°. The $\varepsilon = 90^\circ$ Sky Maps provide for each *DIRBE* wavelength an estimate of the infrared intensity at each pixel based on an interpolation to $\varepsilon = 90^\circ$ of the observations made at various times at ε close to 90°.

2.4 *Galactic Plane Maps*

The Galactic Plane Maps are subsets of the $\varepsilon = 90^\circ$ Sky Maps designed to facilitate studies of the Galaxy. The maps cover Galactic latitudes $|b| < 10^\circ$ at longitudes $30^\circ < \ell < 330^\circ$, and cover $|b| < 15^\circ$ elsewhere. The Galactic Plane Maps file released along with the other products in October 1994 supersedes the original version, which was released in June 1993.

2.5 *Ancillary Data*

Three additional products are intended to facilitate analysis of the sky maps. A Beam Profile Map gives the effective two–dimensional shape of the *DIRBE* beam for each wavelength; normalized system spectral response functions for each *DIRBE* wavelength band are given in tabular form; and a set of Color Correction Tables are provided. The *DIRBE* spectral intensity data, I_ν, are expressed in MJy/sr, assuming the source spectrum is νI_ν =constant. Where this is not true, color corrections must be applied to obtain an accurate intensity. The Color Correction Tables provide the needed correction factors for a variety of source spectral shapes.

Table 2. *FIRAS* Data Products[a]

Product	Description	Number of files	MB/file
Spectral Sky Maps and associated ancillary files	Calibrated, destriped, combined spectra covering 95.4% of the sky; 14 12–MB files, 1 per channel/scan mode.[b] Also, one file of each of the following types for each channel/scan mode: Destriper Correction Spectra; Destriper Stripe Covariances; Destriper Pixel Covariances; Calibration Errors; Calibration Model; Line Profile Models; and Covariance Matrices. Estimates of the possible systematic errors caused by the choice of destriper algorithm are given in 3 Model Errors files. Weight vector and χ^2 files are associated with each of 4 combined–mode Spectral Sky Maps.	95	0.017 – 12
Line Emission Maps	"Maps" of the intensity and the associated uncertainty in each of the following spectral lines: CO (115.27, 230.54, 345.80, 461.04, 576.27, and 691.47 GHz), O_2 (424.75 GHz), [C I] (492.23 and 809.44 GHz), H_2O (556.89 and 1113.3 GHz), [N II] (1461.1 and 2459.4 GHz), [C II] (1900.5 GHz), [O I] (2060.1 GHz), and [Si I] (2311.7 GHz). One file for ν<630 GHz; another for higher frequencies. The high–frequency file includes "maps" of interstellar dust temperature and 1800 GHz optical depth.	2	<1
Continuum Spectrum Maps	Spectral sky maps from which the line emission has been subtracted; 1 file per channel/scan mode[b]	10	12

[a] All of the released products are available from the NSSDC over the network or on tape. A CD–ROM will be prepared as soon as the final product delivery is made. All of the data are in FITS binary tables.

[b] The *FIRAS* instrument had two output ports ("left" and "right") and two frequency bands (60 – 630 GHz and 630 – 2910 GHz), and a Mirror Transport Mechanism (MTM) that could operate at two speeds ("slow" or "fast") and over two scan lengths ("short" or "long"). Some data products represent output port and MTM mode combinations.

3. *FIRAS* DATA PRODUCTS

The Far Infrared Absolute Spectrophotometer surveyed 94.5% of the sky with a 7° beam, measuring absolutely calibrated, moderate resolution spectra at wavelengths ranging from 105 μm to 1 cm. In addition to the Spectral Sky Maps, which contain calibrated spectra from various *FIRAS* channels and scan modes and channel/mode combinations, the high–level data products described below and in Table 2 are available.

3.1 Line Emission Maps

The Line Emission Maps cover the entire sky and give the intensity and the associated uncertainty in each of 16 emission lines, of which the following 9 were detected (Bennett *et al.* 1994): the 158 μm ground state transition of C^+, an important neutral interstellar gas coolant; the N^+ 122 μm and 205 μm transitions, which trace extended low–density ionized gas; the 370 μm and 609 μm lines of C^0, which arise in photodissociation regions; and the CO $J = 2 \rightarrow 1, 3 \rightarrow 2, 4 \rightarrow 3$, and $5 \rightarrow 4$ lines, which trace the molecular gas. As one would expect, the line emission is strongest in the Galactic plane and concentrated in the inner Galaxy.

3.2 Continuum Spectrum Maps and Dust Parameters

At high frequencies, the continuum in a *FIRAS* spectrum is dominated by thermal dust emission; at low frequencies, the cosmic microwave background (CMB) dominates. The Continuum Spectrum Maps cover the whole sky and give spectra from which CMB, zodiacal, and line emission components have been subtracted, leaving behind an interstellar dust signature.

Maps of the interstellar dust temperature and the 167 μm optical depth are included in the "RHSS" Line Emission Map FITS file. A single–temperature dust model was used to make these maps and the far–infrared dust emissivity was assumed to be 1.55. Different models can be used and assumptions made, and corresponding temperature and optical depth maps can be derived straightforwardly from the Continuum Spectrum Maps. Reach *et al.* (1995), for example, report evidence for a ubiquitous cold ($\sim$ 5K) dust component.

REFERENCES

Bennett, C.L., *et al.* 1994, ApJ, 434, 587

Boggess, N.W., *et al.* 1992, ApJ, 397, 420

COBE Diffuse Infrared Background Experiment (DIRBE) Explanatory Supplement, ed. M.G. Hauser, T. Kelsall, D. Leisawitz, and J. Weiland, *COBE* Ref. Pub. No. 95–A (Greenbelt, MD: NASA/GSFC), available in electronic form from the NSSDC

COBE Far Infrared Absolute Spectrophotometer (FIRAS) Explanatory Supplement, ed. J.C. Mather, R.A. Shafer, R.E. Eplee, D.J. Fixsen, J. Gales, R.B. Isaacman, K. Jensen, S. Read, A.R. Trenholme, *COBE* Ref. Pub. No. 95–C (Greenbelt, MD: NASA/GSFC), available in electronic form from the NSSDC

Reach, W.T., *et al.* 1995, ApJ, 450, in press

THE NEAR-INFRARED SCATTERING PROPERTIES OF INTERPLANETARY DUST

G. B. Berriman[1], J. L. Weiland[1], C. M. Lisse[2], W. T. Reach[3], M. G. Hauser[4] & T. Kelsall[5]

[1] *General Sciences Corp., Code 685.3, NASA Goddard Space Flight Center, Greenbelt, MD 20771*
[2] *Hughes STX, Code 685.9, NASA Goddard Space Flight Center, Greenbelt, MD 20771*
[3] *Universities Space Research Association, Code 685, NASA Goddard Space Flight Center, Greenbelt, MD 20771*
[4] *Code 680, NASA Goddard Space Flight Center, Greenbelt, MD 20771*
[5] *Code 685, NASA Goddard Space Flight Center, Greenbelt, MD 20771*

Email ID
berriman@aruba.gsfc.nasa.gov

ABSTRACT

The high photometric quality and full-sky coverage in the COBE DIRBE datasets make possible detailed studies of the interplanetary medium. This paper presents preliminary results of investigations of the infrared scattering properties of interplanetary dust. The purpose of these investigations is to use DIRBE photometry and polarimetry to constrain the composition, size and structure of interplanetary dust grains. The work described is being carried out on two fronts: derivation of the near-infrared scattering phase function and modeling of the run of near-infrared polarization with solar elongation angle.

1. Introduction

The COBE DIRBE has produced the first full-sky survey of the intensity and polarization of the diffuse near-infrared sky at 1.25 μm to 3.5 μm. The principal source of diffuse radiation is the zodiacal light, highly polarized sunlight scattered from interplanetary dust grains. The model of the infrared light of the zodiacal dust cloud presented by Franz *et al.* (1995) and Reach *et al.* (1995) assumed a scattering phase function derived from visual observations. While this phase function (the angular redistribution of light after scattering) indicates that silicates are the predominant scatterers, the infrared color and polarization along one line of sight (Berriman *et al.* 1994), when compared with Mie theory calculations, still allowed for a substantial proportion of carbonaceous material. Clearly, a derivation of the infrared phase function itself is required to validate the model and to probe the nature of the scattering grains. This paper presents preliminary results of the derivation of the scattering phase function in the ecliptic plane at 1.25 μm.

Despite restrictions on the applicability of Mie theory, Frazier *et al.* (1987) have had success in using it to model the elongation dependence of the visible polarization of the zodiacal cloud. Mie theory calculations of the elongation dependence of the near-infrared polarization is clearly warranted, and is an obvious continuation of the initial study of Berriman *et al.* This paper describes the

© 1996 American Institute of Physics

extent to which Mie theory calculations succeed in reproducing the polarization measurements.

The two analyses described here, derivation of the phase function and modeling the polarization properties, are the first steps in efforts to more fully understand the composition, size and structure of interplanetary dust grains. This knowledge will act as the basis for development of physical models of the zodiacal dust cloud.

2. Derivation Of The Intensity Of The Scattered Light

The scattered zodiacal light must be separated from interstellar emission and from emission due to discrete sources. At a given elongation, ϵ and ecliptic latitude, β, the interstellar emission varies continuously throughout the year with the slow precession of the COBE spacecraft in its Sun synchronous orbit. Interstellar emission can therefore be minimized by a judicious selection of time interval. The results described here have been derived from data averaged over the interval 1990 Jan 1-7, when the Galactic Center was behind the line of sight through the Sun and the interstellar emission supplied no more than a few percent of the near-infrared light in the ecliptic plane.

The scattered intensity in the ecliptic plane, $Z_\nu(\epsilon)$, was derived by fitting an analytic function to the observed intensities along lines of constant ecliptic latitude at fixed ϵ. These "slices" were fitted every 5° in elongation to generate the run of $Z_\nu(\epsilon)$ with ϵ. The analytic fit used, described by Hauser (1993), is designed to reproduce the large-scale angular features of the zodiacal light. In practice, the data were fitted with the sum of this empirical function and a Galactic template (to allow the small contribution of the interstellar emission to be taken into account). Bright discrete sources were removed by an interative scheme that weighted residual intensities from the fit inversely with their magnitude. The sources eliminated in this way supply on average some 15% of the total light at 1.25 μm and 2.2 μm, but only 5% at 3.5 μm. The resulting scattered intensities in all three bands decline steadily by roughly 30% across the 64° to 124° elongation range scanned by DIRBE. The faint sources have not been removed and are the principal source of uncertainty in the determination of $Z_\nu(\epsilon)$ in the near-infrared.

3. Derivation of Phase Function

In the visible, moderately strong backscattering seen at scattering angle $\Theta \geq 100°$ distinguishes silicates from carbonaceous grains (Leinert 1975). Because DIRBE observations are limited to $\epsilon \leq 124°$ and backscattering will become important only at $\epsilon \geq 90°$, the crucial issue in analyzing DIRBE data is the accuracy with which backscattering can be investigated.

The phase function $\Phi(\Theta)$ is itself is related to $Z_\nu(\epsilon)$ via:

$$Z_\nu(\epsilon) = \frac{K}{sin^{\nu+1}\epsilon} \int_\epsilon^\pi \Phi(\Theta) sin^\nu \Theta \, d\Theta \qquad (1)$$

where the number density, n, is assumed to vary with distance r (in AU) from the Sun according to a power law with index ν, $n = n_\circ r^\nu$, where $n_\circ$ is the number density at the Earth; here we assume ν=1.3 throughout (Leinert *et al.* 1981). The normalization factor K incorporates n_o and the average grain albedo.

Derivation of Φ involves transforming equation (1) from spatial to scattering coordinates. Two methods widely applied in the visible, inversion and parameterization, have been applied to the DIRBE data.

3.1 The Inversion Method

The phase function is derived simply by differentiating equation (1), so that $\Phi(\Theta)$ is found from measurements of $\partial Z/\partial\epsilon$, itself easily derived from quadratic fits to $Z_\nu(\epsilon)$ *vs.* ϵ. The confusion noise due to faint stars inherent in $Z_\nu(\epsilon)$ limits the accuracy with which $\partial Z/\partial\epsilon$ can be found to $\approx$10%. This imprecision is sufficiently large that it will accommodate phase functions that have strong backscattering and no backscattering. Direct inversion obviously affords no useful constraints on the scattering function.

3.2 Parameterization.

Parametrization is the only practical method applicable to the DIRBE data. We have replicated Hong's (1985) parameterization of the visual zodiacal brightness. The brightness integral in equation (1) is parameterized with the sum of three Henyey Greenstein functions (Hong 1985), which depend upon ϵ and an "asymmetry factor" $g=<\Phi(\Theta)cos\Theta>$; the sum of the relative weights of each function is normalized to unity. The scattered intensities, $Z_\nu(\epsilon)$, are fitted to the parametric expression with the Levenberg-Marquardt non- linear least squares technique, with the measured intensities weighted according to $1/Z_\nu(\epsilon)$.

The results at 1.25 μm are in excellent agreement with those of Hong: the asymmetry factors and weights agree to within a few percent. In particular, the backscattering in the visible is seen in the infrared too, between $\Theta = 100°$ and 124°. This result does, however, require further examination. In Hong's method, the smallest intensities are given highest weight, yet the empirical fits are less reliable as the intensity declines. Moreover, the power law index ν may be significantly larger than 1.3 (Franz *et al.* 1995). Therefore, a more detailed study investigating the sensitivity of phase function to the full allowed range of ν and to different schemes for weighting $Z_\nu(\epsilon)$ must be carried out. Nevertheless, the parameterization method holds promise for deriving the phase function of the scattered infrared light.

4. Modeling the Near-Infrared Polarization

With the aid of Mie theory calculations, Frazier *et al.* (1987) modeled the decline with elongation of the visible polarization with hard, silicate spheres. Berriman *et al.* 's (1994) Mie theory calculations for hard spheres accounted for the near-infrared polarizations at one line-of-sight (ϵ =90°, β=0°), but indicated that scattering from graphite grains may contribute significantly at these wavelengths.

How well do the Mie theory calculations explain the elongation dependence of polarization in the near-infrared? The results of model calculations have been compared with the polarized intensities, calculated at a spatial sampling of 3° at 1.25 μm, 2.2 μm and 3.5 μm. The advantage of polarized fluxes over the more usually quoted degree of polarization is that they are not "diluted" by interstellar or interplanetary emission (which is very likely appreciable at 3.5

μm; Berriman *et al.* 1994). The variation of polarized flux with ϵ specifies the behavior of the interplanetary polarized flux, assuming all other sources of radiation are unpolarized. Comparison of the data with the model calculations (normalized at ϵ=90°) that assume the interplanetary mass spectrum of Grun *et al.* (1985) reveals the inadequacy of the Mie approximation. Whereas the polarized intensities at 1.2 μm, 2.2 μm and 3.5 μm decline by a factor of two from ϵ=64° to 124°, all Mie theory models – whether for hard or porous grains, composed of graphite or silicate grains - have a much steeper elongation dependence, generally of at least a factor or three. The discrepancy is in the sense that the polarized intensities at the lower ϵ's are much smaller than calculated.

These results indicate the importance of the broad wavelength coverage of DIRBE polarimetry. For, despite the fact that data at one wavelength can be accounted for by Mie theory calculations, more sophisticated calculations are needed to account for the visual and infrared polarizations together. The failure of the Mie theory alomost certainly reflects that the grains are not spherical, and it is known that irregular grains of radii tens of microns are common in the solar system (Giese *et al.* 1978). Consequently, we are exploring the application of the Discrete Dipole Approximation, which allows for arbitrary grain shapes, but is limited in its applicability to grain sizes of $\leq$30 μm or so.

5. Conclusions

The parameterization method is a very promising technique for deriving the near-infrared scattering phase function of the interplanetary dust. Preliminary results show the backscattering characteristic of silicate grains. Modeling of the polarization properties of these grains (whatever their actual composition) requires the development of models that are more sophisticated than Mie theory, which is applicable only to smooth spherical or elliptical grains.

Acknowledgements

COBE is supported by NASA's Astrophysics Division. Goddard Space Flight Center (GSFC), under the scientific guidance of the COBE Science Working Group, is responsible for the development and operation of COBE.

References

Berriman, G. *et al.* 1994, ApJLett, 431, L63
Franz, B. *et al.* 1995, this Workshop
Frazier, Bouchet and Mueller 1987, SPIE Volume 819, Infrared Technology, XIII, 2
Giese, R. H. *et al.* 1978, A&A, 65, 265
Grun, E. *et al.* 1985, Icarus, 62, 244
Hauser, M. G. 1993, In "Back To The Galaxy", eds. S. S. Holt, F. Verter, (AIP, New York), 201
Hong, S. S. 1985, A&A, 146, 67
Leinert, C. 1975, Spa. Sci. Rev., 18, 281
Leinert, C. *et al.* 1981, A&A, 103, 177
Reach, W. T. *et al.* 1995, this Workshop

INVERSION OF THE DIRBE PHOTOMETRY TO DERIVE THE THREE DIMENSIONAL DISTRIBUTION OF THE INTERPLANETARY DUST

Bryan A. Franz
Applied Research Corporation
Code 685.3, NASA Goddard Space Flight Center, Greenbelt, MD 20771

William T. Reach
University Space Research Foundation
Code 685, NASA Goddard Space Flight Center, Greenbelt, MD 20771

T. Kelsall
Code 685, NASA Goddard Space Flight Center, Greenbelt, MD 20771

Janet L. Weiland
General Sciences Corp.,
Code 685.3, NASA Goddard Space Flight Center, Greenbelt, MD 20771

ABSTRACT

The Diffuse Infrared Background Experiment (DIRBE) on board NASA's *Cosmic Background Explorer* (*COBE*)[1] satellite has surveyed the entire sky in ten photometric bands covering the wavelength range from 1 to 240 μm at an angular resolution of 0.7°. The spin of the spacecraft, in combination with its orbital motion, allowed the DIRBE instrument to sample redundantly each celestial direction from multiple orientations within the interplanetary dust cloud. The variation in viewing aspect gives rise to time-dependent changes in the observed intensity which are only associated with the changing brightness contributions from the interplanetary dust. We present a three-dimensional, semi-physical model for the emission and scattering of interplanetary dust and describe the techniques employed to optimize the model to match the temporal variations of the infrared sky brightness as observed by DIRBE along selected lines of sight. We then show the accuracy with which the optimized model is able to reproduce the apparent temporal and angular variation of sky brightness associated with the interplanetary dust over the entire sky.

1. INTRODUCTION

Over the period of one day, the DIRBE instrument measures approximately one half the sky at a range of elongation angles between 64° and 124°. Due to the orbital motion of the Earth, this 60° viewing swath shifts by one degree per

[1] *COBE* is supported by NASA's Astrophysics Division. Goddard Space Flight Center (GSFC), under the scientific guidance of the *COBE* Science Working Group, is responsible for the development and operation of *COBE*.

© 1996 American Institute of Physics

day, allowing the DIRBE to sample redundantly each line of sight from multiple orientations within the interplanetary dust (IPD) cloud. The change in viewing aspect gives rise to temporal variations in the dust density and temperature distribution along the line of sight due to variations in distance from the Sun, solar elongation angle, and location of the spacecraft with respect to the peak of the dust density distribution, which is inclined with respect to the ecliptic plane. The inversion procedure exploits these effects to separate the zodiacal light from the brightness contributions associated with stellar and interstellar sources by imposing an analytical form for the dust density distribution, thermal characteristics, and scattering phase function, and then optimizing the model to match the observed temporal variations to selected directions.

2. THE MODEL

The analytical form of the model is fully described by Reach et al. (1995). Briefly, it consists of the integral along the line of sight of the product of an emissivity function and a density distribution function. The emissivity function includes both a thermal component and a scattering component, and the density distribution includes a smooth cloud component (Good 1994), two pairs of asteroidal dust bands (Reach 1992), and a circumsolar dust ring based on the numerical model of Dermott et al. (1994).

3. THE OPTIMIZATION PROCEDURE

The optimization procedure makes use of approximately 200,000 weekly-averaged photometry measurements consisting of about two thousand uniformly distributed lines-of-sight (pixels) in ten wavebands over forty-one weeks. The key to fitting the zodiacal light while ignoring the brightness contributions associated with stellar and interstellar sources is to constrain the fit using only the observed time variation along independent lines-of-sight, while ignoring the pixel-to-pixel variations. This is achieved by including an arbitrary constant along each line-of-sight which forces the mean of the model over all time samples to match the mean of the data. In this way, the optimization procedure only has enough information to match the amplitude and phase of the temporal variation for each pixel. The disadvantage to this approach is that low contrast structure associated with the interplanetary dust (e.g., the asteroidal dust bands) is difficult to fit. This difficulty is overcome by allowing the angular variations to influence the fit in the 12 and 25 μm wavebands at high Galactic latitudes ($|b| > 30°$), where it is reasonable to assume that the pixel-to-pixel variations are dominated by the zodiacal brightness contribution.

4. FINAL RESULTS

Following the optimization procedure, the zodiacal light model was removed from each weekly-averaged photometry map, and the residual maps were combined to form the mission-averaged maps with zodiacal light removed. Figure 1 shows the observed mission-averaged sky brightness at 1.25, 3.5, 12, and 60 μm, and Figure 2 presents the final, mission-averaged maps after zodiacal light removal. The 1.25 μm residual brightness shows no indication of structure with an ecliptic latitude dependence, which suggests that the zodiacal light contribution to the angular variation has been removed correctly. At 3.5 μm, the map does show weak evidence of positive residuals near the ecliptic plane. This waveband contains a mixture of scattered light and thermal emission, and there is reason to believe that it sees a hot dust component which is not supplied by the model (Reach et al. 1995, Berriman et al. 1995). The 12 μm residuals clearly show ecliptic longitudinal banding structure which is most likely associated with shape errors in the band model and/or smooth cloud component. The model tends to oversubtract near the ecliptic plane at 12 μm, while it undersubtracts at 60 μm. This disparity may be a result of the assumption that the particle properties of the dust bands and circumsolar ring are identical to those of the smooth cloud (Reach et al. 1995). In general, the apparent angular variations associated with the IPD in the mission-averaged residual maps are at the few percent level in all bands relative to the observed sky brightness.

REFERENCES

Reach, W. T., *et al.* 1995, this workshop

Berriman, G. B., *et al.* 1995, this workshop

Good, J. 1994, in IRAS Sky Survey Atlas Explanatory Supplement, ed. Wheelock, S. L. et al., JPL Publication 94-11 (Pasadena: JPL)

Reach, W. T. 1992, AJ, 392, 289

Dermott, S. F., Jayraraman, S., Xu, Y. L., Gustafson, B. D. S. & Liou, J. C. 1994, Nature, 369, 719

FIGURE CAPTIONS

Figure 1: Average of 41 weekly-averaged sky brightness maps at 1.25, 3.5, 12, and 60 μm in ecliptic Mollweide projection with longitude zero at center.

Figure 2 Average of 41 weekly-averaged maps with zodiacal light removed at 1.25, 3.5, 12, and 60 μm in ecliptic Mollweide projection with longitude zero at center.

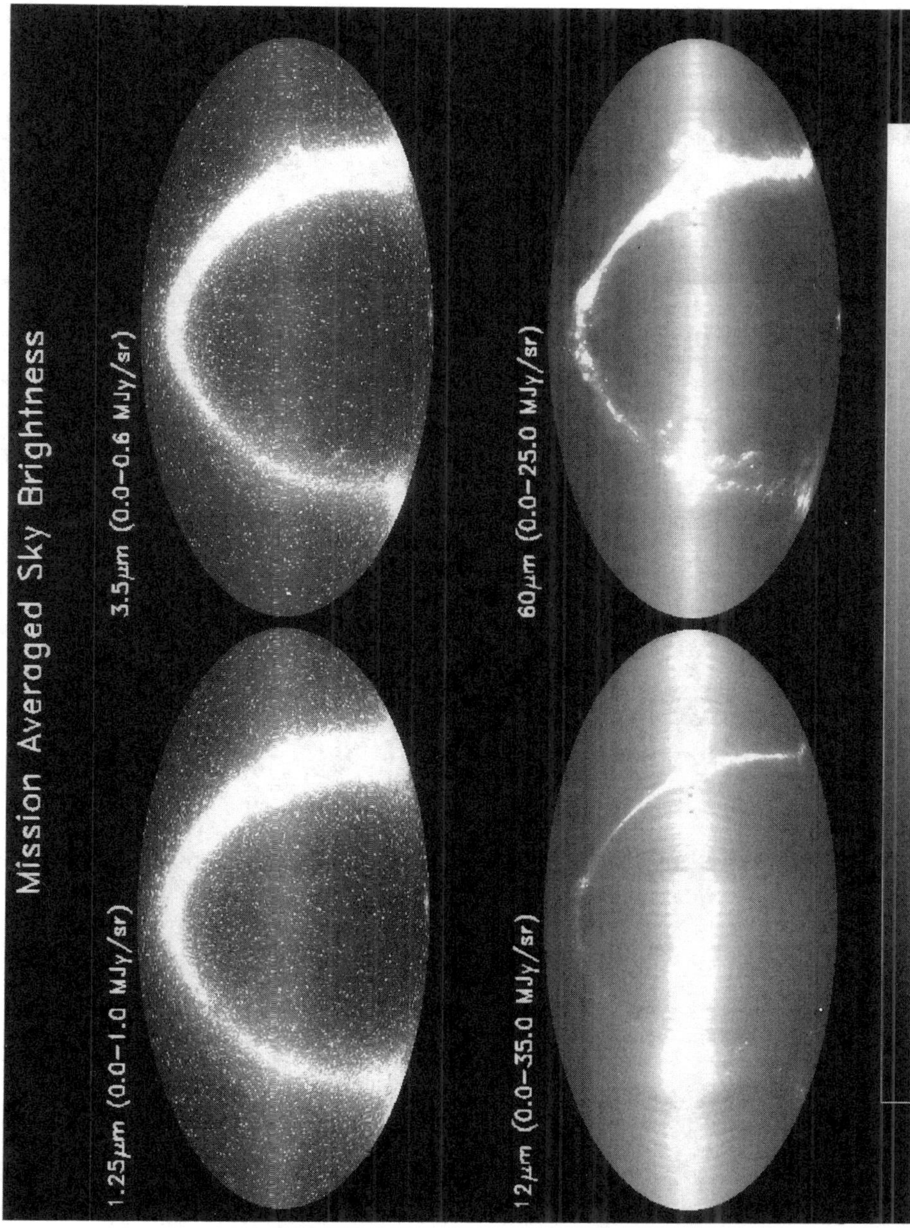
Mission Averaged Sky Brightness
1.25μm (0.0–1.0 MJy/sr)
3.5μm (0.0–0.6 MJy/sr)
12μm (0.0–35.0 MJy/sr)
60μm (0.0–25.0 MJy/sr)

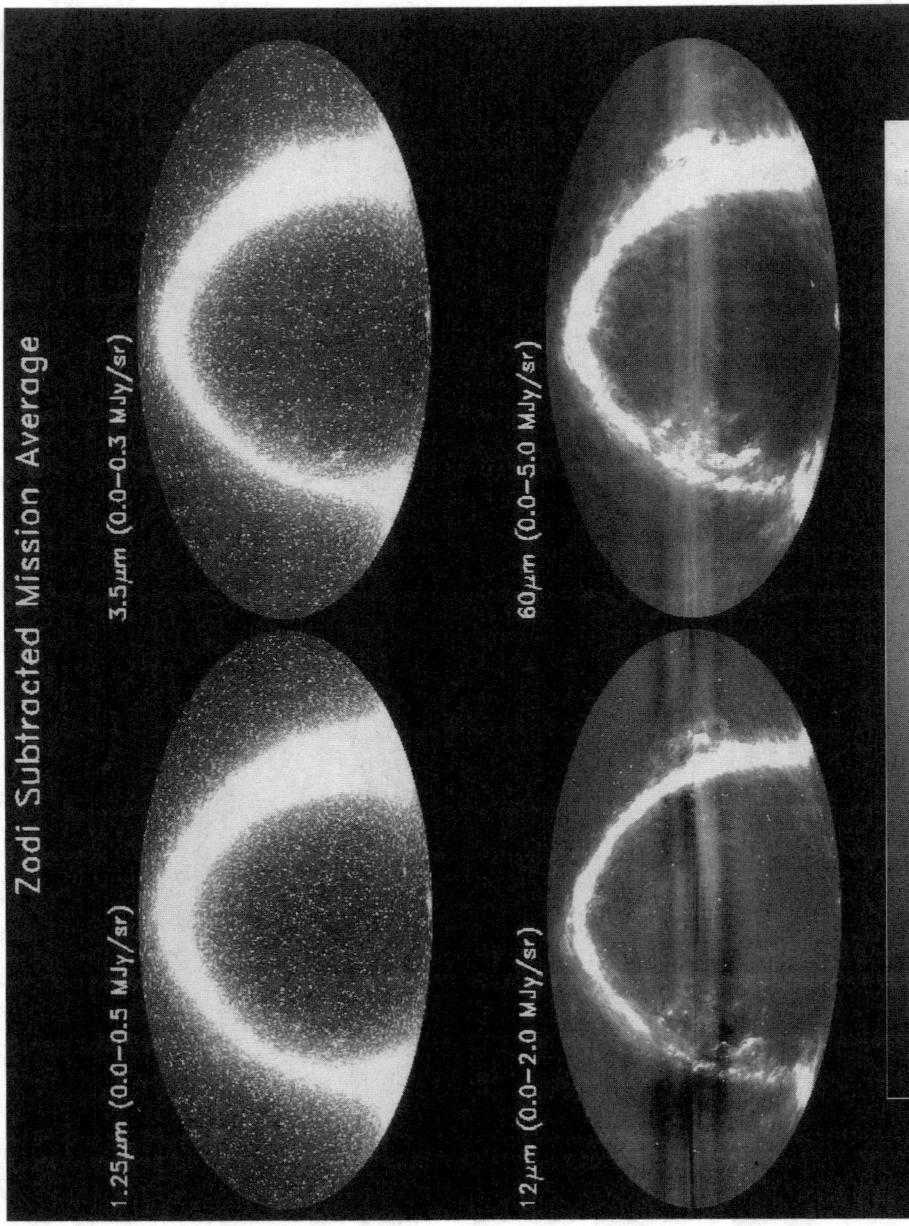
Zodi Subtracted Mission Average
1.25μm (0.0–0.5 MJy/sr)
3.5μm (0.0–0.3 MJy/sr)
12μm (0.0–2.0 MJy/sr)
60μm (0.0–5.0 MJy/sr)

INFRARED POINT SOURCE PHOTOMETRY WITH DIRBE: INSTRUMENT GAIN STABILIZATION AND VARIABLE STAR MONITORING

K. J. Mitchell and G. B. Berriman
General Sciences Corporation, Code 685.3,
NASA/Goddard Space Flight Center, Greenbelt, MD 20771
mitchell@aruba.gsfc.nasa.gov

T. Kelsall
Code 685, NASA/Goddard Space Flight Center, Greenbelt, MD 20771

D. Richardson and J. A. Skard
Hughes/STX, Code 685.3
NASA/Goddard Space Flight Center, Greenbelt, MD 20771

S. W. Stemwedel
Applied Research Corporation, Code 685.3,
NASA/Goddard Space Flight Center, Greenbelt, MD 20771

ABSTRACT

The *COBE/DIRBE* data archive contains a unique record of bright IR point source observations over the entire sky. This paper reviews the method by which point source photometry is calculated from the *DIRBE* time-ordered sky survey data. Also shown are some examples of how this point source photomerty is being used to calibrate the gain of the *DIRBE* photometric system, monitor variable stars, and measure stellar polarization.

1. INTRODUCTION

During its nominal cryogenic-mission lifetime (Dec. 11, 1989 through Sept. 21, 1990) the *COBE/DIRBE* instrument (Hauser *et al.* 1991) observed half of the sky each day in 10 infrared (IR) bands ranging from 1.25 μm to 240 μm. Linear polarization measurements also were obtained in the three shortest bands (1.25, 2.2, and 3.5 μm). Data were recorded simultaneously in all bands every 1/8 second, as *DIRBE* scanned the sky in a helical manner driven by the *COBE* spacecraft spin and orbital motion.

Along with its accurate measurements of the diffuse IR sky, *DIRBE* also obtained a unique set of IR observations of bright discrete sources as they transited through the *DIRBE* (0.7°x 0.7°) beam. Because of the large beam size, *DIRBE* was confusion limited in most bands to observing only the brightest discrete sources (*e.g.* ~2000 sources with $K < 3$ mag. at 2.2 μm for Galactic latitudes $|b| > 20°$). The *DIRBE* mode of operation allowed nearly continuous monitoring of discrete sources with ecliptic latitudes $|\beta| > 60°$, while sources at the ecliptic equator received 60 days of coverage each six months. Point source photometry and polarimetry derived from the *DIRBE* data set can be an important resource for monitoring the instrument gain and for the study of stellar energy distributions, variability, and polarization.

© 1996 American Institute of Physics

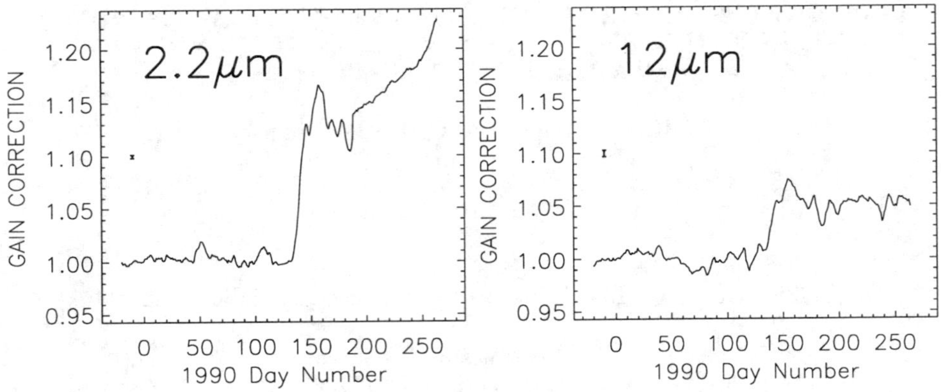

Figure 1. Smoothed versions of the celestial gain corrections (GCs) are plotted as a function of time for the *DIRBE* 2.2 μm and 12 μm bands. The errors bars show the estimated 1σ medium-term (~5-50 days) errors in the GCs. The rise in the GCs after day 130, 1990 mainly reflect an instabiliy in the IRS (see text).

2. CALCULATION OF *DIRBE* POINT SOURCE PHOTOMETRY

DIRBE time-ordered data are used in deriving point source fluxes from individual source passages through the *DIRBE* beam profile. First, a 15-point data string, centered on the source passage and containing both on-source and off-source data, is extracted from the time-ordered data. A background model is then calculated from a quadratic fit to the 5 off-source points on each end of the data string, and the model is subtracted from the original data.

Next, a slice of the normalized *DIRBE* beam profile (*i.e.* beam template) is selected for use based on the source's trajectory through the *DIRBE* beam as derived from the *COBE* fine attitude solution. This beam template is then fit in height and position to the 3 to 4 on-source data points near the center of the data string. The derived instrumental flux of the point source is the multiplicative factor by which the height of the beam template is stretched to fit the amplitude of the on-source data. The typical errors for *DIRBE* point source photometry of bright sources range from ~1% at 1.25 μm to ~10% at 240 μm.

3. CELESTIAL GAIN STABILIZATION OF *DIRBE* PHOTOMETRY

The ability to use *DIRBE* data to accurately model the components of foreground IR radiation and measure the extragalactic IR background depends on the stability of the *DIRBE*'s photometric system. Point source photometry of objects with constant radiative output is being used to produce a relative photometric system in each band which is time-stable throughout the *DIRBE* cryogenic mission. The initial stabilization of *DIRBE* photometry is accomplished through use of frequent observations (at 5-20 min. intervals) of the *DIRBE* Internal Reference Source (IRS). The IRS has proven to be very stable over time scales of days, allowing for removal of short time scale instrumental gain variations. However, over longer timescales celestial calibration is needed to stabilize the relative gain of the IRS-corrected photometry. (Note: the IRS correction was not applied at 140 μm and 240 μm.)

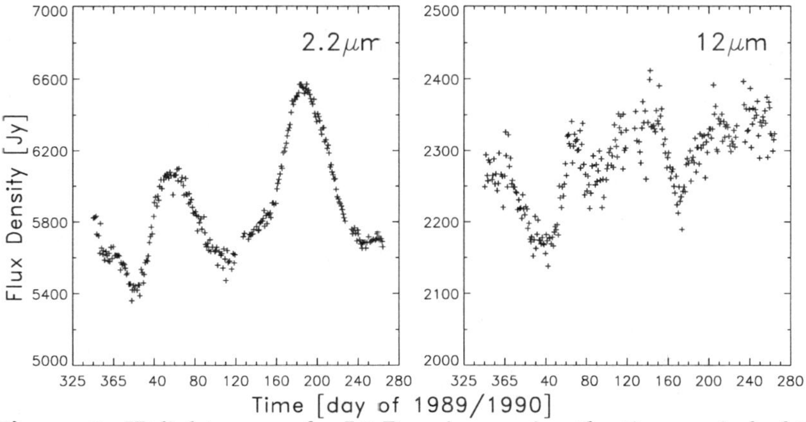

Figure 2. IR light curves for L2 Puppis covering the time period of the *DIRBE* cryogenic mission. The plotted data are daily averages of point source photometry with an approximate absolute calibration applied.

A self-defined celestial photometric system is formed in each *DIRBE* band by deriving a self-consistent set of photometric standard values (PSVs) for the calibration objects used in that band. The point source photometry for the calibration objects is divided by their PSVs, putting observations of all calibrators on the same normalized scale. Time-binned averages of normalized point source photometry are used as the celestial gain coefficients (GCs). Figure 1 shows smoothed versions of the GCs as a function of time in the 2.2 μm and 12 μm bands, which map changes in the gain of the IRS-based *DIRBE* photometry. The final *DIRBE* photometric system is based on the application of these GCs.

4. VARIABLE STAR LIGHT CURVES IN THE IR

The DIRBE temporal and multiwavelength coverage of the sky in the IR is unique. Point source photometry derived from *DIRBE* observations will be an important resource for monitoring the IR light curves of many bright, variable stars over the *DIRBE* mission lifetime. As an example, Figure 2 shows the 2.2 μm and 12 μm light curves of the variable star L2 Puppis.

L2 Puppis is classified as an M5 emission line giant of type SRb, with semi-regular variability. The V-band light curve in Petit (1987) shows evidence of a secondary variation with a period slightly less than the primary period of 140 days. The *DIRBE* light curves in Figure 2 cover two full periods of the L2 Puppis primary cycle which is clearly seen at 2.2 μm. However, the amplitude of the secondary mode increases in strength relative to the primary mode at 12 μm. This wavelength-dependent behavior is most likely attributable to the presence of a dust shell surrounding the star (*e.g.* Judge and Stencel 1991).

5. POINT SOURCE POLARIZATIONS IN THE IR

The construction and operation of the *DIRBE* polarimeter at 1.25, 2.2, and 3.5 μm are detailed in a separate paper at this workshop by Berriman *et al.* At each of the three wavelengths, the point source photometry measurements in the polarization channels give the Stokes parameters Q and U through

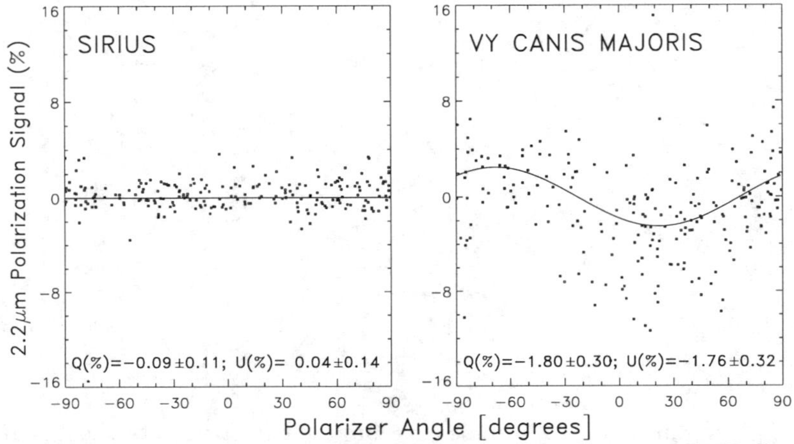

Figure 3. Polarization observations at 2.2 μm of Sirius and VY Canis Majoris. Also shown are the Stokes parameters derived from fitting eqn. 1 to the data.

an iterative, outlier-resistant least-squares fit of the form:

$$\text{POLARIZATION SIGNAL} = Q\ cos(2\theta) + U\ sin(2\theta) \qquad (1)$$

where POLARIZATION SIGNAL is the fractional difference between the point source fluxes measured in the polarization channel and the total-intensity channel, with the fluxes calibrated to the same relative photometric scale. The polarizer orientation, θ, is derived from the position of the object and the instantaneous *DIRBE* scan direction on the sky.

Polarization measurements at 2.2 μm of the stars Sirius and VY Canis Majoris are shown in Figure 3. The data for Sirius indicates that no significant polarization is measured to a 1σ accuracy of 0.18%, as expected for a nearby, early type star. This demonstrates that instrumental polarization in *DIRBE* is inconsequential. The polarization data for VY Canis Majoris are clearly modulated with polarizer orientation, with a peak signal of 2.5%. The derived IR polarizations presented here agree with the highest values reported by McCall and Hough (1981). These measurements suggest that *DIRBE* polarization observations will be valuable in modeling the atmospheres of very late type stars.

Acknowledgements

COBE is supported by NASA's Astrophysics Division. Goddard Space Flight Center, under the scientific guidance of the *COBE* Science Working Group, is responsible for the development and operation of *COBE*.

References

Hauser, M.G. et al. 1991, AIP Conf. Proceed. 222, 161
Judge, P.G. and Stencel, R.E. 1991, ApJ, 371, 357
McCall and Hough 1981, A&A Suppl., 42, 141
Petit, M. 1987, "Variable Stars", [John Wiley & Sons]

SUBTRACTION OF GALACTIC STELLAR EMISSION FROM *COBE* [1] / DIRBE OBSERVATIONS

Richard G. Arendt
Applied Research Corp.
Code 685.3, NASA Goddard Space Flight Center, Greenbelt, MD 20771

Henry T. Freudenreich
Hughes/STX
Code 685.9, NASA Goddard Space Flight Center, Greenbelt, MD 20771

Eli Dwek
Code 685, NASA Goddard Space Flight Center, Greenbelt, MD 20771

ABSTRACT

The expected faint extragalactic emission at near-infrared wavelengths is observed through foreground emissions from interplanetary dust particles, Galactic stars, and the Galactic interstellar medium (ISM). The stellar emission component can be removed from the *COBE* Diffuse Infrared Background Experiment (DIRBE) data by using a model for the 3-D distribution of stars and dust within the Galaxy (the "SKY" model of Wainscoat *et al.* 1992 and Cohen 1993, 1994). This model has been constrained by source counts from the mid-IR through the UV. We use this model to account for the emission of the many faint stars which blend into a smooth unresolved foreground when observed with the 0.7° DIRBE beam. The regions of the brightest distinguishable point sources are simply blanked from the sky maps and excluded from further analysis.

Comparison between the model and DIRBE maps which have had zodiacal emission from interplanetary dust (IPD) removed (Reach *et al.* 1995, Franz *et al.* 1995) show that the model is a good representation of the actual sky. However, the comparison also shows that there is room for improvement in some aspects of the modeling.

1. THE BRIGHT STARS (1.25 - 25 μm)

Rather than modeling and removing the emission of each of the detected near-IR point sources individually, we simply excluded regions with bright foreground stars (and other compact sources). Bright sources are blanked at thresholds of 15, 15, 15, 15, 85, and 110 Jy at 1.25, 2.2, 3.5, 4.9, 12 and 25 μm, respectively. Blanking removes the center pixel of a source and the four adjacent pixels. This exclusion rejects 37, 35, 18, 10, 2.6, and 1.8% of the sky above $|b| = 30°$ at wavelengths from 1.25 to 25 μm. The remaining faint sources are smoothly distributed and must be represented by a model (the "Faint Source Model").

[1] *COBE* is supported by NASA's Astrophysics Division. Goddard Space Flight Center (GSFC), under the scientific guidance of the *COBE* Science Working Group, is responsible for the development and operation of *COBE*.

© 1996 American Institute of Physics

2. THE FAINT SOURCE MODEL

The Faint Source Model used to subtract the remaining near-IR Galactic discrete source emission consists of the integrated sky brightness calculated using the SKY source count model of Wainscoat *et al.* (1992), and Cohen (1993, 1994, priv. comm.). The SKY model, and our Faint Source Model both include 87 source types, with individual scale heights, densities, and intrinsic spreads in absolute magnitude. The sources are distributed among 5 structural components: disk, arms, ring, bulge, halo.

Extinction from an exponential disk of dust with separate radial and vertical scale heights mimics the gross properties of Galactic extinction and creates a visible dust lane in the inner Galaxy at 1.25 and 2.2 μm. We assume R(sun) = 8.5 kpc, but deviate from current SKY models by including a solar offset from the Galactic disk of z(sun) = 18 pc (*e.g.* Lynga 1982, Pandey *et al.* 1990). [Cohen (1995) has very recently used an updated version of the SKY model to derive z(sun) = 15 pc from IRAS source counts, and a less well determined value of 18 pc from K band star counts at the NGP.]

Aside from calculating surface brightness instead of source counts, the Faint Source Model is also tailored to the DIRBE data by excluding stars with an apparent magnitude brighter than the bright source thresholds indicated above, and calculating the model at the same resolution as the standard DIRBE maps ($\sim$0.32° x 0.32° pixels). Our Faint Source Model matches the mean intensities calculated from the SKY model to within $\sim$5% at high latitudes. Differences can be larger at low latitude, but low latitude regions are not expected to be useful for measurements of faint backgrounds.

3. MAPS

The effects of subtracting the Faint Source Model from the near-IR DIRBE observations are illustrated by the Mollweide projection maps in Figure 1. Each row in the figure depicts one wavelength. The first column shows images of the sky at 1.25 to 4.9 μm after subtraction of the zodiacal light (Reach *et al.* 1995, Franz *et al.* 1995). These maps are displayed on a logarithmic scale from 0.01 to 100 MJy/sr. Images of the Faint Source Model on the same intensity scale as the preceding maps are in the center column. The third column shows the sky after subtraction of the zodiacal light model and the Faint Source Model, and blanking the bright sources (including the Magellanic Clouds). Because of the greatly reduced dynamic range in these residual maps, they are displayed on a linear scale from -0.1 to 0.1 MJy/sr. The regions where bright sources have been removed appear black, as do some regions where the foreground subtraction has produced negative intensities.

Maps are not shown for 12 and 25 μm because faint discrete Galactic sources contribute very little to the sky brightness at these wavelengths.

4. SUMMARY

In order to remove the contribution of Galactic starlight to measurements of the infrared sky brightness:

1) At all wavelengths, we disregard directions containing bright stars, and other extended bright sources of emission.

2) At 1.25, 2.2, 3.5, 4.9, 12, and 25 μm, we subtract a model of the

Galactic emission based on source count models by Wainscoat *et al.* (1992) and Cohen (1993, 1994) which are constrained by multi-wavelength comparisons with observed source counts.

Having subtracted the zodiacal light (Reach *et al.* 1995, Franz *et al.* 1995) and the Galactic stellar emission:

* The resulting maps of residual emission at 1.25 - 4.9 μm, show greatly reduced gradients of intensity as a function of Galactic latitude.

* Maps at 12 and 25 μm are only slightly changed by the subtraction of Galactic starlight, because of much stronger emission from the ISM (Weiland et al. 1995).

* Accounting for the offset of the Sun from the midplane of the disk, $z(sun) = 12$ - 24 pc, eliminates systematic differences between the near-IR residual intensities in the North and South Galactic hemispheres.

* An asymmetry of the residuals at 1.25 - 4.9 μm in the region of the Galactic bulge suggests that a better model should include a prolate rather than the oblate bulge used in this model (*e.g.* Blitz & Spergel 1992, Weiland et al. 1994, Dwek *et al.* 1995)

* Other obvious problems at low latitude ($|b| < 20°$) may be caused by inaccuracies in our Faint Source Model, which does not match the SKY model as well as at high latitudes.

* Removal of Galactic stellar emission makes relatively small deficiencies in the model for the zodiacal light much more apparent.

REFERENCES

Blitz, L. & Spergel, D. N. 1991, ApJ, 379, 631

Cohen, M. 1993, AJ, 105, 1860

Cohen, M. 1994, AJ, 107, 582

Cohen, M. 1995, ApJ, 444, 874

Dwek, E., *et al.* 1995, ApJ, 445, 716

Franz, B., *et al.* 1995, this workshop

Lynga, G. 1982, A&A, 109, 213

Pandey, A. K., Bhatt, B. C., & Mahra, H. S. 1990, A&A, 234, 128

Reach, W. T., *et al.* 1995, this workshop

Wainscoat, R. J., Cohen, M., Volk, K., Walker, H. J., & Schwartz, D. E. 1992, ApJS, 83, 111

Weiland, J. L. *et al.* 1994, ApJ, 425, L81

Weiland, J. L. *et al.* 1995, this workshop

FIGURE CAPTION

Figure 1. (next page) These maps show a) the near-IR sky after subtraction of the zodiacal light model, b) the Faint Source Model, and c) the residual intensity after subtraction of the Faint Source Model and bright source blanking. (See text for details.)

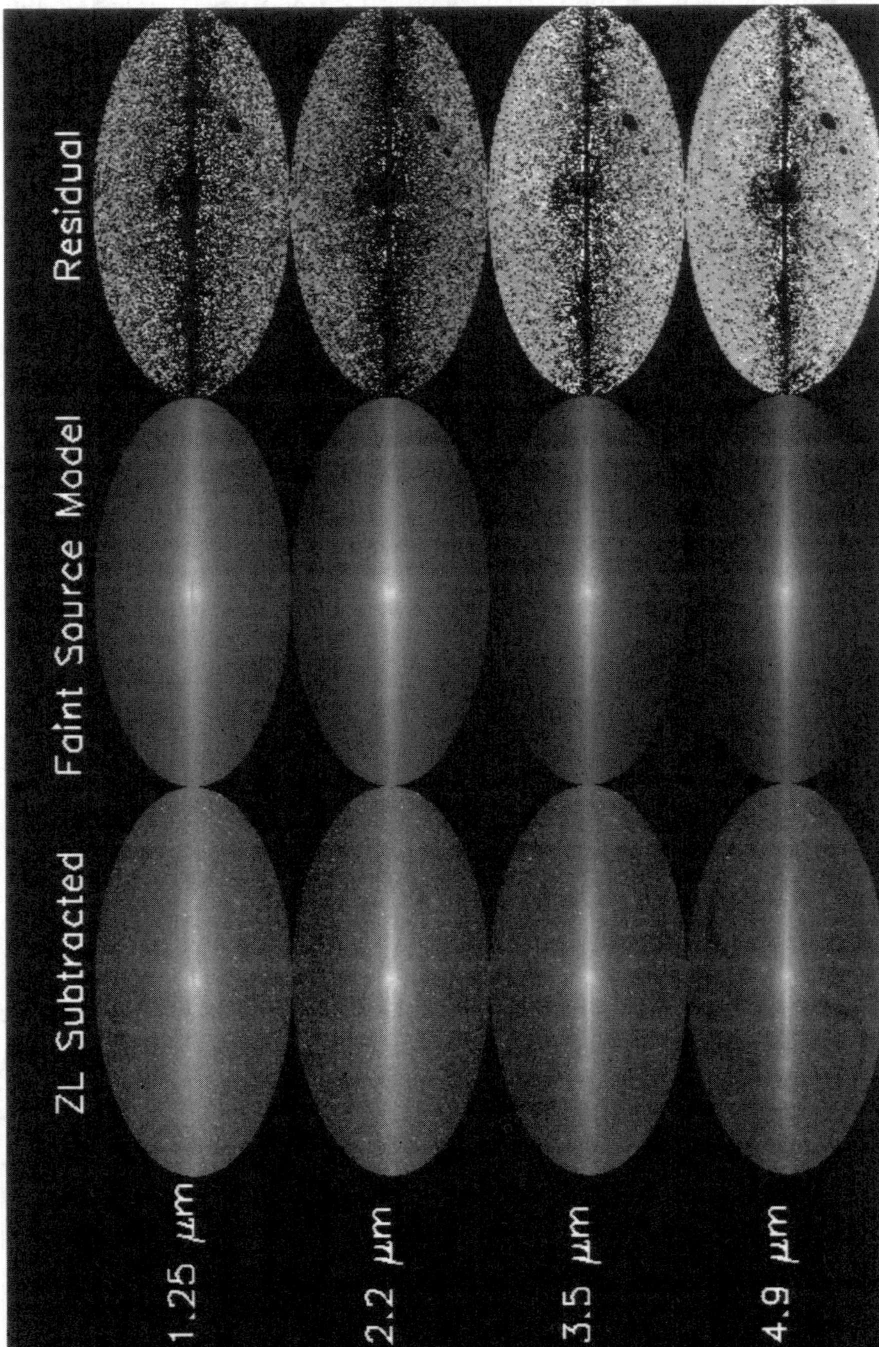
ZL Subtracted
Faint Source Model
Residual
1.25 μm
2.2 μm
3.5 μm
4.9 μm

A THREE-DIMENSIONAL DECOMPOSITION OF THE INFRARED EMISSION FROM DUST IN THE MILKY WAY

T. J. Sodroski and R. G. Arendt
Applied Research Corporation, Code 685.3,
NASA/Goddard Space Flight Center, Greenbelt, MD 20771

N. Odegard and J. L. Weiland
General Sciences Corporation, Code 685.3,
NASA/Goddard Space Flight Center, Greenbelt, MD 20771

and

E. Dwek, M. G. Hauser, and T. Kelsall
Code 680, NASA/Goddard Space Flight Center, Greenbelt, MD 20771

ABSTRACT

We have constructed a three-dimensional model of the Galactic large-scale infrared emission from dust associated with the molecular (H_2), neutral atomic (H I), and extended low-density ionized (H II) gas phases of the interstellar medium. The model incorporates a three-dimensional map of the molecular and neutral atomic hydrogen gas distributions, derived from available ^{12}CO and H I surveys by using the radial velocity information in the spectral lines as a distance indicator, and available 5 and 19 GHz radio-continuum surveys to trace the column density of ionized gas. We have used the model to decompose the *COBE* Diffuse Infrared Background Experiment (DIRBE) 12 - 240 μm observations of the Galactic plane region ($|b| < 5°$), from which the zodiacal-light and stellar emission have been subtracted, into distinct emission components associated with each gas phase within selected ranges of galactocentric distance. An interstellar dust model was fit to the resulting far-infrared spectra to derive the large-scale properties of the dust associated with each gas phase. The implications of our results for proposed models of interstellar dust and the current picture of the Galactic large-scale infrared emission are discussed.

1. Comparison With Previous Decomposition Methods

Two distinct approaches for decomposing the infrared observations of the Galactic plane region into the respective contributions from the different gas phases of the ISM have previously been presented. Sodroski et al. (1989, 1994) derived longitude profiles of the average line-of-sight emissivities of the dust associated with each of the three most massive (H_2, H I, and H II) gas phases by applying a linear matrix inversion to IRAS 60 and 100 μm observations, and DIRBE 140 and 240 μm observations, within subregions of longitude and latitude. Bloemen et al. (1990) correlated IRAS 60 and 100 μm maps with maps of velocity-integrated ^{12}CO and H I emission corresponding to various galactocentric distance intervals to derive the radial distributions of the dust emissivities within the H_2 and H I gas phases. Giard et al. (1994) applied this

© 1996 American Institute of Physics

inversion technique to IRAS 12 and 25 μm maps. The approach of Bloemen et al. (1990) more accurately accounts for expected radial variations in dust emissivity for the H_2 and H I gas phases than the approach of Sodroski et al. (1989). However, Bloemen et al. (1990) made the assumption, based on *a priori* argument, that dust within extended low-density and compact H II regions contributes negligibly to the Galactic infrared luminosity. The decomposition method presented here is a significant improvement over these earlier approaches in that it accurately acccounts for radial variations in dust emissivity, and also includes the contribution from dust associated with H II regions.

2. Infrared and Radio Maps of the Galactic Plane Region

Maps of the 12, 25, 60, 100, 140, and 240 μm emission from the Galactic plane region ($|b| < 5°$) were obtained from the DIRBE data averaged over 10 months of cryogenic operation. A model for the zodiacal emission (Reach et al. 1995; Franz et al. 1995) was subtracted from each of the maps, and a model for the contribution of point sources (Arendt et al. 1995) was also subtracted from the 12 and 25 μm maps. Using data from Dame et al. (1987), Weaver & Williams (1973, 1974a,b), and Kerr et al. (1986), maps of velocity-integrated ^{12}CO and H I emission corresponding to the four galactocentric distance intervals 1.5 kpc $< D <$ 3 kpc, 3 kpc $< D <$ 6 kpc, 6 kpc $< D <$ 8.5, and $D >$ 8.5 kpc were derived assuming the Galactic rotation curve of Brand & Blitz (1993) and a Sun-to-Galactic Center distance of 8.5 kpc (Kerr & Lynden-Bell 1986). A map of the expected 5 GHz radio-continuum emission from the Galactic plane region was derived using the Haynes et al. (1978) 5 GHz radio-continuum survey for $280° < l < 40°$, $|b| < 1.5°$, and the 19 GHz radio-continuum survey (converted to expected flux at 5 GHz by assuming purely thermal emission) of Boughn et al. (1991) elsewhere. For comparison, all of the above maps were degraded to the DIRBE angular resolution ($\sim 0.7°$) within the region $280° < l < 40°$, $|b| < 1.5°$, and to the angular resolution ($\sim 3°$) of the Boughn et al. (1991) survey elsewhere.

3. Decomposition of the Galactic Infrared Emission

The maps of the Galactic 12 - 240 μm emission were decomposed as follows. We assumed that the emissivities per H atom for the molecular and neutral atomic hydrogen gas are constant within each of the 4 galactocentric distance intervals (i.e., rings) and that the emissivity per H atom for the ionized gas phase is constant within 2 rings, one inside the solar circle, and one outside the solar circle. Maps of the H atom column density of each gas phase within each ring were obtained from the velocity-integrated ^{12}CO and H I maps, and expected 5 GHz radio-continuum map (see Sodroski et al. (1989)). The maps of the H atom column density of the ionized gas were derived by assuming that all of the 5 and 19 GHz radio-continuum emission from the region $270° < l < 90°$, $|b| < 5°$ arises from within the solar circle. For each line of sight, the emission at each infrared wavelength is then given by the emissivity per H atom for each gas phase within each ring times the H atom column density of that gas phase within the ring, summed over all rings and the 3 gas phases. We thus have a linear matrix equation with an H-atom column-density matrix of 1889 columns (i.e, the number of lines of sight) by 10 rows (i.e, the number of unknown emissivities).

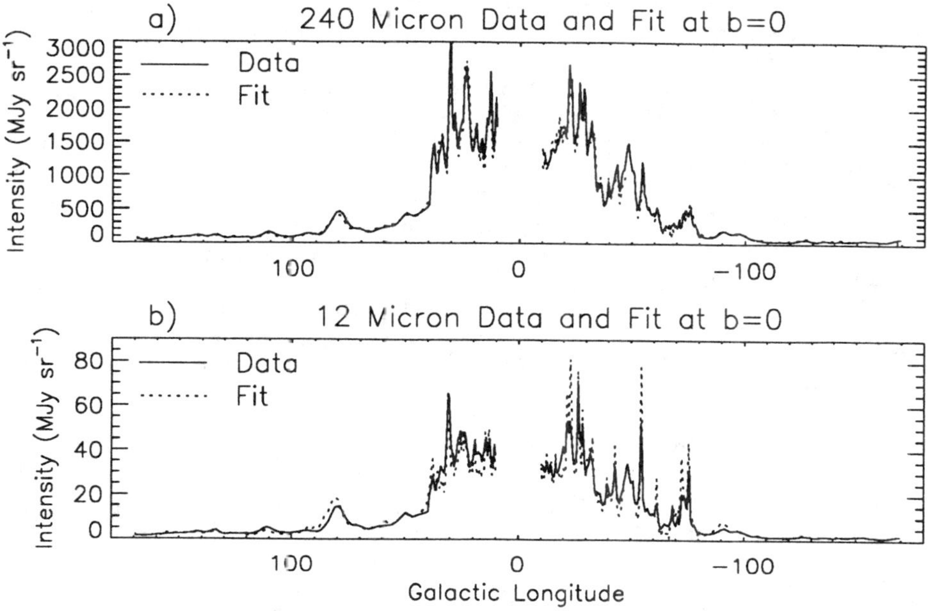

Figure 1 – (a) Longitude profile of the observed 240 μm Galactic plane emission (solid line), and fit to observed profile from decomposition algorithm (dotted line). (b) Same as (a), for 12 μm emission.

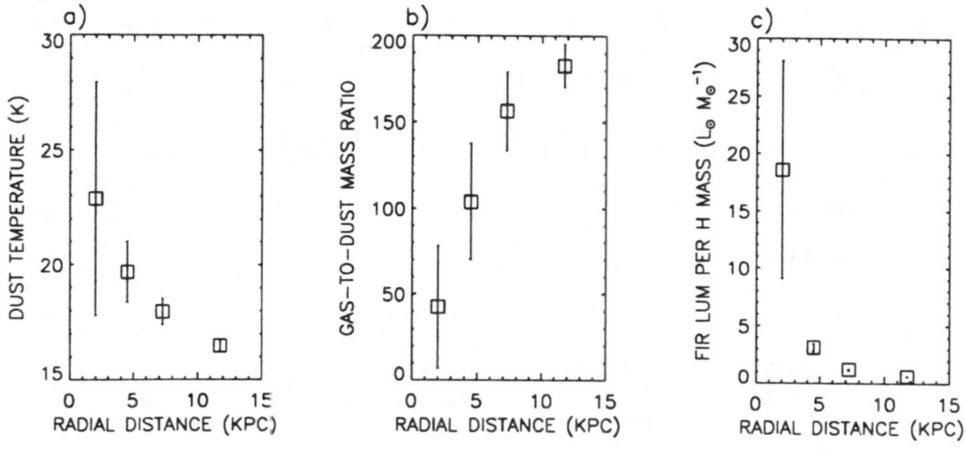

Figure 2 – The radial distributions of the derived physical conditions for the H I gas phase: (a) dust temperature; (b) gas-to-dust mass ratio; (c) far-infrared luminosity per hydrogen mass

At each wavelength between 60 and 240 μm, the linear matrix equation was inverted to obtain the unknown emissivities and the best fit to the observed intensities by computing the generalized inverse of the H-atom column-density matrix (Golub & Reinsch 1970). Longitude profiles of the observed and calculated intensities at $b = 0°$ for the 240 μm band are shown in Figure 1. The physical conditions of the large dust grain component within each gas phase were then determined by fitting a modified blackbody with an adopted emissivity index of n = 2 to the derived 140 and 240 μm emissivities (see, for example, Hauser et al. 1984). The value of X, the ratio of H_2 column density to ^{12}CO intensity, for each of the 4 rings was obtained by requiring that the value of the dust-to-gas mass ratio derived for the H_2 gas phase equals that derived for the H I gas phase. From the values of the dust-to-gas mass ratio for the H I gas phase and the ratio of H_2 column density to ^{12}CO intensity, extinction correction factors at 12 and 25 μm were estimated. The 12 and 25 μm observations were then similarily decomposed (Figure 1).

4. Derived Dust Physical Conditions and Model Parameters

The radial distributions of dust temperature, gas-to-dust mass ratio, and total FIR luminosity per hydrogen mass for the H I gas phase are shown in Figure 2. The distributions of dust temperature and FIR luminosity per hydrogen mass for the H I gas phase exhibit steep systematic increases with decreasing galactocentric distance, consistent with the radial gradients expected for dust heated by the Galaxy's ambient interstellar radiation field (ISRF; Cox et al. 1986). Based on these distributions we estimate that the energy density of the ISRF increases by a factor of $\sim$ 6 between the solar circle and a galactocentric distance of 2 kpc. The radial distributions of dust temperature and FIR luminosity per hydrogen mass for the H_2 gas phase exhibit similar gradients as the corresponding quantities for the H I gas phase. This result is consistent with the notion that dust associated with the H_2 gas is heated in part by the ISRF and in part by nearby and embedded OB stars (Leisawitz & Hauser 1988; Sodroski et al. 1989), since the star formation rate per unit mass of gas is also believed to increase sharply with decreasing galactocentric distance (see, for example, Sodroski 1991).

The gas-to-dust mass ratio for the H I gas shows a steep systematic decline toward the Galactic center. The radial gradient in the dust-to-gas mass ratio is equivalent, within the uncertainties, to the metallicity gradient in the Galactic disk (Panagia & Tosi 1981; Shaver et al. 1983; Wilson & Rood 1994). Our estimates for X within each ring are consistent with the finding of Sodroski et al. (1995) that X increases by more than an order of magnitude from the Galactic center to a galactocentric distance of 13 kpc.

The 12 - 240 μm spectra of the dust associated with the H I gas are consistent with the expected spectra based on the model of Désert et al. (1990). This suggests that very small transiently-heated dust grains are the dominant contributor to the mid-infrared ($\lambda < 40$ μm) luminosity from the molecular ring (D $\sim$ 5 kpc) to a galactocentric distance of at least 12 kpc.

Acknowledgements

COBE is supported by NASA's Astrophysics Division. Goddard Space Flight Center (GSFC), under the scientific guidance of the *COBE* Science Working Group, is responsible for the development and operation of *COBE*.

References

Arendt, R. G., Freudenreich, H. T., & Dwek, E. 1995, this workshop
Bloemen, J. G. B. M., Deul, E. R., & Thaddeus, P. 1990, A&A, 233, 437
Boughn, S. P., Cheng, E. S., Cottingham, D. A., & Fixsen, D. J. 1991, in After the First Three Minutes (AIP Conf. Proc. 222), ed. S. S. Holt, C. L. Bennett, & V. Trimble (New York: AIP), 107
Brand, J., & Blitz, L. 1993, A&A, 275, 67
Cox, P., Krügel, E., & Mezger, P. G. 1986, A&A, 155, 380
Dame, T. M., et al. 1987, ApJ, 322, 706
Désert, F. X., Boulanger, F., & Puget, J. L. 1990, A&A, 237, 215
Franz, B. A., et al. 1995, this workshop
Giard, M., Lamarre, J. M., Pajot, F., & Serra, G. 1994, A&A, 286, 203
Golub, G. H., & Reinsch, C. 1970, Numer. Math., 14, 403
Hauser, M. G., et al. 1984, ApJ, 285, 74
Haynes, R. F., Caswell, J. L., & Simons, L. V. S. 1978, Australian J. Phys., Astrophys. Suppl., 45, 1
Kerr, F. J., Bowers, P. F., Jackson, P. D., & Kerr, M. 1986, A&AS, 66, 373
Kerr, F. J., & Lynden-Bell, D. 1986, MNRAS, 221, 1023
Leisawitz, D., & Hauser, M. G. 1988, ApJ, 332, 954
Panagia, N., & Tosi, M. 1981, A&A, 96, 306
Reach, W. T., et al. 1995, this workshop
Shaver, P. A., McGee, R. X., Newton, L. M., Danks, A. C., & Pottash, S. R. 1983, MNRAS, 204, 53
Sodroski, T. J. 1991, ApJ, 366, 95
Sodroski, T. J., Dwek, E., Hauser, M. G., & Kerr, F. J. 1989, ApJ, 336, 762
Sodroski, T. J., et al. 1994, ApJ, 428, 638
Sodroski, T. J., et al. 1995, in press
Weaver, H., & Williams, D. R. W. 1973, A&AS, 8, 1
———. 1974a, A&AS, 17, 1
———. 1974b, A&AS, 17, 251
Wilson, T. L., & Rood, R. T. 1994, ARA&A, 32, 191

100 μm FLUCTUATIONS OF A HIGH GALACTIC LATITUDE COMPLEX

Alain Abergel and Francois Boulanger
Institut d'Astrophysique Spatiale
CNRS - Univ. Paris XI,Bât. 121, 91405 Orsay Cedex, France

Email ID
abergel@iaslab.ias.fr, boulanger@iaslab.ias.fr

ABSTRACT

The local correlations of the fluctuations at different angular scales of the 100 μm IRAS emissions of a large high galactic interstellar complex are analysed. First a wavelet analysis is applied to locally quantify the brightness fluctuations at different scales, then a map of the spectral indexes is built and compared with the emission one.

1. INTRODUCTION

Observations of the interstellar medium (ISM) show intricated features at all scales. Statistical properties of the ISM structure have been analysed by several groups using connected objects (Bazell & Désert 1988; Dickman et al. 1990; Vogelaar and Wakker 1994; Falgarone et al. 1991) or "clumps" extracted from a wavelet analysis (Langer et al. 1993). However, all results based on the analysis of extracted objects are questionable since, whatever the method to identify these objects is, there is no proof that these objects correspond to physical entity in the clouds.

Gautier et al. (1992) have computed the spectral index of the power law used to fit the two-dimensional power spectra for the 100 μm emission of five different complexes. Their approach is independent of any doubtful assumption, but the spatial information is lost.

In this paper we analyse the spatial variations of the spectral index of the 100 μm emissions of an interstellar complex, by computing for each pixel and different scales the amplitude of the brightness fluctuations using the wavelet transform, and then for each pixel the local spectral index. We present results obtained with the field number 8 of the first release of the IRAS Sky Survey Atlas (ISSA) at 100 μm (Fig. 1). The method and results obtained at 60 and 100 μm are detailed in Abergel et al. (1995).

2. DATA PROCESSING

2.1 Wavelet analysis

The wavelet transform has been introduced by Morlet (1983) to analyse seismic data. We have used isotropic analysing wavelets, so that the frequency domain is scanned varying only one scale factor a. The wavelet chosen for our study is the so-called "Mexican hat". The initial image is convolved with the analysing wavelet progressively dilated. Following Slezak et al. (1990), the whole frequency domain is scanned by mean of a dyadic analysis, where each

© 1996 American Institute of Physics

scale factor a is related to the following one by a factor of 2: a= 2^n, with n in the range 0-5.

2.2 Spectral indexes map

A fluctuation of the input image at the given scale 2^n introduces in the wavelet transform a fluctuation around zero at the same location and at the corresponding scale 2^n. Therefore we have computed a map of rms fluctuation inside circular window of radius 2^n. Then, for each pixel, the variations of the amplitude of the fluctuations with the spatial frequency is fitted with the power law: $F(\nu) = F_0\,(\nu/\nu_0)^{-\beta}$.

The computed values of β are significant if the log(F) - log(ν) relation does not strongly deviate from a linear law. Therefore we have eliminated pixels with a standard deviation of the log(F) - log(ν) relation from a linear law greater than 0.4.

Fig. 2 shows the map of the computed values of the spectral index β from the 100 μm map of Fig. 1. Dark pixels have a standard deviation of the log(F) - log(ν) relation from a linear law above 0.4 (17% of the pixels). A least square fitted gaussian curve (dotted line, $\beta_{cen} = 0.63$ at 100 μm , σ=0.18) is superimposed to the histogram (continuous line). Both the sky brightness fluctuations and the noise fluctuations contribute on computed fulctuations maps. The noise contribution is quantified by Abergel et al. (1995). They have shown that the histogram of β is shifted by around -0.05 because of the noise contribution.

3. ANALYSIS OF THE RESULTS

3.1 Scale variations of the amplitudes of fluctuations

The mean statistical fluctuations at 100 μm for a given angular scale θ in units of arcmin is fitted with the formula: $F(\theta) = 0.12\,(\theta/1.5)^{0.67}$ MJy sr^{-1}. This formula has been established from fluctuations computed at angular scales within the range 1.5-48 arcmin, and we suggest to use it to estimate very roughly the value of the mean statistical fluctuations inside our field at 100 μm for smaller or larger scales. For θ= 1", we obtain a mean statistical fluctuation of 6 10^{-3} MJy sr^{-1}.

3.2 Spatial variations of the correlation of the amplitude of fluctuations

For most of the analysed map, the relation log(F)-log(ν) can be considered as linear. This result shows that within most of the interstellar matter, the amplitudes of fluctuations at different angular scales in the range 1.5-48 arcmin are correlated. The map of the spectral indexes β shows spatial variations not correlated with the 100 μm brightness. It does not contain any coherent or organised feature, and looks significantly more homogeneous that the brightness image.

For dark pixels of Fig. 2, there is no visible correlation between the amplitude of fluctuations at different scales, which should indicate a change in the physics of the cloud hierarchy from the largest to the smallest observed structures. Small coherent structures corresponding to persistent zones of high vorticity are observed in turbulent flows (e.g. Douady et al. 1991) and correlate with zones where the dissipation energy is concentrated (Falgarone & Puget 1995). Can we identify regions with no coupling between fluctuations at different scales to coherent high vorticity structures observed in turbulent flows? The absence of velocity information prevents in this study the testing of this idea.

3.3 Interpretation of β values

The distribution of β has a gaussian shape with σ=0.18 and centered to β_{cen} = 0.63, value which shifts to 0.68 to take into account the noise contribution. Moreover, as the 100 μm emission is known to trace all the matter illuminated by the radiation field of the galaxy (out of very dense regions), the relation $F(\nu) \; \alpha \; (\nu)^{-\beta}$ can also be written: $\Delta N_{col}(l) \; \alpha \; l^{\beta}$ where $\Delta N_{col}(l)$ is the fluctuation of the total column density of the line of sight for a scale l.

Since Larson (1981), several authors have shown from different molecular observations that the internal density n of molecular clouds or condensations varies with size according to: $n(l) \; \alpha \; l^{p}$. This relation can also be written $N_{col}(l) \; \alpha \; l^{p+1}$. As reviewed by Scalo (1987), most of these studies indicate p= -1.4 to -1.1, and therefore that N_{col} slowly decreases with l.

The value of p+1 is obviously outside the distribution of the spectral index β which is centered on 0.68 (with the noise correction). This result is not necessarily surprising because the molecular and 100 μm emissions do not trace the same component of matter. IRAS images at 100 μm trace the interstellar medium independently of chemical composition and density (except for very high densities where the dust is not sufficiently heated to emit at 100 μm), while the molecular emission regions which are dense enough for the molecules to exist and to be excited by collisions. Molecular observations are strongly influenced by the small scale structure of the observed clouds. Moreover the small fractions of the sky which are covered are frequently biased toward high extinction regions.

REFERENCES

Abergel A., Boulanger F., Delouis J.M. , Dudziak G. , Steindling S 1995, Submitted to A&A

Bazell D., Désert FX 1988, ApJ 333, 353

Désert F. X., Boulanger F., Puget J. L. 1990, A&A 237, 215

Dickman R.L., Horvath M.A., Margulus M. 1990, ApJ 365, 58

Douady S;, Couder Y., Brachet M.E. 1991, PRL 67, 983

Falgarone E., Phillips T. G.,J1990, ApJ 359, 344

Falgarone E., Phillips T. G., Walker C. K. 1991, ApJ 378, 186

Falgarone E., Puget J.L. 1995, A&A 293, 840

Gautier III T.N., Boulanger F., Prault M., Puget J.L. 1992, AJ 103 (4), 1313

Larson R.B., 1981, MNRAS 194, 809

Langer W.D., Wilson R.W., Anderson C.H. 1993, ApJ 408,L5

Morlet J. 1983. In: Chen C.H. (ed.) Issues on Acoustic Signal/Image Processing and Recognition. Springer, Berlin.

Scalo J.M. 1987. In: Hollenbach D.J., Thronson H.A. Jr.(eds.) Proc. Instellar processes Symp., Instellar processes. Reidel, Dordrecht, p.J349

Slezak E., Bijaoui A., Mars G. 1990, A&A 227, 301Green, W. 1993, ApJ, 666, 666

Vogelaar M.G.R., Wakker B.P., A&A 291, 557

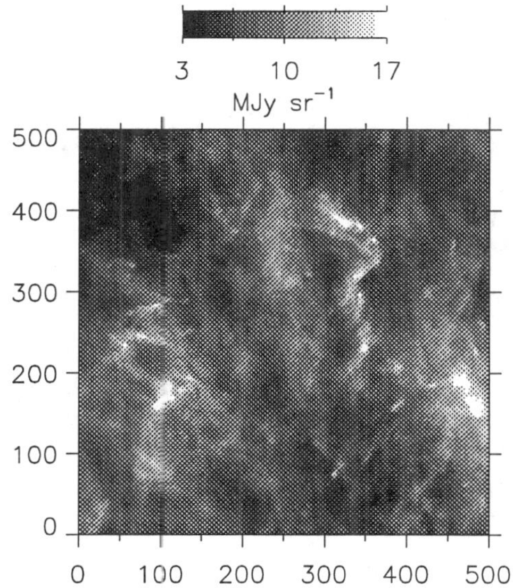

Figure 1: Field number 8 of the IRAS Sky Survey Atlas (ISSA) at 100 μm lying to the West of the Chameleon complex (α = 18h 00m, δ = -80°). Total dimensions: 500 x 500 pixels, i.e. 12.5° x 12.5°.

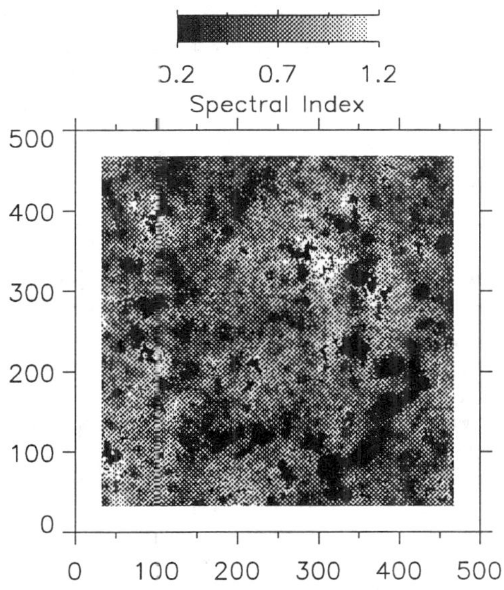

Figure 2: Map of the spectral index β computed from the map of Fig. 1.

A DIRBE 12-240 MICRON SURVEY FOR EXTERNAL GALAXIES

Sten Odenwald[1,2], Jeffrey Newmark[1,2] and George Smoot[3]
[1]Code 685, NASA Goddard Space Flight Center, Greenbelt, MD 20771
[2] Applied Research Corporation, Landover, MD 20785
[3] University of California, Berkeley CA
odenwald@tahiti.gsfc.nasa.gov

ABSTRACT

We describe the first results from a search for external galaxies in the COBE Diffuse Infrared Background Experiment (DIRBE) all-sky survey. Excluding the Large and Small Magellanic Clouds, as many as 56 galaxies were detected, with 7 detected at $\lambda > 100$ μm.

1. Introduction

Many galaxies are strong sources of far-IR radiation (Soifer, Houck and Neugebauer, 1987), particularly at 60 and 100 μm. The nature of galaxian far-IR emission among late-type spirals is widely acknowledged to be radiation reprocessed by dust grains present in a variety of sizes and temperatures in the interstellar medium (Cox and Mezger, 1989). Detection of sub-millimeter dust emission at $\lambda > 100$ μm is a particularly important measure of the quantity of cold ($T_d < 25$ K) dust in a galaxy. In the Milky Way, this dust is known to be pervasive, and probably the dominant source of far-IR luminosity (Sodroski et al., 1994). Although not specifically designed for galaxy surveys, the large-beam and photometric bands at 140 and 240 μm make the DIRBE instrument capable of providing important information on the integrated quantity of cold dust in nearby galaxies.

2. Observations

The DIRBE instrument on board NASA's COBE satellite has surveyed the entire sky in 10 photometric bands from 1.25 to 240 μm. A detailed description of the DIRBE instrument and its calibration is given in the DIRBE Explanatory Supplement (Hauser et al., 1995). The instantaneous DIRBE beam profile in each band is constrained by an internal field stop to 0.7° x 0.7°, however scan motion broadens the beam profile so that the beam size can be somewhat larger than this especially in the in-scan direction. For the purpose of galaxy detection, this large beam size implies that all but the nearest galaxies such as M-31 and the Magellanic Clouds will remain unresolved. Since the DIRBE all-sky survey is pixelized at a scale of 20′, all of the detectable flux from external galaxies will fall within individual 3x3 pixel patches centered on the nominal galaxy position. The positions to be searched were obtained from the IRAS Catalog of Extragalactic Objects (Fullmer and Lonsdale, 1989) which lists 11,444 galaxies detected at 100 μm. We also used the CfA Galaxy Survey (Huchra et al., 1992) to search for new IR galaxies in the 4% of the sky not surveyed by IRAS.

Sky backgrounds at the angular scales of the DIRBE beam are generally complex due to the presence of interstellar 'cirrus' emission (Low et al., 1989),

© 1996 American Institute of Physics

and to zodiacal emission. The background emission near each galaxy was determined by fitting a plane to a succession of nested annuli with radii of 1.0, 1.5 and 2.0°. The fits do not include emission from the pixels within 0.5°of the galaxy. The background-subtracted peak flux of the galaxy was then compared with the residual background noise after plane fitting to determine the S/N of the galaxy emission in each of the three annuli.

To be classified as a detection, the far-IR emission must occur as a correlated deviation in at least two bands with S/N $>$ 4, or S/N $>$ 5 in a single band. A study of the statistical properties of this detection algorithm was performed by analyzing a 256x256 pixel sky patch in each band. Although the initial histogram of pixel intensities was non-Gaussian, after local background subtraction, the resulting pixel intensity distribution was essentially Gaussian. This implies that the false positive frequency may be calculated from Gaussian statistics, and that a threshold based on a S/N criterion provides a reasonable discriminant between real and spurious detections. In addition to meeting this S/N requirement, the candidate must also be readily distinguishable from obvious background clutter. We have reduced this source of false positives by visually inspecting each candidate field in the DIRBE sky data, and eliminating candidates with obvious cirrus and stellar contamination; the latter primarily due to planetary nebulae and star-forming regions.

3. Results

We were able to identify 56 galaxies that appear to be sufficiently bright to be seen by DIRBE. This list represents an upper limit to the number of DIRBE-detected bright galaxies since some of the unresolved detections may be cirrus-related. A more rigorous identification procedure involving IRAS and DIRBE-derived cirrus flags will be discussed in a forthcoming paper.

Several well-known far-IR galaxies were detected including M-82, NGC 253, Arp 220, and IC 342. The Andromeda galaxy, M-31, was not only detected by DIRBE, but also resolved. DIRBE also appears to have detected 4 new far-IR emitting galaxies, M-106, NGC 92, IC 751 and CfA0004-4144 not previously surveyed by IRAS. These were found at a total S/N between 6-10. The first three are classified in the CfA Catalog as Sb, Sa and Sc/Irr respectively. No classification is available for CfA0004-4144. The distribution of galaxy candidates with galaxy type is shown in Table 1. Boldfaced entries indicate galaxies for which emission at 140 or 240 μm has been detected. These galaxies are: M31, M33, M82, M83, NGC 253 and NGC 4945.

4. Summary

We have identified as many as 56 galaxies in the DIRBE extragalactic survey. In a forthcoming paper we will refine our detection list, and use the DIRBE photometry at 12 - 240 μm to constrain the properties of very cold dust emission from the detected galaxies.

Acknowledgements

COBE is supported by NASA's Astrophysics Division. Goddard Space Flight Center, under the scientific guidance of the COBE Science Working Group, is responsible for the development and operation of COBE.

References

Soifer, B.T., Houck, J.R. & Neugebauer, G. 1987, ARAA 25, 187.
Cox, P. & Mezger, P.G. 1989, Astro. & Astrop. Rev. 1, 49.
Sodroski, T. et al. 1994, ApJ 428, 638.
Hauser, M. et al. 1995, "DIRBE Explanatory Supplement", COBE Ref. Pub. No. 95-A (Greenbelt, MD: NASA/GSFC)
Fullmer, L. & Londsdale, C. 1989, NSSDC Archive Number 7113.
Huchra, J.P. et al. 1992, NSSDC Archive Number 7144.
Low, F. et al, 1989, ApJ (Letter) 278, L19.

Table 1: DIRBE Galaxy Detections and Types

S0,a	Sb	Sbc	Sc,d	SA,B	Seyfert	Liner	Misc.
			M33				
			M61				
			M101				
			IC694				
			IC342				
			IC751				
			IC5179	**M83**			
			N7793	**N253**			
			N**4945**	N613			
			N4559	N986			
			N4038	N1313			
			N3621	N2403	M51		**M82**
M94		**M63**	N3556	N4030	N1097		N838
N92	**M31**	M106	N1448	N4631	N1365		N1072
N2798	M66	N134	N1385	N7552	N7582		N3256
N4490	N891	N1808	N300	N7625	Arp220	N3079	Mk273
N5128	N4102	N3310	N55	Mk201	Mk231	N5005	C0004

THE LONG WAVELENGTH ITERATION SCHEME

G. L. Comer
Department of Science and Mathematics
Parks College of Saint Louis University, Cahokia, IL 62206

Abstract. The long wavelength iteration scheme is a technique that can be used to obtain fairly generic solutions of the Einstein equations for cosmologies which are inhomogeneous and anisotropic. At the heart of the scheme is the assumption that spatial gradients of the gravitational and matter fields are small. For instance, the gravitational field solution is obtained as an expansion in terms of a given "seed" metric, where the first-order term in the expansion is the "seed," the next order contains two derivatives of the "seed," and so on. When the scale of spatial variation is much bigger than the Hubble radius, then only the first-order term is needed to describe scales smaller than the Hubble radius. The higher order terms are required once the scale of spatial variation is on the order of, or smaller than, the Hubble radius.

1. Introduction

New windows, which ultimately look out on the distribution of matter in the universe, are being opened using the data obtained by the COBE satellite. Peering through one window, we see anisotropies in the cosmic microwave background radiation; through another (opened by the DIRBE experiment), we expect new insights on the cosmic infrared background radiation. For the gravitation theorist pursuing applications in cosmology, the main goal is to look through windows like these, take note of the landscape, and then determine how the lay of the land—with its valleys and hills—affects, and is affected by, the gravitational field. Unfortunately, this goal of describing the "hills" and "valleys" (anisotropies and inhomgeneities) in a generic way is hard to realize because the Einstein equations being a set of ten coupled nonlinear second order partial differential equations, are exactly solvable only when most of the degrees of freedom of the gravitational field are "frozen" out, that is, when a large degree of symmetry for the underlying spacetime is assumed to exist. For instance, the Friedmann-Robertson-Walker (FRW) metric is an exact solution of the Einstein equations, having a small number of unspecified parameters, which models spacetimes that are homogeneous and isotropic. Despite the complications, progress on a general analysis has been made, but only for those cases where the linearized theory of perturbations of a FRW background can be applied. The purpose here is to briefly outline the long wavelength iteration scheme, which is a way to analyze solutions to the Einstein equations that are anisotropic and inhomogeneous and go beyond the regime covered by linear perturbation theory.

This approach was first introduced by Lifschitz and Khalatnikov (see the discussion in (1), and the references contained therein). In more recent times, two different groups have employed the same expansion: Salopek and collaborators (see (2), and references therein), who use it in their analysis of the Hamilton-Jacobi equation for General Relativity; and Deruelle and collaborators (see (3), and references therein, (4), and (5)), who, among other things, discuss how the form of the equation of state for the over-all matter content of the universe affects the growth or decay of cosmological inhomogeneities.

© 1996 American Institute of Physics

2. The Iteration Scheme

We start by constructing the characteristic length and time for which the metric varies, and thereby determine the conditions under which the long wavelength approximation, where *all* spatial gradients are neglected, is particularly accurate. This also simultaneously determines the starting point where the iteration scheme must be implemented. For a thorough discussion of the details, see Comer *et al.* (3).

The metric is written in the synchronous coordinate system, where time is the proper length along the geodesics perpendicular to an initial spacelike surface, *i.e.*,

$$ds^2 = -dt^2 + \gamma_{ij}(t, x^k) dx^i dx^j \quad , \quad (i,j = 1,2,3) \; . \tag{1}$$

Assuming a perfect fluid stress-energy and using the metric written above results in the following for the Einstein equations:

$${}^{(3)}R^j_i + \frac{1}{2\sqrt{\gamma}} \frac{\partial}{\partial t} \left(\sqrt{\gamma} K^j_i \right) = \frac{2\dot{K} + K^l_k K^k_l}{4(2 - 3\Gamma - 2\Gamma u^m u_m)} \left(2\Gamma u^j u_i + \delta^j_i (2 - \Gamma) \right) \; , \tag{2}$$

$$8\pi G \rho = \frac{2\dot{K} + K^l_k K^k_l}{2(2 - 3\Gamma - 2\Gamma u^m u_m)} \; , \tag{3}$$

and

$$8\pi G \rho u_i = -\frac{1}{2\sqrt{1 - u^k u_k}} D_j \left(K^j_i - K \delta^j_i \right) \; , \tag{4}$$

where an overdot signifies $\partial/\partial t$. In these equations we have the following terms: ${}^{(3)}R_{ij}$ is the Ricci tensor formed out of γ_{ij}; $K_{ij} = \dot{\gamma}_{ij}$ and $K = K^i_i$; ρ is the density; the equation of state defining the pressure is $p = (\Gamma - 1)\rho$, $0 \leq \Gamma \leq 2$; u^j is the fluid three-velocity; $\gamma \equiv \det(\gamma_{ij})$; and D_j in Eq. 4 denotes a covariant derivative with respect to γ_{ij}. All spatial indices are lowered and raised with γ_{ij}.

Let L be the characteristic length on which the metric varies spatially, so that

$$\partial_i \gamma_{jk} \sim L^{-1} \gamma_{jk} \; . \tag{5}$$

The Hubble time H^{-1} (where $H \equiv \dot{a}/a$ and $a^2 \equiv \gamma^{1/3}$) is the characteristic time on which the metric varies temporally, so that

$$\dot{\gamma}_{jk} \sim H \gamma_{jk} \; . \tag{6}$$

The long wavelength approximation, where ${}^{(3)}R^j_i$, $D_j K^j_i$, and $D_i K$ are neglected in the Einstein equations, becomes accurate when the characteristic scale L of spatial variation is much bigger than the Hubble radius $R_H \equiv (aH)^{-1}$, *i.e.*, $L \gg R_H$.

So long as $L \gg R_H$, it thus suffices to neglect *all* spatial gradients in the Einstein equations. However, the Hubble radius, in general, depends on time and eventually the condition $L \gg R_H$ can be violated. Thus, neglecting all spatial gradients is only a starting point. To get more accurate (and "longer-lived") results, we expand γ_{ij} in terms of ever increasing spatial gradients of a "seed" metric h_{ij} which does not depend on the time parameter t. Since each term of the expansion must be a tensor, the gradients of the "seed" enter via all possible contractions and derivatives (leaving two free indices) of the Riemann tensor formed from the "seed." For instance, one of the special cases considered by Comer *et al.* (3) (which turns out to be an attractor for a more general class of solutions) uses

$$\gamma_{ij} = A_o(t)h_{ij} + A_2(t)Rh_{ij} + B_2(t)R_{ij} + ... \tag{7}$$

(where R_{ij} is the Ricci tensor and R the Ricci scalar) for the metric and

$$\rho = \rho_o(t) + \rho_2(t)R + ... \tag{8}$$

and

$$u_i = u_3(t)R_{,i} + ... \tag{9}$$

for the matter variables. These are put into the Einstein equations, which are then rearranged until all "like" (*e.g.*, in Eq. 2, terms proportional to either h_{ij}, Rh_{ij}, or R_{ij}, *etc.*) terms are gathered. Note that the correcting terms in Eqs. (7), (8), and (9) are small in the sense that the Riemann tensor formed from the "seed" contains two spatial gradients, for instance, we have for the Ricci tensor

$$R_{ij} \sim L^{-2}h_{ij} \ . \tag{10}$$

The iteration part (of 'the long wavelength *iteration* scheme') enters when determining the time dependent coefficients (*i.e.*, $A_o(t)$, $A_2(t)$, $B_2(t)$, $\rho_o(t)$, $\rho_2(t)$, $u_3(t)$, *etc.*): First, one solves the Einstein equations by neglecting all spatial gradients; this is equivalent to the long wavelength approximation and determines $A_o(t)$ and $\rho_o(t)$, the three-velocity u^j having no first-order contribution. Next, one considers contributions due to the terms containing two spatial gradients, and thereby determines the third-order coefficients $A_2(t)$, $B_2(t)$, $\rho_2(t)$, and $u_3(t)$. To get enough equations for these coefficients, one takes into account the independence of the functions R_{ij} and Rh_{ij}. This yields a set of second-order coupled ordinary differential equations for the time-dependent coefficients alone. The time parameter t enters these equations explicitly via the coefficients $A_o(t)$ and $\rho_o(t)$, which exist as both source terms and as coefficients multiplying the various time derivatives of $A_2(t)$, $B_2(t)$, $\rho_2(t)$, and $u_3(t)$. Using the results for the first- and third-order coefficients, one then continues in the same way to the next order, *i.e.*, terms with four spatial gradients, and so on until all orders have been determined. At each iteration, the time-dependent coefficients will depend on the coefficients determined from previous iterations.

Note that in the iteration process the "seed" metric h_{ij} does not get determined; it is thus freely specified in the problem. At face value, the "seed" represents a set of six functions of space. However, by a suitable change of coordinates, this number can always be reduced to three. It is through the specification of these functions that inhomogeneity is introduced. (Anisotropy can enter at the first-order, through an integration constant that results when determining $A_o(t)$. See Comer *et al.* (3).) As a check, one can even specify the "seed" to be the maximally symmetric three-metric of the FRW solution, modulo the expansion factor. The long wavelength scheme then reproduces, order-by-order, the FRW solution.

3. Future Work

We have discussed above that the purpose of the long wavelength iteration scheme is to provide a general framework for analyzing inhomogeneous and anisotropic cosmological solutions to the Einstein equations that go beyond the linear regime. That being done, we would now like to know what restrictions must be imposed so that these solutions are not in conflict with the data obtained, for instance, by COBE. As a first step, Comer *et al.* (6) have determined, to the third-order, light-ray paths by solving the null geodesic equations. These have been used to obtain the long wavelength version of the Sachs-Wolfe effect.

ACKNOWLEDGMENTS

I wish to thank Nathalie Deruelle and David Langlois for many fruitful discussions on the long wavelength iteration scheme, C. A. Fuchs, J. Potvin, and W. Thacker for their careful reading of this manuscript, and Parks College of Saint Louis University for partial support.

REFERENCES

1. Landau, L. D. and Lifschitz, E. M., *The Classical Theory of Fields*, 4th ed., New York: Pergamon Press, 1975, ch. 14, pp. 358-397.
2. Croudace, K. M., Parry, J., Salopek, D. S., and Stewart, J. M., *Ap. J.*, **423**, pp. 22-32 (1994).
3. Comer, G. L., Deruelle, N., Langlois, D., and Parry, J., *Phys. Rev. D*, **49**, pp. 2759-2768 (1994).
4. Deruelle, N. and Goldwirth, D., *Phys. Rev. D*, **51**, pp. 1563-1568 (1995).
5. Deruelle, N. and Langlois, D., *Phys. Rev. D*, in press.
6. Comer, G. L., Deruelle, N., and Langlois, D., in preparation.

A STRATOSPHERIC BALLOON OBSERVING FACILITY: MILO BASE

M.C. Falvella
Agenzia Spaziale Italiana
Base Lancio Palloni Stratosferici, S.S.113 C.DA MILO I-91100 TRAPANI

ABSTRACT

In a favourable geographic position (38° 01′N, 12° 35′E) Milo Base guarantees long duration flights on Mediterrean Sea (at least 20 hours) at atmosphera limit (40 km of altitude) from Sicily to Spain. With a low X ray background it represents a special point for astrophysical observation. Programs for the next years, collaboration feasibility, work facilities and obtained results will be presented and discussed.

1. Introduction

Every year having experimental comparison with astrophysical theory becomes more important and data have to be continuosly up-dated by vangard technologies giving better reliability. Not always satellite data, also if very interesting, can be satisfactory because of the technical time scale between the experiment proposal and the data availability (about 15 years). Moreover planning a satellite missions involves costs not always compatible with scientific budgets. On the other side demand for experiments at atmosphere limit increases every year. Infact stratospheric balloon experiments in a time scale of a pair of years let today to up-date and improve data. Italian Space Agency (ASI) offers a modern facility for stratospheric ballooning by Milo Base (Sicily). Milo Base is in an extremely favorable geographic position (38°N, 12°E) which not only guarantees long duration (at least 20 hours of plafond) flights on the Mediterranean Sea from Sicily to Spain during westerly stratospheric flow but also provides low X ray background and calm of wind in anticlyclonic condition during day to night and night to day breezes variation. Besides Milo other two telemetry and acquisition stations, located respectively in Palma de Mallorca and Arenosillo (Spain), are involved in balloon flight managing to guarantee a complete and continuos link.

2. MILO BASE

Milo base is a last world war cast-off airport turned into a stratospheric balloon launching base. With a 500 mt diameter launching area, entirely enlightened for nocturnal flights, and a launching vehicle up to 2500 Kg gondola weight, the Base can hosts in its hangars for integration up to 6 scientific payloads at the same time with the possibility to use two workshop and electronic laboratories, a gondola suspending amagnetic system and several good precision pivots (20′) for gondola pointing system. The flight success is assured by telemetry and telecommand ground station and on board equipment,flight control and balloon tracking system,real time digital and analog data acquisition and recording, data elaboration center network connected, meteorological sounding station and telex, telefax and radio facilities. Data transmission during balloon flight is made by a telemetry system. Actually it is an IRIG FDM (Frequency Division Multiplex) system working in UHF band (400 and 444 MHz). The system

© 1996 American Institute of Physics

is based on redundancy devices whether on acquisition and transmission or on telecommand decoding chain. Ground station configuration allows reception and decoding of all the signals encoded on board. Many devices are projected to allow an easy managing of the H/K parameters and a decoding of scientific data. The ground station is also provided with a GPS receiver for U.T. synchronization better than 1 microsec; the NASA 36 BIT code or IRIG B code is generated for real time recording. The Base provides a receiving chain for data acquisition, processing and recording in real time. Two analog recorders (RACAL Storehorse) of the standard IRIG permit a recording of video mixer FDM, the universal time (U.T.) and other service information on 1 inch magnetic tape with 14 channels covering 2 MHz of bandwidth. In the standard configuration acquisition, recording and quick-look are performed by three different systems: LORAL 500 for data acquisition and pre-treatment, MicroVAX 3800 for data recording and VAXstation 3100 for data handling and quick-look. The workstation 3100 is connected in LAVC to MicroVAX 3800 to provide data visualization and analysis on scientific teams request. During flight it is possible displaying and printing on-line data. Software is prepared before basing on specific request (it is possible to print, plot and display data). Analog and digital data will be stored by Base up to 5 years since flight. For flights with more than one experiment on board we can guarantee data property with special procedures of data package. Data will be available to their owners on different tapes or cartridges as soon as possible. There is a dedicated 9600 b/s link between Palma de Mallorca and Milo Base to quick-look data collected by Palma station. MicroVAX 3800 is connected by Decnet (ASITP 38.271) and Internet (193.204.124.1) protocols and a modem connected to PC is also available (39-923-553432).

3. PAST ACTIVITY

In 1975 a collaboration memorandum between ASI and NASA gave the green light to project of a stratospheric balloon launch facilities in Sicily with the first transoceanic flight and autumn turn rounds. Besides a cooperation between ASI, CNES (FR) and INTA (E), named ODISSEA, gave since 1977 the opportunity to perform transmediterrean flights during summer months. More than 40 experiments have been successfully launched from Milo inside Odissea collaboration.

ASTROPHYSICAL EXPERIMENTS LAUNCHED FROM MILO

- MILO1 '75, Cosmic particles, Bristol University
- MILO2 '75, X-Sources and C.B.R.,CNR and Washington Univ.
- MILO3 '75, GAMMA-Astronomy, FCTR, CNR and MAX Plank Inst.
- ODISSEA1 '77, Cosmic particles, Conie Spain
- ODISSEA2 '77, GAMMA-Atronomy, FCTR, GIFCO and TESRE CNR
- PAF '78, Radio Astronomy, IROE/GIFCO CNR
- CAESAR/CLEOPATRA '78, GAMMA-Astronomy, CERS/CEN CNES
- CELIMENE '78, X-Astronomy, IAS CNR
- AGLE '78, IR-Astronomy,CERS/CEN CNES
- CELIMENE '79, X-Astronomy, IAS CNR
- CIRCE '80, X-Astronomy, IAS CNR
- ULISSE '80, IR-Astronomy, IROE CNR
- ENEA '81, X-Astronomy, IAS CNR
- ULISSE '81, IR-Astronomy, IROE CNR

- POKER '81, X-Astronomy, IAS CNR
- MINIZEBRA '81, X-Astronomy, CNR and Southampton Univ.
- PHOSWICH '82, X-Astronomy, TESRE Univ.
- ARGO '82, IR-Astronomy, Rome Univ.
- ELENA '83, GAMMA-Astronomy, TESRE CNR
- TELEMACO '83, IR-Astronomy, IROE CNR
- POKER '84, X-Astronomy, IAS CNR and Tubingen Univ.
- TELEMACO '84, IR-Astronomy, IROE CNR
- POKER '85, X-Astronomy, IAS CNR and Tubingen Univ.
- FIGARO '86, GAMMA-Astronomy, CERS/CEA CNES and IFCAI/IAS CNR
- PALLAS '86, X-Astronomy,IAF/IAS CNR and Southampton Univ.
- PALLAS '87,X-Astronomy,IAF/IAS CNR and Southampton Univ.
- ARGO '87, IR- Astronomy, IROE/IFA CNR
- AROME '87,IR-Astronomy, CNES
- MINITIR '89,IR-Astronomy,Rome Univ. and IROE/CAISMI/IAS
- AROME '90,IR-Astronomy, CNES
- TRIP '90,IR-Astronomy, TESRE CNR
- LAPEX '92,X-Astronomy, TESRE/IAS CNR and CESR
- AROME '92, IR-Astronomy, CESR/LPSP CNES, and Meudon Obs.

On 4th August 1993 a new cooperation between ASI and INTA allowed the first launch with total italian responsability recovered by spanish team about a day after. ARGO experiment was flown from Trapani to Spain: it was a very successfull experiment and gave high scientific quality results. ARGO is a mm-wave telescope, optimized for the measurement of CMB anisotropies at the degree angular scale, in four wavelength bands at 2.0, 1.2, 0.8 and 0.5 mm. High throughput bolometric detectors cooled at 0.3 K were used, featuring a sensitivity of the order of 100 $\mu K/\sqrt{Hz}$ in the 2.0 and 1.2 mm bands. The beamsize was 0.86^o FWHM, the beamthrow 1.8^o peak to peak; sine modulation and square demodulation were used for the detection of sky brightness gradients. The ARGO experiment was developed by Università di Roma La Sapienza (P. de Bernardis, P.I.), ENEA Frascati and IROE-CNR Firenze. The weight of the payload was 1500 Kg. The experiment was launched at 18:00 U.T. on August 4, 1993, and landed in Spain 20 hours later. The use of a $10^6 m^3$ balloon allowed to reach a float altitude of 42 Km. During the night several high Galactic latitude regions were scanned by the experiment (Hercules, Aries, Taurus). The very high altitude reached and the optimal performance of the pointing system and of the detection system allowed a clear detection of CMB anisotropy, well above the detector noise. A 63 independent points data set (the largest degree-scale anisotropy data set to date), features the expected CMB spectral signature in the 2.0 and 1.2 mm bands (de Bernardis et al. 1994a,b). The level of the measured CMB anisotropy is $\Delta T_{rms} = (60 \pm 20)\mu K$. Cirrus and Galactic plane dust emission were also measured in all the four bands allowing the measurement of the dust temperature (20 K) and of the spectral index of emissivity (1.5). The data from the second part of the flight (a 154 data set at moderate Galactic latitudes) are currently under analysis.

4. FUTURE PROGRAM

Programs at near and medium term are strongly encouraged while several payloads are ready to fly from Milo Base starting from 1995 transmediterranean campaign. Moreover several groups require to test some instrumentations for future application on satellites or big balloon-borne experiments. In many cases the single equipment weight is less than 100 kg and it works at low bit rate. It is planned for the future to include in each transmediterranean campaigns at least one payload "General purpose" containing up to 6 small compatible experiments.

FROM 1995 CAMPAIGN

- A.R.D. (Tecnological flight) Atmospheric Reentry Demostrator, an ESA program devoted to qualify applied reetry technologies.
- S.Q.M. (X-Astronomy) Strange Quark Matter, a cosmic rays experiment (P.I. O. Savedra)
- PALLAS (X-Astronomy) 20-300KeV experiment devoted to NGA study (P.I. F. Perotti)
- T.I.R. (IR-Astronomy) Cosmic background anisotropy experiment (P.I. F. Melchiorri)
- GIZERO (Local flight) High sensitive scientific instruments test in environment without gravity and dynamic pertubations (P.I. F. Fuligni)
- A.R.C.A. (General purpose flight)
  It hosts the following tests:
    - Solar cells (FIAR)
    - Limited Streamer Tube for FIGARO experiment (P.I. B. Sacco)
    - Accelerometer for GIZERO experiment
    - Inclinometers for GIZERO experiment
    - Solar sensor for BOOMERANG experiment (P.I. P. de Bernardis)

Acknowledgements

It is a pleasure to acknowledge the fundamental contribution of Prof. P.de Bernardis and Dr. S.Masi for the fruitful discussion and information about ARGO program.

References

-G.Rum, O.Cosentino, A.Leonardi, M.C.Falvella, E. Morelli and D.Spoto, COSPAR '92, Washington U.S.A.
-O.Cosentino, M.C.Falvella, A.Leonardi, E.Morelli, D.Spoto, I.J.S.G.E. '94, Huelva Spain
-M.C. Falvella, F.C.S.T. '95 ,accepted, Tehran Iran
-P.de Bernardis, L.Amicone, A.De Luca, M.De Petris, M.Epifani, M.Gervasi, G.Guarini, S.Masi, F.Melchiorri, V.Natale, A.Boscaleri, G.Natali, F.Pedichini, Ap.J.Lett. 360, L31, (1990)
-S. Masi, P. de Bernardis, M. De Petris, M. Epifani, M. Gervasi, G. Guarini, Ap.J.Lett. 366, L51-L55, (1991)
-P.de Bernardis, E. Aquilini, A. Boscaleri, M. De Petris, G. D' Andreta, M. Gervasi, E. Kreysa, L. Martinis, S. Masi, P. Palumbo, F. Scaramuzzi, Ap.J.Lett. 422, L33, (1994)
-P.de Bernardis, E. Aquilini, A. Boscaleri, M. De Petris, M. Gervasi, L. Martinis, S. Masi, V. Natale, P. Palumbo, F. Scaramuzzi, IL NUOVO CIMENTO, 16, 721, (1993)
-P.de Bernardis, G.De Gasperis, S.Masi, N.Vittorio, Ap.J.Lett., 433, L1, (1994)

TRIP: A COOLED TELESCOPE FOR MEASUREMENTS OF THE NEAR INFRARED COSMOLOGICAL BACKGROUND AT BALLOON ALTITUDE

G. Ventura, M. Malaspina, G. Morigi, M.R. Attolini,
A. Brighenti, P. Cazzola, P. Calzolari, S. Cortiglioni,
G. Giovannini, N. Mandolesi, E.Palazzi
I.TE.S.R.E./C.N.R., Via Gobetti, 101, 40129 Bologna, Italy

A. Boscaleri
I.R.O.E./C.N.R., Via Panciatichi, 64, 50127 Firenze, Italy

Email ID
trip@botes1.tesre.bo.cnr.it

ABSTRACT

A cooled telescope (250 mm, F/2) has been designed and prepared for measurements of both local and cosmological Near Infrared backgrounds in the waveband 1-5 μm at balloon altitudes. The separation of various foreground contributions should be possible by applying observing strategies based on multiple sky scans with different filters. The whole apparatus will be tested in flight by using a linear array of 32 InSb detectors cryogenically cooled and thermally stabilized at 52 K. The instrument sensitivity will allow the measurement of most of the local backgrounds. A description is given of the flight instrumentation and of the expected performances based on laboratory test results.

1. INTRODUCTION

The main scientific aim of the TRIP experiment (TRIP is the acronym for the Italian equivalent of Cooled IR Telescope for Balloon flights) is the detection and measurement of the Near Infrared Cosmological Background (NICB) in the waveband 1-5 μm. The best fit with the scientific goals would be attained with space missions using satellites, which would permit long observing times without foreground by the atmosphere. Flying at balloon altitude is a compromise between costs and scientific capabilities in a field in which few data have been collected. Atmospheric contamination is not all that must be dealt with; in fact, the total NIRB flux is a combination of the extragalactic cosmological background and other "local" backgrounds such as: Zodiacal Emission (ZE), resulting from the diffuse component of the Zodiacal Light (ZL) and the thermal component of the Interplanetary Dust (IPD), integrated Star Light (SL) and the atmospheric emission (ATM). Estimates for the local backgrounds are available from both observations and models developed by different authors on the basis of existing data (Murdock & Price, 1985, for ZE; Bahacall & Soneira, 1980, Mamon & Soneira, 1982, Franceschini et al, 1991, for SL; Mc Clatchey et al, 1972 & 1973, for the atmospheric emission; Vallance Jones, 1973, for OH line emission).

The TRIP is designed to fly at ballon altitudes and to detect the various IR emissions at different elevations and hour angles in order to have sufficient sky coverage at different wavelengths: this should permit to evaluate the "local" emissions (ATM, ZL, IDP, SL) and to extract the NICB (for a more detailed discussion on the local backgrounds see Greco, 1989; Morroi, 1989).

© 1996 American Institute of Physics

The actual flight configuration, with a linear 32 elements InSb detector, is mainly devoted to testing and calibrating the most critical parts of the experiment and to verify the capability of the adopted measurement strategies to separate the various components of the 1-5 μm total emission. A detailed description of the experiment can be found in Ventura et al, 1995.

2.PAYLOAD CONFIGURATION

The telescope, the optics and the detection assembly

The cryostat (Figure 1), is a two-gap dewar which houses the optics, the focal plane detector with the filter wheel and the baffles. To avoid ice formation, the entrance of the telescope is protected by a shutter which is opened only at float altitude. To observe the IR background in the wavelength range 1-5 μm at about 40 km altitude a cooled apparatus is necessary to minimize the thermal emission from the mirrors and the telescope mounting. The cooling is obtained by flowing liquid He from an external dewar through a transfer line which enters the telescope in the upper part of the inner gap.

The telescope is a double Cassegrain system (250 mm, F/2) with an angular scale at the focal plane of about 7 arcmin/mm and a PFOV of 1.4 arcmin (total FOV = 51 arcmin).

Developed by Cincinnati Corp. (USA), the focal plane detector is a 32 pixel InSb linear array working in the range 1-5 μm: more details about these detectors can be found in the technical literature (Cincinnati Corp. data; Nyblack, 1985; Bird, 1988). The noise behaviour of InSb arrays is also widely treated in the literature (for example, see Davies and Nyblack, 1986). The measured readout noise is $\approx 3500\ e^-$ rms per pixel, which means the experiment will suffer from "kTC" noise limitation . Future upgrades of the detection system will depend on the results of the first flight which should confirm the feasibility of the measurement.

The cryogenic apparatus

The temperature control of the detection system is performed by forcing LHe into the cryostat. The liquid He flow is regulated both by controlling the power to the heating resistors inside the storage dewar and by controlling the differential pressure between the cryostat and the storage dewar itself. This leads to a control of the temperature of the inner part of the telescope (1st mirror) within ± 1 K (around 40 K), while the detector temperature is stabilized at a higher value (52 ± 0.1 K) by a separate servo control system (Cortiglioni et al, 1986).

The attitude control systems.

The telescope elevation is independently controlled by the main experiment CPU which can power the zenith dc-motor in a servo-loop mode. The status is determined by a 12-bit binary absolute encoder mechanically coupled to the zenith rotation axis (position resolution $\approx$ 5 arcmin). The azimuth pointing system operates from the signal of a horizontal magnetometer to position the gondola in the desired direction within ± 10 arcmin.

The azimuth pointing direction can be varied either by remote commands (TCs) or by the on-board software by powering a motor which drives an independent platform carrying the magnetometer. The pointing system CPU is connected to the experiment CPU via a standard RS232 communication link which permits the transfer of the pointing system data to the experiment CPU for telemetry data transmission and the reception of TCs for the remote control from the experiment CPU (Boscaleri et al, 1990, 1994).

The service electronics (SE) and the data handling system (DHS)

The SE includes the batteries, the dc/dc converters and the housekeeping module which collects the data from the sensors which are distributed over the payload to monitor the working status of the various subsystems during the flight (Ventura et al, 1992).

The DHS is built around two CPUs, the *main* CPU and the *detector* CPU

The *main* CPU controls all the payload subsystems, the data flow to the PCM encoder, the data collection from the *detector* CPU, the data transmission to/from the pointing system, the parallel TCs reception from the flight decoder and the TCs distribution to the users.

The *detector* CPU runs under dedicated-ROM resident software and is fully reserved to the control of the InSb detector, i.e. it sets the exposure time and controls the detector data storage before a complete detector image is collected for transmission by the main CPU.(Ventura et al, 1992, 1995).

3. OBSERVING STRATEGIES AND EXPECTED IN-FLIGHT PERFORMANCES

Figure 2 represents the detection system overall sensitivity together with the measured total fluxes data from Matsumoto (1988), as compared with the emissions expected in the observed sky directions for a June 1995 flight; future possible improvements are also indicated by considering the use of bidimensional 64 x 64 InSb arrays with typical readout noise of $\approx$ 500 $e^{-}/pixel$. The TRIP should be able to detect the local emissions at their maximum (ZL_{max}, SL_{max}) and the ATM by using the measurement routines based on the "multiple sky scan" techniques which should permit the extraction of individual components.

The measurements are performed both by positioning the gondola to various azimuth angles and pointing the telescope at 6 elevation angles for each filter (7). The center wavelengths of the filters are: 2.025, 2.15, 2.28, 2.39, 2.5, 3.54, 4.61 μm; filter 0 is a stop position used to perform internal calibrations and dark current measurements.

Any parameter can be set either by the on-board software or by TCs and any change to the measurement sequence is possible by issuing TCs from the on-ground equipment. At each zenith observing angle and for any filter the multiple exposures (n) should give redundancy in collected data so reducing the statistical error by a factor of $\sqrt{n}$. The measurement routines will take place only after the completion of the in-flight calibration procedure which will comprise both absolute and internal calibrations.

Absolute in-flight on-source calibrations are necessary to perform a calibration of the whole instrument; on-ground absolute calibrations are in fact impossible as the lid of the cryostat cannot be opened when the telescope is cold.

The large sky coverage obtained by using this scan procedure should permit, during the off line analysis, to arrange the data as a function of any coordinate system, depending on the background to be extracted. In the first flight sky regions near to the galactic plane and to the ecliptic plane (where the SL and ZE are strongest) will be privileged.

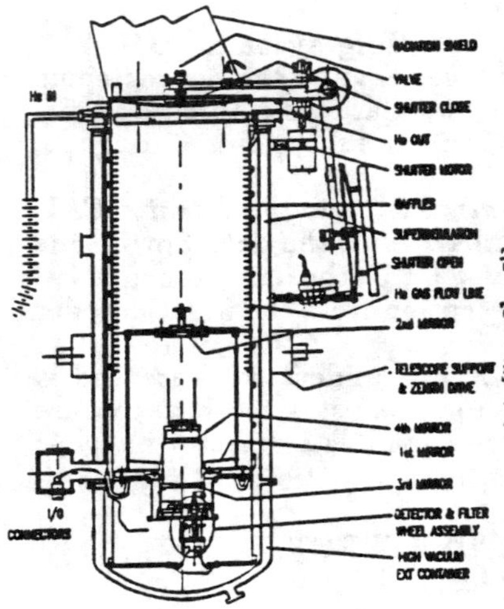

Figure 1. Cross sectional view of the telescope and of the cryostat.

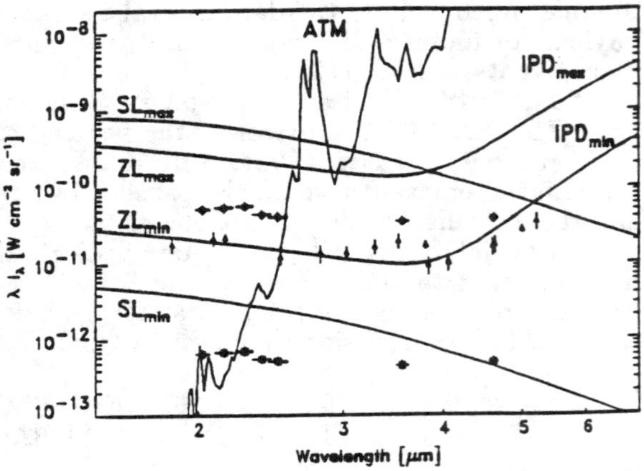

Figure 2. Estimated **TRIP** sensitiviy (•) for 7 filters as compared to the expected local backgrounds for a June '95 flight.Matsumoto's data (Δ) and InSb 64x64 detector array sensitivity (•) are also reported.

REFERENCES

Bahacall, J.N., Soneira, R.M. 1980, ApJ Suppl, 44 , 73

Bird, M.C. 1988, PhD Thesis, Univ. of Edimborough

Boscaleri, A. et al 1990, SPIE Proceedings, 127

Boscaleri, A. et al 1994, Meas. Sci. Technol., 5, 190

Cortiglioni, S. et al 1986, "The TRIP pressure control system", CNR Int.Report N. 119

Davies, R.M., Nyblack, C.A. 1986, SPIE Proceedings, 627 , 438

Franceschini, A. et al 1991, AA Supp., 89, 285

Greco, C. 1989, Thesis, Univ. of Bologna

MacClatchey, R.A. et al 1972, AFCRL-TR-55-213

MacClatchey, R.A. et al 1973, Agard Conf. Proc. 183

Mamon, A.G., Soneira, R.M. 1982, ApJ, 255 , 181

Matsumoto, T. et al 1988, ApJ, 332 , 575

Morroi, R. 1989, Thesis, Univ. of Bologna

Murdock, T.L., Price, S.D. 1985, AJ, 90 , 375

Nyblack, C.A. 1985, Proc. of the 28th Midwest Symposium on circuits and Systems Conference, Louisville , 4

Vallance Jones, A. 1973, Sp. Sci. Rev., 15, 355

Ventura, G. 1992, "The TRIP Service Electronics", CNR Int.Report N. 159

Ventura, G. et al 1994, "The TRIP experiment", CNR Int.Report N. 169

Ventura, G. et al 1995, Exp. Astr., submitted

Progress Report on the Las Campanas/Hubble Space Telescope Extragalactic Background Light Program (3000-8000 Å)

Rebecca A. Bernstein
Caltech 105-24, Pasadena, CA, 91125

Wendy L. Freedman
OCIW, 813 Santa Barbara St., Pasadena, CA 91101

Barry F. Madore
NED/IPAC, Caltech 100-22, Pasadena, CA, 91125

ABSTRACT

We present a status report on results of an ongoing experiment to measure both the broadband spectral energy distribution and the power spectrum of spatial fluctuations of the Extragalactic Background Light at 5 wavelengths covering the range from 3000 to 8000 Å. The results of coordinated, simultaneous CCD imaging of several high-Galactic-latitude fields using the Las Campanas duPont 2.5m and Swope 1.0m telescopes are discussed, and plans for scheduled HST observations in the ultraviolet, optical and near-infrared are presented.

BACKGROUND AND MOTIVATION

The extragalactic background light (EBL) is the integrated flux from all extragalactic sources. The surface brightness, the spectrum of spatial fluctuations, and the spectral energy distribution of the EBL are each strongly dependent on the assumed epoch of galaxy formation, the nature of galaxy evolution, and the assumed cosmology (Ω_0,H_0,Λ). Clearly, one measurement of the EBL cannot constrain all of the involved parameters; however, measurements of the absolute flux and power spectrum of the EBL over a *range of wavelengths* will isolate different subsets of the variables, and can together greatly limit the range of acceptable models.

For example, the EBL at ~3000Å will be dominated by redshifts $z < 2$, where the differences between cosmological models are not pronounced, but different star formation rates have a strong effect on the total EBL flux (Martin & Bowyer 1989). At ~8000Å, the EBL includes light from objects at $z < 7.5$ and is, therefore, more sensitive to cosmological models. Even at 5000Å, the EBL is predicted to differ by a factor of two for $\Omega_0 = 0.1$ and $\Omega_0 = 1$, due to smaller volume elements at high redshift for $\Omega_0 = 1$ (Yoshii & Takahara 1988).

The EBL is also a powerful probe of ultra-faint galaxies which are too faint to be resolved individually. The steep slope of observed number counts indicates that the greatest contribution to the EBL is from the faintest sources. If the number counts continue to diverge, the EBL will be dominated by a population of objects not yet directly observed. The power spectrum (the Fourier transform of the autocorrelation function) of the EBL is a measure of both the luminosity density and the angular (and spatial) correlation function of these galaxies. It is thus a direct measurement of the spatial clustering of these faint objects.

© 1996 American Institute of Physics

MEASURING THE EBL

While the EBL is composed of light from galaxies, the integrated light of the night sky is dominated by airglow (atmospheric line emission), zodiacal light, and Galactic star light at levels up to 100 times that expected for the EBL itself. In the last 15 years, several attempts have been made to measure the EBL, the most recent and noteworthy of these attempts have been reviewed by Mattila (1990). The majority of these have employed techniques to isolate the foreground without measuring the constituent components independently. Unfortunately these experiments have been unsuccessful.

While it is a much more complicated experiment than most previous attempts, our method involves the direct measurement and subtraction of the foreground components. Dube, Wickes and Wilkinson (1979) are the only other group to have tried this technique. Using a photoelectric photometer they adopted a sec$z$ approximation to determine the airglow component, applied aperture masks to block out stars brighter than 20 mag, and used narrow–band filters centered on solar Fraunhofer lines to determine the contribution from zodiacal light (which has a solar spectrum, Beggs 1963). Their greatest source of difficulty was the time variability of the sky (extinction and emission). Our experiment utilizes methods which build on these techniques, but also substantially improves the accuracy to which each of these foreground components can be identified and removed using both dramatically improved technology and new techniques.

THE DATA: LAS CAMPANAS OBSERVATORY AND HST

Because this is a delicate experiment prone to systematic errors and requiring careful calibration, two independent data sets are being used. The first is comprised of images and spectra obtained simultaneously using the 1m and 2.5m telescopes at LCO; the second data set will be comprised of images and spectra from HST. Each of the foreground components can be redundantly determined using these data sets. Furthermore, the three different telescopes provide independent checks of the photometry and object identification, and external estimates of systematic errors.

The images will be used to measure the EBL using wide bandpass filters for maximum sensitivity. The spectra, taken simultaniously with the images, are used strictly to monitor the zodiacal light.

FOREGROUND REMOVAL: ATMOSPHERIC EMISSION

In ground–based observations, the airglow at ~5000Å is roughly 100 times the anticipated intensity of the EBL. To correct for airglow, we take advantage of the fact that both atmospheric emission and extinction are functions of the pathlength through the atmosphere. A point source of light (i.e., a star) appears exponentially fainter as the line-of-sight pathlength through the atmosphere increases. At the same time, the airglow brightens along that same line of sight, also as a function of pathlength. Both effects are dependent upon the total pathlength through the atmosphere, but they are independent probes of that quantity. To identify the airglow component in our ground–based images, we can therefore use the stars present in each frame as tracers of the instantaneous, effective pathlength through the atmosphere. This gives a much more accurate estimate of airmass than the usual, analytic sec(z) approximation, and allows us to extrapolate the sky intensity as a function of the instantaneous, effective airmass of each exposure and to thereby obtain the intensity of the background above the atmosphere.

Using all the stars in a field which is observed roughly every twenty minutes from zenith to sec(z) ≈ 2, we can accurately calculate the effective extinction at the time each image was taken, and therefore the effective pathlength through the atmosphere. The average sky value for the frame (the average of all pixels not containing detectable objects) is then plotted as a function of this effective pathlength, and the sky value is extrapolated to give the background level above the Earth's atmosphere (the sky at zero effective pathlength). One such plot is depicted in Figure 1.

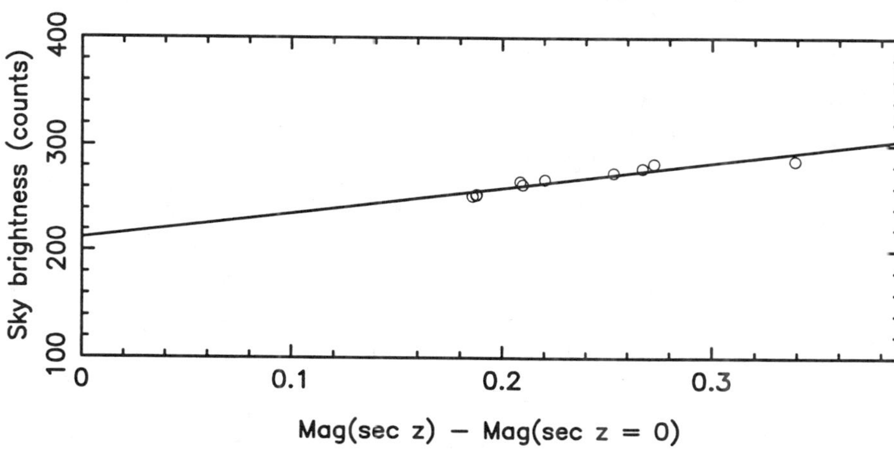

Figure 1: Airglow extrapolation using data taken in one night with the 2.5m telescope at LCO, Gunn-g filter. The RMS scatter is 2.17 counts, or $< 1\%$.

FOREGROUND REMOVAL: ZODIACAL LIGHT

In the wavelength range 3000Å to 8000Å, the zodiacal light has been very well studied, and it is well established that the zodiacal light reproduces the solar spectrum in both continuum shape and in the depth and width of individual lines (Leinert 1975). In addition, it has been determined that the particles responsible for scattering the sunlight are not influenced by solar activity, flares, meteor streams, comets, interplanetary plasma or magnetic fields (Leinert 1982). This being the case, it is possible to identify the flux level of the zodiacal continuum using the known equivalent widths of the strong absorption features (Fraunhofer lines) in the spectrum of the Sun (see Figure 2).

In practice, the airglow complicates the analysis of the ground-based spectra. For this reason, a single star was included in the spectrum of each of our target fields. By comparing the spectrum of the star, to the spectrum of the sky for a series of exposures, we can remove the airglow as a function of the star's extinction in 10Å bins, in the same manner as was outlined above for the images. Once the atmospheric contribution is removed, the zodiacal spectrum remains. The spectrum obtained from HST will be centered on "empty" regions in our fields, and will be dominated by the zodiacal light.

HST will also take spectra in parallel with imaging, so we will have independent measurements of the zodiacal component in each of our fields each time observations are conducted. Note that the HST spectra will be free of atmospheric extinction and emission, simplifying the reduction.

As a final check of the zodiacal light removal, the geometry of the dust layer causing the scattered zodiacal light is well known. Based on this geometry, we can predict the percentage variation in the zodiacal contribution to our fields depending on the time of year (position in the Earth's orbit) of our three observation epochs. This will provide an independent check on our spectral solution from both LCO and HST.

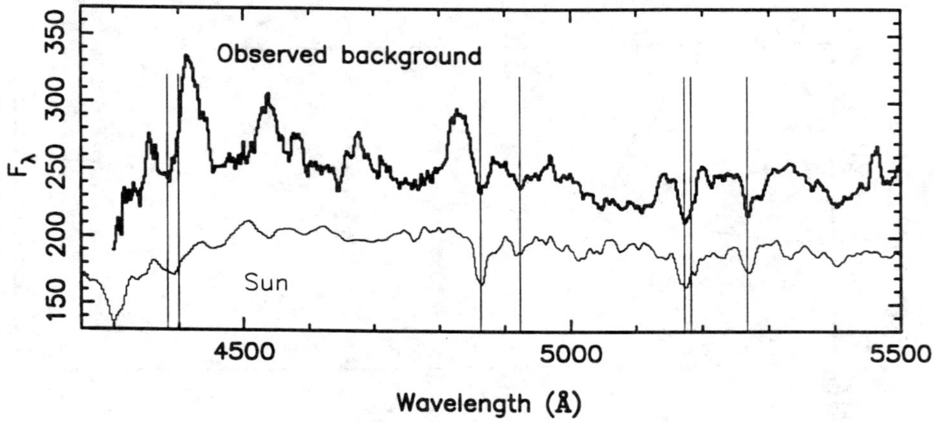

Figure 2: The solar spectrum (blue, with flux in units of 10^{-6}ergs/sec/cm^2/Å) compared to the spectrum of the background (EBL + zodiacal light + airglow, with flux in 10^{-18} ergs/sec/cm^2/Å) in one program field. The following solar features are marked: 4384 Fe I & 4405 Fe I, blended; 4861 Hβ; 4920 Fe I; 5173 Mg I; 5183 Mg I; 5269 Fe I.

REFERENCES

Beggs, D.W., Blackwell, D.E., Dewhirst, D.W., Wolstencroft, R.D., 1963, MNRAS, 127(5), 329

Dube, R.R., Wickes, W.C., & Wilkinson, D.T. 1979, ApJ, 215, L51

Leinert, Ch. 1975, Space Science Reviews, 18, 281

Leinert, Ch. 1982, AA, 110, 115

Martin, C. & Bowyer, S. 1989, ApJ, 338, 677

Mattila, K. 1990, in Proc IAU 139, Galactic and Extragalactic Background Radiation, ed. S. Bowyer & Ch. Leinert (Dordrecht: Dluwer Academic Publisher)

Tinsley, B.M. 1978, ApJ, 220, 816

Yoshii, Y. & Takahara, F. 1988, ApJ, 326, 1

PRONAOS submillimeter Observations of the Orion clouds

I.Ristorcelli[1], A.De Luca[2], M.Giard[1], F. Pajot[2], J.P.Torre[2], G.Serra[1], J.M.Lamarre[2]
[1] C.E.S.R., CNRS-UPS, BP 4346, F31029, Toulouse cedex, France.
[2] IAS, bât 120, Campus d'Orsay, 91405 Orsay CEDEX, France.

Email ID
ristorce@cesr.cnes.fr

ABSTRACT

Submillimeter observations have been performed by the french balloon-borne experiment PRONAOS, during its first flight in September 1994. The instrument consisted mainly of a two-meter telescope and a multiband photometer (from 180 to 1100 μm). Despite bad altitude and attitude controls, a map of an extended area (10'x50') containing the M42 nebula, has been drawn up in the four photometric bands, with angular beams from 2' to 3.5'. The flux values measured in the four bands, on the brightest source associated to the BN/KL objects, are compatible with a thermal spectrum emitted by dust with an emissivity power law $Q_{em}\alpha\nu^{1.5}$. Around this source, three other brightness peaks are observed, two of them coincident with enhancements in the 100 μm IRAS map. The third one is about $5' \times 5'$ in size, and can be interpreted as the dust emission of a very cold (about $10 \pm 2K$) molecular cloud . The in-flight "Noise Equivalent Brightness" is $1.7MJysr^{-1}s^{1/2}$ in the longer wavelength band. The pointing stability is better than 3 arcsec and the observation of BN/KL objects shows that the telescope presents an image quality compatible with a diffraction limitation in all bands. These characteristics make PRONAOS-SPM a very accurate instrument to perform cosmological observations such as the Sunyaev-Zeldovitch effect measurement.

1. INSTRUMENT CHARACTERISTICS

PRONAOS is a french balloon-borne experiment devoted to submillimeter astronomy. It consists of four main sub-systems : a two-meter Cassegrain telescope, a stabilized gondola and two focal instruments, a Multiband Photometer (SPM) and a heterodyne Spectrometer (SMH), planned to fly alternatively on the experiment. The SPM instrument has been designed to realise simultaneous measurements in four bands from 180 to 1100 μm and its characteristics have been described in Lamarre et al. (1994). It uses He^3 cooled bolometers (300mK) and dichroic filters inside a cryostat; the warm optics ensures, in particular, the beam switching and the in-flight calibration by means of two blackbodies. The absolute accuracy of this calibration is better than 10%. The sensitivities reached during the flight as well as the field of view in each photometric band are given in table 1. The electrical signal provided by each bolometer is amplified, filtered, demodulated and synchronously transmitted by telemetry to the ground station. The telescope (manufactured by Matra Marconi Space-Toulouse) is made of a 2m diameter primary mirror, consisting of six identical panels in carbon fiber and covered by a gold coating. A servo-controlled loop system between capacitive sensors and actuators on the panels allows, all along

© 1996 American Institute of Physics

the flight, to correct any misalignment introduced by thermal or gravity effects, and maintains the nominal alignment of the panels. The payload pointing is provided by the gondola using a two-stage system. The first one ensures the azimuthal stabilization within a relative precision of 30 arcseconds, thanks to a magnetometer and a gimbal servoed with gyrometers and inertial wheels. The fine pointing is obtained with an inertial platform which drifts are corrected thanks to the pointing of a star sensor that can be oriented off-axis from the telescope. The final relative pointing accuracy is better than 5 arcseconds. The PRONAOS-SPM experiment was launched on September 17, 1994, from the NASA launching site of Fort-Sumner (New Mexico, USA). Unfortunately, the failure of the star sensor during daytime, together with a much too low ceiling altitude at night, did not allow to perform the observations planned. Only very partial observations were possible. We report here the results obtained in the region of the Orion M42 nebula. The observations have been done with a cross-elevation scanning at a constant speed of $20 arcsec/s$ and with a beam switching amplitude of 5.35 arcmin. The covered map corresponds to $50 \times 10 arcmin^2$.

Table 1:

Band	1	2	3	4
$\Delta\lambda(\mu m)$	180-240	240-340	340-540	540-1100
FOV(arcmin)	2	2	2.5	3.5
NET($mKs^{1/2}$)	1.2	0.9	0.5	0.35
NEB($Wm^{-2}sr^{-1}s^{1/2}$)	1.1e-6	3.3e-7	5.1e-8	6.8e-9

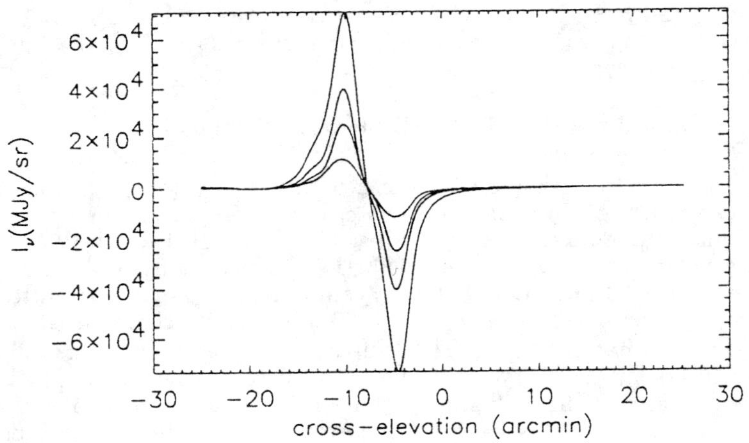

Figure 1: Measured signal on the BN/KL objects in the four photometric bands. The scaling factors applied on the bands 3 and 4 are respectively : x2 and x6.

2. RESULTS

Figure 1 shows the peak of the measured signal in the four photometric bands, along the scan containing the BN/KL objects. The contours map of the signal measured in the second photometric band, for instance, is given in figure 2. Four brightness peaks are visible and have been identified to the following sources: 1) The brightest peak corresponds to the BN/KL objects.
2) An extended emission region over around $5' \times 20'$, 15' SE from BN/KL coincides with an IRAS 100 μm band flux enhancement and a CO emission peak; it has been identified by Castets et al.(1990) with a compressed gas area due to a shock driven into the neutral gaz by the expansion of the HII region.
3) The third brightness peak in the PRONAOS map is located at $10'W \times 4'N$ from BN/KL. It is associated with the edge of an ionized region (which emission peak at 100 μm is about 20'W from BN/KL).
4) An extended peak emission (over $5' \times 5'$) is clearly observed at 30'NW from BN/KL, particularly in the second photometric band (240 - 340 μm). This area coincides neither with a brightness enhancement of the CO nor with the 100 μm IRAS band emission.

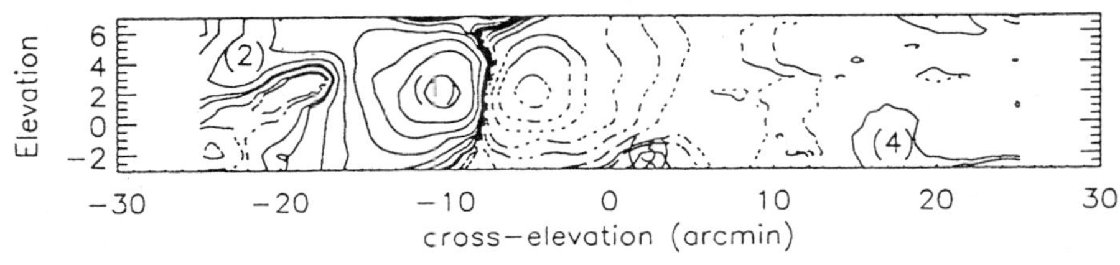

Figure 2: Contours of the measured signal in the second photometric band. Large negative signal (dashed contour) is due to the beam switching

The emission spectrum associated with each of these sources is displayed on figure 3. It can be noticed that, except for the fourth one, the general shape is similar. It is compatible with a dust emission spectrum, modeled with a single dust temperature, and a power law for the dust emissivity $Q_{em} \alpha \nu^{\beta}$ We have determined the temperature and spectral index values by a least square method. The corresponding fit $F_\nu = A\nu^{\beta} B_\nu(T) = \tau_\nu B_\nu(T)$ (where $B_\nu(T)$ is the blackbody Planck function) is shown on figure 3, and the values of T, β and the resulting optical depth in each band (τ_i), are given in table 2.

Table 2:

Source	T(K)	β	τ_1	τ_2	τ_3	τ_4
BN/KL objects	87 ± 10	1.5 ± 0.14	1.8e-2	1.4e-2	8.5e-3	3.3e-3
source (2)	27 ± 3	2.4 ± 0.2	4.e-3	2.1e-3	8.7e-4	3.e-4
source (3)	37 ± 11	1.8 ± 0.2	7.8e-4	4.e-4	2.1e-4	1.0e-5

The fourth source spectrum presents a narrow peak in the 240-340 μm band. It cannot be explained as a simple dust emission, even using a distribution law for the grains temperature and size. The spectrum can be described using two components : i) a dust emission (dashed line in Fig.3) corresponding to a mean very cool temperature of $10 \pm 2K$ ($\beta \sim 2$), and ii) an additive flux contribution in a narrow band ($\Delta\lambda \sim 100\,\mu$m), around $\lambda \sim 300\,\mu$m.

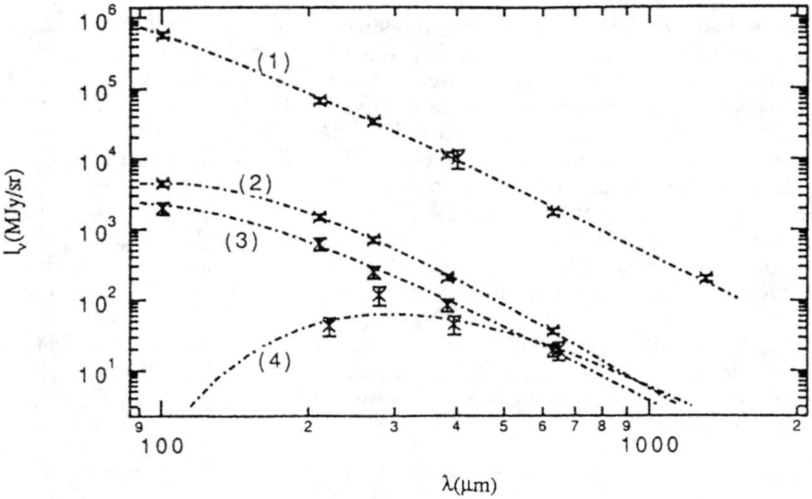

Figure 3: Emission spectrum of the identified sources. The fluxes measured with PRONAOS-SPM are given at effective wavelengths determined in each photometric band in accordance with the spectrum. We have overplotted on the BN/KL spectrum, the data of Pipher et al.1978, Keene et al.1982 and Mezger et al.1990, at 100, 400 and 1300 μm respectively. The fluxes deduced from IRAS data at 100 μm have also been added to the spectrum of sources (2) and (3). The dashed curves correspond to the fit described in the text.

ACKNOWLEDGEMENTS

We are indebted to the technical staff who contributed to the design, the manufacture and the tests of the PRONAOS-SPM instrumentation, mainly the CNES team (with the PRONAOS Project Manager, François Buisson) and the CNRS-IAS and CNRS-CESR engineers and technicians, lead by Guy Guyot. We are grateful to the NSBF-NASA for the balloon launching operations.

REFERENCES

Castets, A., Duvert, G., Bally, J., Langer, W.D., Wilson, R.W., 1990, A&A, 234, 469

Keene, J., Hildebrand, R.H., Whitcomb, S.E., 1982, ApJ, 215, L11

Lamarre, J.M., Pajot, F., Torre, J.P., Guyot, G., Bernard, J.P., De Luca, A., Giard, M., Mangin, J., Peytureaux, R., Puget, J.L., Serra, G., Boulanger, F., Desert, F., Ristorcelli, I., Jegoudez, G., Lagardere, H., Leblanc, L., Pons, R., Recouvreur, G., Barthelemy, M., Bourguignon, C., Crussaire, J.P., Dambier, G., Delille, L., Korczak, E., Lepeltier, J.P., Leriche, B., Lizambert, C., Narbonne, J., Plaisant, C., Reigner, J., Renault, J.C., Rioux, C., 1994, Infrared Phys. Technol., Vol. 35, No. 2/3, 277

Mezger, P.G., Wink, J.E., Zylka, R., 1990, A&A, 228, 95

Pipher, J.L., Duthie, J.G., Savedoff, M.P., 1978, ApJ, 219, 494

APPENDIX A: LIST OF PARTICIPANTS

Alain Abergel
Institut d'Astrophysique Spatiale
Batiment 121
F-91405 Orsay Cedex FRANCE

Nabila Aghanim
Institut d'Astrophysique Spatiale
Batiment 121 Universite Paris XI
F-91405 Orsay Cedex FRANCE

Thomas Ake
CSC
Mail Code 681
NASA/GSFC
Greenbelt, MD 20771

Richard G. Arendt
Applied Research Corp.
Code 685.3
NASA/GSFC
Greenbelt, MD 20771

Anthony J. Banday
Code 685
NASA/GSFC
Greenbelt, MD 20771

James Bartlett
Observatoire de Strasbourg
Universite' Louis Pasteur
11, rue de l Universite'
67000 Strasbourg
FRANCE

Stephen Benka
AIP
One Physics Ellipse
College Park, MD 20740

Charles Bennett
Mail Code 685
NASA/GSFC
Greenbelt, MD 20771

Peter Berczik
Main Astronomical Observatory
Ukranian National Academy of Sciences
Dept. for the Physics of Star and Galaxies
Golosiiv Kiev-22, UKRAINE

Jean-Philippe Bernard
Institut d'Astrophysique Spatiale
IPAC, MS100-22
Pasadena, CA 91125

Rebecca Bernstein
Caltech 105-24
Pasadena, CA 91125

Bruce Berriman
GSC
Mail Code 685.3
NASA/GSFC
Greenbelt, MD 20771

Francois N. Bouchet
Institut d'Astrophysique Spatiale de Paris, CNRS
98 bis Boulevard Arago
F-75014 Paris FRANCE

Francois Boulanger
Institut d'Astrophysique Spatiale
Batiment 121 Universite Paris XI
F-91405,Orsay FRANCE

Sean Casey
Code 683/HSTX
NASA/GSFC
Greenbelt, MD 20771

Emmanuel Caux
CESR-CNRS/UPS
31029 Toulouse Cedex
FRANCE

Hong-Yee Chiu
Mail Code 910.4
NASA/GSFC
Greenbelt, MD 20771

Martin Cohen
Radio Astronomy Laboratory
601 Campbell Hall
University of California-Berkeley
Berkeley, CA 94720

Greg Comer
Parks College of St. Louis University
Dept. of Sciences & Mathematics
Caholkia, IL 62206

Daniel Crenshaw
Mail Code 681
NASA/GSFC
Greenbelt, MD 20771

Stan Dermott
University of Florida
Dept. of Astronomy
224 SSRB
Gainesville, FL 32611

Francois-Xavier Desert
Institut d'Astrophysique Spatiale
Batiment 121 Universite Paris XI
F-91405, Orsay Cedex FRANCE

Jacques Delabrouille
Institut d'Astrophysique Spatiale
Batiment 121 Universite Paris XI
F-91405 Orsay Cedex FRANCE

Gianfranco De Zotti
Osservatorio Astronomico di Padova
Vicolo del' Osservatorio, 5
I-35122 Padova, ITALY

Eli Dwek
Mail Code 685
NASA/GSFC
Greenbelt, MD 20771

Nicholas Epchtein
Observatoire de Paris
Dept. de Recherche Spatiale
5, Place Jules Janssen
F-92195 Meudon Prinicipal Cedex
FRANCE

Michael Fall
Space Telescope Science Institute
3700 San Martin Drive
Baltimore, MD 21218

Maria Cristina Falvella
Italian Space Agency
A.S.I.-.Base Lancio Palloni
Statosferici
S.S. 113- C.DA Milo
I-91100 Trapani, ITALY

Giovanni G. Fazio
Harvard Smithsonian Center
for Astrophysics
CFA, MS/65
60 Garden Street
Cambridge, MA 02138

Hume A. Feldman
Princeton University
Physics Department
Jadwin Hall
P.O. Box 708
Princeton, NJ 05480-0708

James E. Felten
Code 685
NASA/GSFC
Greenbelt, MD 20771

Alberto Franceschini
Padova Astronomical Obs.
Vicolo Osservatorio 5
I-35122 Padova, ITALY

Bryan Franz
Code 685
NASA/GSFC
Greenbelt, MD 20771

Jonathan Gardner
University of Durham
Department of Physics
South Road
Durham, DH1 3LE,
UNITED KINGDOM

Martin Giard
CNRS/CESR
BP 4346, F-31029
Toulouse Cedex FRANCE

Richard Gispert
Institut d'Astrophysique Spatiale
Batiment 121 Universite Parix XI
91405 Orsay Cedex, FRANCE

Karl D. Gordon
Ritter Astrophysical Research Center
The University of Toledo
Toledo, OH 43606

Krzysztof Gorski
Code 685
NASA/GSFC
Greenbelt, MD 20771

Gian Luigi Granato
Dipartmento di Astronomia
Vicolo del ' Osservatorio, 5
I-35122 Padova, ITALY

Bruno Guiderdoni
Institut d' Astrophysique de Paris,
CNRS
98 bis Boulevard Arago
F-75014 Paris, FRANCE

Shaul Hanany
University of California-Berkeley
Center for Particle Astrophysics
301 LeConte Hall
Berkeley, CA 94720

Michael Hauser
Code 680
NASA/GSFC
Greenbelt, MD 20771

Sally Heap
Code 681
NASA/GSFC
Greenbelt, MD 20771

Sumita Jayaraman
University of Florida
Dept.of Astronomy
201, SSRB
Gainesville, FL 32611-2055

Alexander Kashlinsky
NORDITA
Blegdamsvej 17
Copenhagen DK-2100
DENMARK

Thomas Kelsall
Code 685
NASA/GSFC
Greenbelt, MD 20771

Alexander Kutyrev
Code 685
NASA/GSFC
Greenbelt, MD 20771

Christoph Leinert
Max-Planck-Institut Fur Astronomie
Kon'gstuhal 17
D69117 Heidelberg
GERMANY

David Leisawitz
Code 631
NASA/GSFC
Greenbelt, MD 20771

Carol Lonsdale
IPAC
Caltech
Pasadena, CA 91125

Gary Mamon
Institut d' Astrophysics de Paris
98 bis Boulevard Arago
F-75014 Paris, FRANCE

John Mather
Code 685
NASA/GSFC
Greenbelt, MD 20771

Kenneth Mitchell
Code 685.3
NASA/GSFC
Greenbelt, MD 20771

Harvey Moseley
Code 685
NASA/GSFC
Greenbelt, MD 20771

Sten Odenwald
Applied Research Corp.
Cosmology Data Analysis Center
7601 Ora Glen Drive
Greenbelt, MD 20770

Leonid Ozernoy
Code 685
NASA/GSFC
Greenbelt, MD 20771

Jim Peebles
Joseph Henry Lab/Jadwin Hall
Princeton University
P. O. Box 708
Princeton, NJ 08544-0708

Jean-Loup Puget
Institu d'Astrophysique Spatiale
Bat 121, Universite Paris Sud
F-91405 Orsay FRANCE

Denis Puy
SISSA-ISAS
Via Beirut n 2/4
34013 Trieste ITALY

William Reach
Code 685
NASA/GSFC
Greenbelt, MD 20771

Antonio Regaldo,
SCIENCE AAAS
1333 H Street, NW
Washington, D.C. 20005

Paul Richards
University of California-Berkeley
Department of Physics
366 LeConte Hall #7300
Berkeley, CA 94720-7300

Brigitte Rocca-Volmerange
Institut d'Astrophysique de Paris
98 bis Boulevard Arago
F-75014 Paris FRANCE

Michael Rowan-Robinson
Blackett Laboratory
Imperial College
Prince Consort Road
London, SW7 2BZ
UNITED KINGDOM

Stephaine Ruphy
Observatorie de Paris-Meudon
5, Place Jules Janssen
F-92195 Meudon Principal Cedex
FRANCE

Rick Shafer
Code 685
NASA/GSFC
Greenbelt, MD 20771

Joe Silk
Astronomy Department
University of California-Berkeley
601 Campbell Hall
Berkeley, CA 97020

Robert Silverberg
Code 685
NASA/GSFC
Greenbelt, MD 20771

Randall Smith
University of Wisconsin-Madison
Department of Physics
1150 University Avenue
Madison, WI 53706

George Smoot
University of California-Berkeley
LBL 50/351
Berkeley, CA 94720

Thomas Sodroski
ARC/GSFC
Mail Code 685.3
Greenbelt, MD 20771

Floyd Stecker
Code 660
NASA/GSFC
Greenbelt, MD 20771

Rich Talcott
Assistant Editor
Astromony Magazine
P.O. Box 1612
Waukesha, WI 53187

Gary Toller
Code 685.3
NASA/GSFC
Greenbelt, MD 20771

Virginia Trimble
University of California-Irivin
Physics Department
Irvine, CA 92717

Frank Varosi
Mail Code 685
NASA/GSFC
Greenbelt, MD 20771

Shoba Veeraraghaven
NRC
Code 685
NASA/GSFC
Greenbelt, MD 20771

Jan Vrtilek
Mail Code SZA
NASA HQ
Washington, D.C. 20546-0001

Janet Weiland
Code 685.3
NASA/GSFC
Greenbelt, MD 20771

Michael Werner
233-303 JPL
4800 Oak Grove Dr.
Pasadena, CA 91109

Edward Wishnow
LLNL, L401
P.O. Box 808
Livermore, CA 94550

Adolf N. Witt
Ritter Observatory
The University of Toledo
2801 W. Bancroft
Toledo, OH 43606

APPENDIX B: SUBJECT INDEX

Page numbers refer to the *first* page of the contribution in which the subject appears

A

B

C

D

E

F

G

N

O

P

Q

R

S

T

U

V

W

X

Y

Z

Author Index

A

B

C

D

E

F

G

H

J

K

AIP Conference Proceedings

	Title	L.C. Number	ISBN
No. 215	X-Ray and Inner-Shell Processes (Knoxville, TN 1990)	90-84700	0-88318-790-6
No. 216	Spectral Line Shapes, Vol. 6 (Austin, TX 1990)	90-06278	0-88318-791-4
No. 217	Space Nuclear Power Systems (Albuquerque, NM 1991)	90-56220	0-88318-838-4
No. 218	Positron Beams for Solids and Surfaces (London, Canada 1990)	90-56407	0-88318-842-2
No. 219	Superconductivity and Its Applications (Buffalo, NY 1990)	91-55020	0-88318-835-X
No. 220	High Energy Gamma-Ray Astronomy (Ann Arbor, MI 1990)	91-70876	0-88318-812-0
No. 221	Particle Production Near Threshold (Nashville, IN 1990)	91-55134	0-88318-829-5
No. 222	After the First Three Minutes (College Park, MD 1990)	91-55214	0-88318-828-7
No. 223	Polarized Collider Workshop (University Park, PA 1990)	91-71303	0-88318-826-0
No. 224	LAMPF Workshop on (π, K) Physics (Los Alamos, NM 1990)	91-71304	0-88318-825-2
No. 225	Half Collision Resonance Phenomena in Molecules (Caracas, Venezuela 1990)	91-55210	0-88318-840-6
No. 226	The Living Cell in Four Dimensions (Gif sur Yvette, France 1990)	91-55209	0-88318-794-9
No. 227	Advanced Processing and Characterization Technologies (Clearwater, FL 1991)	91-55194	0-88318-910-0
No. 228	Anomalous Nuclear Effects in Deuterium/ Solid Systems (Provo, UT 1990)	91-55245	0-88318-833-3
No. 229	Accelerator Instrumentation (Batavia, IL 1990)	91-55347	0-88318-832-1
No. 230	Nonlinear Dynamics and Particle Acceleration (Tsukuba, Japan 1990)	91-55348	0-88318-824-4
No. 231	Boron-Rich Solids (Albuquerque, NM 1990)	91-53024	0-88318-793-4
No. 232	Gamma-Ray Line Astrophysics (Paris-Saclay, France 1990)	91-55492	0-88318-875-9
No. 233	Atomic Physics 12 (Ann Arbor, MI 1990)	91-55595	088318-811-2
No. 234	Amorphous Silicon Materials and Solar Cells (Denver, CO 1991)	91-55575	088318-831-7

	Title	L.C. Number	ISBN
No. 235	Physics and Chemistry of MCT and Novel IR Detector Materials (San Francisco, CA 1990)	91-55493	0-88318-931-3
No. 236	Vacuum Design of Synchrotron Light Sources (Argonne, IL 1990)	91-55527	0-88318-873-2
No. 237	Kent M. Terwilliger Memorial Symposium (Ann Arbor, MI 1989)	91-55576	0-88318-788-4
No. 238	Capture Gamma-Ray Spectroscopy (Pacific Grove, CA 1990)	91-57923	0-88318-830-9
No. 239	Advances in Biomolecular Simulations (Obernai, France 1991)	91-58106	0-88318-940-2
No. 240	Joint Soviet-American Workshop on the Physics of Semiconductor Lasers (Leningrad, USSR 1991)	91-58537	0-88318-936-4
No. 241	Scanned Probe Microscopy (Santa Barbara, CA 1991)	91-76758	0-88318-816-3
No. 242	Strong, Weak, and Electromagnetic Interactions in Nuclei, Atoms, and Astrophysics: A Workshop in Honor of Stewart D. Bloom's Retirement (Livermore, CA 1991)	91-76876	0-88318-943-7
No. 243	Intersections Between Particle and Nuclear Physics (Tucson, AZ 1991)	91-77580	0-88318-950-X
No. 244	Radio Frequency Power in Plasmas (Charleston, SC 1991)	91-77853	0-88318-937-2
No. 245	Basic Space Science (Bangalore, India 1991)	91-78379	0-88318-951-8
No. 246	Space Nuclear Power Systems (Albuquerque, NM 1992)	91-58793	1-56396-027-3 1-56396-026-5 (pbk.)
No. 247	Global Warming: Physics and Facts (Washington, DC 1991)	91-78423	0-88318-932-1
No. 248	Computer-Aided Statistical Physics (Taipei, Taiwan 1991)	91-78378	0-88318-942-9
No. 249	The Physics of Particle Accelerators (Upton, NY 1989, 1990)	92-52843	0-88318-789-2
No. 250	Towards a Unified Picture of Nuclear Dynamics (Nikko, Japan 1991)	92-70143	0-88318-951-8
No. 251	Superconductivity and its Applications (Buffalo, NY 1991)	92-52726	1-56396-016-8
No. 252	Accelerator Instrumentation (Newport News, VA 1991)	92-70356	0-88318-934-8
No. 253	High-Brightness Beams for Advanced Accelerator Applications (College Park, MD 1991)	92-52705	0-88318-947-X
No. 254	Testing the AGN Paradigm (College Park, MD 1991)	92-52780	1-56396-009-5

	Title	L.C. Number	ISBN
No. 255	Advanced Beam Dynamics Workshop on Effects of Errors in Accelerators, Their Diagnosis and Corrections (Corpus Christi, TX 1991)	92-52842	1-56396-006-0
No. 256	Slow Dynamics in Condensed Matter (Fukuoka, Japan 1991)	92-53120	0-88318-938-0
No. 257	Atomic Processes in Plasmas (Portland, ME 1991)	91-08105	0-88318-939-9
No. 258	Synchrotron Radiation and Dynamic Phenomena (Grenoble, France 1991)	92-53790	1-56396-008-7
No. 259	Future Directions in Nuclear Physics with 4π Gamma Detection Systems of the New Generation (Strasbourg, France 1991)	92-53222	0-88318-952-6
No. 260	Computational Quantum Physics (Nashville, TN 1991)	92-71777	0-88318-933-X
No. 261	Rare and Exclusive B&K Decays and Novel Flavor Factories (Santa Monica, CA 1991)	92-71873	1-56396-055-9
No. 262	Molecular Electronics—Science and Technology (St. Thomas, Virgin Islands 1991)	92-72210	1-56396-041-9
No. 263	Stress-Induced Phenomena in Metallization: First International Workshop (Ithaca, NY 1991)	92-72292	1-56396-082-6
No. 264	Particle Acceleration in Cosmic Plasmas (Newark, DE 1991)	92-73316	0-88318-948-8
No. 265	Gamma-Ray Bursts (Huntsville, AL 1991)	92-73456	1-56396-018-4
No. 266	Group Theory in Physics (Cocoyoc, Morelos, Mexico 1991)	92-73457	1-56396-101-6
No. 267	Electromechanical Coupling of the Solar Atmosphere (Capri, Italy 1991)	92-82717	1-56396-110-5
No. 268	Photovoltaic Advanced Research & Development Project (Denver, CO 1992)	92-74159	1-56396-056-7
No. 269	CEBAF 1992 Summer Workshop (Newport News, VA 1992)	92-75403	1-56396-067-2
No. 270	Time Reversal—The Arthur Rich Memorial Symposium (Ann Arbor, MI 1991)	92-83852	1-56396-105-9
No. 271	Tenth Symposium Space Nuclear Power and Propulsion (Vols. I–III) (Albuquerque, NM 1993)	92-75162	1-56396-137-7 (set)
No. 272	Proceedings of the XXVI International Conference on High Energy Physics (Vols. I and II) (Dallas, TX 1992)	93-70412	1-56396-127-X (set)

	Title	L.C. Number	ISBN
No. 273	Superconductivity and Its Applications (Buffalo, NY 1992)	93-70502	1-56396-189-X
No. 274	VIth International Conference on the Physics of Highly Charged Ions (Manhattan, KS 1992)	93-70577	1-56396-102-4
No. 275	Atomic Physics 13 (Munich, Germany 1992)	93-70826	1-56396-057-5
No. 276	Very High Energy Cosmic-Ray Interactions: VIIth International Symposium (Ann Arbor, MI 1992)	93-71342	1-56396-038-9
No. 277	The World at Risk: Natural Hazards and Climate Change (Cambridge, MA 1992)	93-71333	1-56396-066-4
No. 278	Back to the Galaxy (College Park, MD 1992)	93-71543	1-56396-227-6
No. 279	Advanced Accelerator Concepts (Port Jefferson, NY 1992)	93-71773	1-56396-191-1
No. 280	Compton Gamma-Ray Observatory (St. Louis, MO 1992)	93-71830	1-56396-104-0
No. 281	Accelerator Instrumentation Fourth Annual Workshop (Berkeley, CA 1992)	93-072110	1-56396-190-3
No. 282	Quantum 1/f Noise & Other Low Frequency Fluctuations in Electronic Devices (St. Louis, MO 1992)	93-072366	1-56396-252-7
No. 283	Earth and Space Science Information Systems (Pasadena, CA 1992)	93-072360	1-56396-094-X
No. 284	US-Japan Workshop on Ion Temperature Gradient-Driven Turbulent Transport (Austin, TX 1993)	93-72460	1-56396-221-7
No. 285	Noise in Physical Systems and 1/f Fluctuations (St. Louis, MO 1993)	93-72575	1-56396-270-5
No. 286	Ordering Disorder: Prospect and Retrospect in Condensed Matter Physics: Proceedings of the Indo-U.S. Workshop (Hyderabad, India 1993)	93-072549	1-56396-255-1
No. 287	Production and Neutralization of Negative Ions and Beams: Sixth International Symposium (Upton, NY 1992)	93-72821	1-56396-103-2
No. 288	Laser Ablation: Mechanismas and Applications-II: Second International Conference (Knoxville, TN 1993)	93-73040	1-56396-226-8
No. 289	Radio Frequency Power in Plasmas: Tenth Topical Conference (Boston, MA 1993)	93-72964	1-56396-264-0

	Title	L.C. Number	ISBN
No. 290	Laser Spectroscopy: XIth International Conference (Hot Springs, VA 1993)	93-73050	1-56396-262-4
No. 291	Prairie View Summer Science Academy (Prairie View, TX 1992)	93-73081	1-56396-133-4
No. 292	Stability of Particle Motion in Storage Rings (Upton, NY 1992)	93-73534	1-56396-225-X
No. 293	Polarized Ion Sources and Polarized Gas Targets (Madison, WI 1993)	93-74102	1-56396-220-9
No. 294	High-Energy Solar Phenomena A New Era of Spacecraft Measurements (Waterville Valley, NH 1993)	93-74147	1-56396-291-8
No. 295	The Physics of Electronic and Atomic Collisions: XVIII International Conference (Aarhus, Denmark, 1993)	93-74103	1-56396-290-X
No. 296	The Chaos Paradigm: Developments an Applications in Engineering and Science (Mystic, CT 1993)	93-74146	1-56396-254-3
No. 297	Computational Accelerator Physics (Los Alamos, NM 1993)	93-74205	1-56396-222-5
No. 298	Ultrafast Reaction Dynamics and Solvent Effects (Royaumont, France 1993)	93-074354	1-56396-280-2
No. 299	Dense Z-Pinches: Third International Conference (London, 1993)	93-074569	1-56396-297-7
No. 300	Discovery of Weak Neutral Currents: The Weak Interaction Before and After (Santa Monica, CA 1993)	94-70515	1-56396-306-X
No. 301	Eleventh Symposium Space Nuclear Power and Propulsion (3 Vols.) (Albuquerque, NM 1994)	92-75162	1-56396-305-1 (Set) 156396-301-9 (pbk. set)
No. 302	Lepton and Photon Interactions/ XVI International Symposium (Ithaca, NY 1993)	94-70079	1-56396-106-7
No. 303	Slow Positron Beam Techniques for Solids and Surfaces Fifth International Workshop (Jackson Hole, WY 1992)	94-71036	1-56396-267-5
No. 304	The Second Compton Symposium (College Park, MD 1993)	94-70742	1-56396-261-6
No. 305	Stress-Induced Phenomena in Metallization Second International Workshop (Austin, TX 1993)	94-70650	1-56396-251-9
No. 306	12th NREL Photovoltaic Program Review (Denver, CO 1993)	94-70748	1-56396-315-9

	Title	L.C. Number	ISBN
No. 307	Gamma-Ray Bursts Second Workshop (Huntsville, AL 1993)	94-71317	1-56396-336-1
No. 308	The Evolution of X-Ray Binaries (College Park, MD 1993)	94-76853	1-56396-329-9
No. 309	High-Pressure Science and Technology—1993 (Colorado Springs, CO 1993)	93-72821	1-56396-219-5 (Set)
No. 310	Analysis of Interplanetary Dust (Houston, TX 1993)	94-71292	1-56396-341-8
No. 311	Physics of High Energy Particles in Toroidal Systems (Irvine, CA 1993)	94-72098	1-56396-364-7
No. 312	Molecules and Grains in Space (Mont Sainte-Odile, France 1993)	94-72615	1-56396-355-8
No. 313	The Soft X-Ray Cosmos ROSAT Science Symposium (College Park, MD 1993)	94-72499	1-56396-327-2
No. 314	Advances in Plasma Physics Thomas H. Stix Symposium (Princeton, NJ 1992)	94-72721	1-56396-372-8
No. 315	Orbit Correction and Analysis in Circular Accelerators (Upton, NY 1993)	94-72257	1-56396-373-6
No. 316	Thirteenth International Conference on Thermoelectrics (Kansas City, Missouri 1994)	95-75634	1-56396-444-9
No. 317	Fifth Mexican School of Particles and Fields (Guanajuato, Mexico 1992)	94-72720	1-56396-378-7
No. 318	Laser Interaction and Related Plasma Phenomena 11th International Workshop (Monterey, CA 1993)	94-78097	1-56396-324-8
No. 319	Beam Instrumentation Workshop (Santa Fe, NM 1993)	94-78279	1-56396-389-2
No. 320	Basic Space Science (Lagos, Nigeria 1993)	94-79350	1-56396-328-0
No. 321	The First NREL Conference on Thermophotovoltaic Generation of Electricity (Copper Mountain, CO 1994)	94-72792	1-56396-353-1
No. 322	Atomic Processes in Plasmas Ninth APS Topical Conference (San Antonio, TX)	94-72923	1-56396-411-2
No. 323	Atomic Physics 14 Fourteenth International Conference on Atomic Physics (Boulder, CO 1994)	94-73219	1-56396-348-5

	Title	L.C. Number	ISBN
No. 324	Twelfth Symposium on Space Nuclear Power and Propulsion (Albuquerque, NM 1995)	94-73603	1-56396-427-9
No. 325	Conference on NASA Centers for Commercial Development of Space (Albuquerque, NM 1995)	94-73604	1-56396-431-7
No. 326	Accelerator Physics at the Superconducting Super Collider (Dallas, TX 1992-1993)	94-73609	1-56396-354-X
No. 327	Nuclei in the Cosmos III Third International Symposium on Nuclear Astrophysics (Assergi, Italy 1994)	95-75492	1-56396-436-8
No. 328	Spectral Line Shapes, Volume 8 12th ICSLS (Toronto, Canada 1994)	94-74309	1-56396-326-4
No. 329	Resonance Ionization Spectroscopy 1994 Seventh International Symposium (Bernkastel-Kues, Germany 1994)	95-75077	1-56396-437-6
No. 330	E.C.C.C. 1 Computational Chemistry F.E.C.S. Conference (Nancy, France 1994)	95-75843	1-56396-457-0
No. 331	Non-Neutral Plasma Physics II (Berkeley, CA 1994)	95-79630	1-56396-441-4
No. 332	X-Ray Lasers 1994 Fourth International Colloquium (Williamsburg, VA 1994)	95-76067	1-56396-375-2
No. 333	Beam Instrumentation Workshop (Vancouver, B. C., Canada 1994)	95-79635	1-56396-352-3
No. 334	Few-Body Problems in Physics (Williamsburg, VA 1994)	95-76481	1-56396-325-6
No. 335	Advanced Accelerator Concepts (Fontana, WI 1994)	95-78225	1-56396-476-7 (Set) 1-56396-474-0 (Book) 1-56396-475-9 (CD-Rom)
No. 336	Dark Matter (College Park, MD 1994)	95-76538	1-56396-438-4
No. 337	Pulsed RF Sources for Linear Colliders (Montauk, NY 1994)	95-76814	1-56396-408-2
No. 338	Intersections Between Particle and Nuclear Physics 5th Conference (St. Petersburg, FL 1994)	95-77076	1-56396-335-3
No. 339	Polarization Phenomena in Nuclear Physics Eighth International Symposium (Bloomington, IN 1994)	95-77216	1-56396-482-1

	Title	L.C. Number	ISBN
No. 340	Strangeness in Hadronic Matter (Tucson, AZ 1995)	95-77477	1-56396-489-9
No. 341	Volatiles in the Earth and Solar System (Pasadena, CA 1994)	95-77911	1-56396-409-0
No. 342	CAM -94 Physics Meeting (Cacun, Mexico 1994)	95-77851	1-56396-491-0
No. 343	High Energy Spin Physics Eleventh International Symposium (Bloomington, IN 1994)	95-78431	1-56396-374-4
No. 344	Nonlinear Dynamics in Particle Accelerators: Theory and Experiments (Arcidosso, Italy 1994)	95-78135	1-56396-446-5
No. 345	International Conference on Plasma Physics ICPP 1994 (Foz do Iguaçu, Brazil 1994)	95-78438	1-56396-496-1
No. 346	International Conference on Accelerator-Driven Transmutation Technologies and Applications (Las Vegas, NV 1994)	95-78691	1-56396-505-4
No. 347	Atomic Collisions: A Symposium in Honor of Christopher Bottcher (1945-1993) (Oak Ridge, TN 1994)	95-78689	1-56396-322-1
No. 348	Unveiling the Cosmic Infrared Background (College Park, MD, 1995)	95-83477	1-56396-508-9
No. 349	Workshop on the Tau/Charm Factory (Argonne, IL, 1995)	95-81467	1-56396-523-2
No. 350	International Symposium on Vector Boson Self-Interactions (Los Angeles, CA 1995)	95-79865	1-56396-520-8
No. 351	The Physics of Beams Andrew Sessler Symposium (Los Angeles, CA 1993)	95-80479	1-56396-376-0
No. 352	Physics Potential and Development of $\mu^+ \mu^-$ Colliders: Second Workshop (Sausalito, CA 1994)	95-81413	1-56396-506-2
No. 353	13th NREL Photovoltaic Program Review (Lakewood, CO 1995)	95-80662	1-56396-510-0
No. 355	Eleventh Topical Conference on Radio Frequency Power in Plasmas (Palm Springs, CA 1995)	95-80867	1-56396-536-4
No. 357	10th Topical Workshop on Proton-Antiproton Collider Physics (Batavia, IL 1995)	95-83078	1-56396-543-7
No. 358	The Second NREL Conference on Thermophotovoltaic Generation of Electricity	95-83335	1-56396-509-7